Chicago Public Library

REFERENCE

Form 178 rev. 1-94

DISCARD

D0927968

SITE
REMEDIATION
Planning and Management

J. Andy Soesilo
Stephanie R. Wilson

LEWIS PUBLISHERS

Boca Raton   New York   London   Tokyo

Acquiring Editor:	Joel Stein
Project Editor:	Albert W. Starkweather, Jr.
Marketing Manager:	Greg Daurelle
Direct Marketing Manager:	Arline Massey
Cover design:	Denise Craig
PrePress:	Kevin Luong
Manufacturing:	Sheri Schwartz

Library of Congress Cataloging-in-Publication Data

Soesilo, J. Andy
    Hazardous waste site remediation / J. Andy Soesilo and Stephanie R. Wilson
       p.   cm.
    Includes bibliographical references and index.
    ISBN 1-56670-207-0 (alk. paper)
      1. Hazardous waste site remediation.   I. Wilson, Stephanie R. II. Title.
    TD1052.S63 1997
    628.5′2—dc20   96-38473
                                                                         CIP

This book contains information obtained from authentic and highly regarded sources. Reprinted material is quoted with permission, and sources are indicated. A wide variety of references are listed. Reasonable efforts have been made to publish reliable data and information, but the author and the publisher cannot assume responsibility for the validity of all materials or for the consequences of their use.

© 1997 by CRC Press, Inc.
Lewis Publishers is an imprint of CRC Press

No claim to original U.S. Government works
International Standard Book Number 1-56670-207-0
Library of Congress Card Number   96-38473
Printed in the United States of America   1 2 3 4 5 6 7 8 9 0
Printed on acid-free paper

ABOUT THE AUTHORS

J. Andy Soesilo, Ph.D., currently is the manager of the Hazardous Waste Section at the Arizona Department of Environmental Quality. He is responsible for the management of the state's hazardous waste program, which encompasses permitting, site remediation and corrective action, emergency response, inspections, enforcement, planning, and data management. He is also an adjunct professor at the Center for Environmental Studies at Arizona State University. Dr. Soesilo received his B.S. (1972) in civil engineering from Petra Christian University, his M.S. (1976) in urban planning from the Asian Institute of Technology, and his Ph.D. (1986) in geography from Arizona State University.

Stephanie R. Wilson currently is working in the Office of Customer Service and External Affairs at the Arizona Department of Environmental Quality (ADEQ). Previous jobs included Acting Assistant Director for the ADEQ waste programs, lead planner for the ADEQ Waste and Air Quality programs, manager of the Underground Storage Tank (UST) program and consulting. Ms. Wilson received her B.S. in zoology and M.S. in biology from Kent State University.

ACKNOWLEDGMENTS

The authors would like to acknowledge thanks for reviewing parts of the manuscript to Dan Marsin, Manager of Emergency Response; Roddy Miller, Manager of Laboratory/Quality Control Service Section; Moses Olade, Ph.D., Manager of Site Assessment Hydrology Unit; and Bill Ruddiman, Hydrologist; who are all with the Arizona Department of Environmental Quality. Our appreciation for manuscript review is also given to Will Humble, Epidemiology Program Supervisor with the Arizona Department of Health Services. Finally, a thankful recognition is given to Professor Donald McTaggart of the Department of Geography and to Professor Alvin Mushkatel of the School of Public Affairs, Arizona State University, who provided insight and valuable comments on the structure and content of the text.

PREFACE

This book describes the management of remediation of contaminated sites from the planning perspective. It discusses the problems, scope, process, and trends of site remediation. Specific discussion is focused on the statutory and regulatory provisions, site assessment requirements, and remediation planning processes. The book covers environmental sampling, site characterization, risk assessment, cleanup criteria, technology and technology screening, and public participation.

The features of the book allow the reader to comprehensively grasp the problems and issues related to site contamination and to quickly and easily read and understand the legal provisions pertinent to site remediation. The book provides the reader with an understanding of the complex site remediation planning process through the discussion of environmental sampling, site characterization, human health and ecological risk assessment, and the development of site remediation standards. It also provides the reader with information on available site remediation technologies, the procedure to screen appropriate technologies and the trends of site remediation.

The book is intended for individuals responsible for the management of site remediation programs for hazardous waste, Superfund, and underground storage tank sites. It is also intended for environmental planners, elected officials, citizens, and environmental management students who wish to learn more about the integrated aspects of site remediation.

To satisfy this purpose, the book was written in as detailed and comprehensive a manner as possible, but with simplicity as its main feature. It systematically compiles various sources of documentation and blends them into an integrated planning text. This allows the reader to grasp the nature and scope of site remediation in a single, easy-to-read book.

CONTENTS

PART IV
REMEDIAL ACTION PLANNING

PART V
CONCLUSION

PART I
INTRODUCTION

The objective of this introduction is to inform readers about the problems and issues associated with site remediation. Because of its far reaching consequences, site remediation represents one of the most complex and controversial issues in the various environmental programs. The first chapter of the book describes the problems extensively. It deals with the diverse regulatory provisions related to site remediation, the different levels of remediation standards, the scientific uncertainties built into the program, the liability and equity issues, and the linkage between site remediation and urban redevelopment programs.

The second chapter of the book is intended to introduce the readers to the mission, scope, approach, and processes of planning. It discusses the relationship of site remediation planning with the broader concepts of environmental planning and management. The chapter also outlines the contents of remedial planning practices. Included in this discussion is the grouping of the chapters into a topical classification to help the readers achieve a better understanding of the site remediation process.

1   PLANNING ISSUES IN SITE REMEDIATION

Among the major environmental concerns in the United States, hazardous waste site remediation represents one of the most complex and controversial issues. The debate embraces very sensitive aspects of societal concerns. It includes the issues of health and safety, liability and equity, trust and fear, the reliability of remediation technology, and the need for urban economic development. Probst et al.[1] observe, for example, that although everyone complains about the site remediation program, regulatory provisions that some would like to see relaxed are the same provisions that others believe need to be strengthened. Additionally, although some experts believe that the current human health and ecological risks at many hazardous waste sites are negligible or even nonexistent, the public consistently regards such sites as posing a more serious threat than any other environmental problem.

Site remediation has also become an environmental justice issue on the part of minority groups complaining of the perceived inequities of the past two decades of U.S. environmental policy. As Probst et al. indicate, the minority groups maintain that waste disposal facilities have been and continue to be disproportionately sited in minority neighborhoods, and allege that minorities are given short shrift when decisions are made about which contaminated site to remediate and when. Clearly, the existence of contaminated sites is a result of past mismanagement of hazardous waste. In order to better appreciate site remediation issues, it is necessary to trace the problems back through the history of the United States' hazardous waste management which is tied to the country's progress in economic development.

The U.S. began its economic surge with the industrial revolution in the latter part of the 19th century. Rapid growth continued through the early 20th century, fueled by the defense demands of World War II. After the war, the nation's revitalized industrial base shifted from armaments to the production of consumer goods. The production of both military and consumer goods generated major quantities of municipal refuse and industrial waste. The ability and concern to manage waste in an environmentally protective manner were lacking; garbage, refuse, sludge, and other discarded materials generated by Americans in their daily lives made its way into the environment, posing a serious threat to ecological systems and public health. This situation prompted Congress to pass the Solid Waste Disposal Act (SWDA) in 1965 which provided funding to states and local governments for the development of safe garbage disposal methods.

Although the 1965 Act expressed concerns over hazardous waste management practices, it was passed with the primary purpose of improving solid waste disposal practices. The SWDA describes hazardous waste as a solid waste or combination of solid wastes, which, because of its amount, concentration, or physical, chemical, or infectious characteristics, may contribute to adverse health effects. When a hazardous waste is improperly treated, stored, transported, disposed of, or otherwise managed, it may pose hazards to human health and the environment.

The 1976 amendment to the SWDA, called the Resource Conservation and Recovery Act (RCRA), greatly expanded the provisions pertaining to the management of hazardous waste. The RCRA program is aimed at assuring that hazardous waste is handled in a manner that protects human health and the environment. In doing so, RCRA regulations specify wastes that are hazardous and establish administrative requirements for the generators of hazardous waste, the transporters, and the owners or operators of hazardous waste treatment, storage, and disposal (TSD) facilities.

RCRA is primarily a prevention-oriented program. In order to minimize the risk to human health and the environment, its focus is on preventing the generation of hazardous waste and, if a hazardous waste is generated, preventing the release of hazardous waste into the environment. Although RCRA creates a framework for the proper management of wastes, the Act does not address the problems of hazardous waste found at inactive or abandoned sites or those spills that require emergency response. This realization occurred in the late 1970s, when national attention was focused on the discovery of a large number of abandoned, leaking hazardous waste dumps that were contaminating the environment and threatening human health. These chemical dumps, which represent a series of "toxic time-bombs waiting to go off", are scattered across the country, i.e., New Jersey, Delaware, Michigan, Iowa, Tennessee, Louisiana, California, Oregon, and Washington (Brown, 1980).[2]

The first major incident was Love Canal in Niagara Falls, NY, where people were evacuated from their homes after hazardous waste buried for more than 25 years by a chemical company seeped to the surface and into basements. The canal, originally designed to transport water, was capped during the 1940s with clay and soil after being used for the burial of hazardous waste. In the early 1950s, a community was built around the canal. Over the years, the canal leaked and by 1978 the danger to the residents had become so great that President Jimmy Carter was forced to declare the canal area a national disaster.[3]

In 1988, a Love Canal habitability study was completed after six years of research. The $16 million study found that, after being cleaned up, two thirds of the Love Canal neighborhood would be suitable for residential use, the remainder could be used for industrial or commercial purposes. Based on the study's findings, the Love Canal Area Revitalization Agency developed a master plan. Residential resettlement started in September 1990.[4,5]

Times Beach, MO, illustrates another outstanding story of past hazardous waste mismanagement. In this case, oil contaminated with a hazardous waste called dioxin was used to pave local roads. The hazardous waste subsequently contaminated the soil and groundwater in the community.[6]

Residents of Times Beach were relocated by the federal government in 1983 at a cost of $33.5 million. As of 1990, an estimated $52.5 million had been spent to clean up Times Beach. Syntex Agribusiness was named as one of several potentially responsible parties. On June 20, 1990, the company agreed to continue the cleanup project, estimated to cost between $90 and $110 million, in addition to reimbursing the federal government $10 million of the estimated $86 million it had spent. The remediation plan included the incineration of dioxin-contaminated soil. The remediation is expected to be completed in 1997.[7–9]

The discovery of the series of hazardous waste contaminated sites in the 1970s resulted in awareness that a disaster of this type and magnitude was outside existing environmental laws. A Congressional committee survey found that one third of the 3,383 waste disposal sites used since 1950 by the 53 largest U.S. chemical companies were not subject to federal regulations.[10] In response to the need for the cleanup of hazardous waste in abandoned sites, the Comprehensive Environmental Response, Compensation and Liability Act (CERCLA), popularly known as "Superfund", was enacted in 1980. The Act authorized a tax to be levied on the chemical industry. The Superfund program helped to clean hazardous waste sites by removing contaminated soils from the sites, and/or remediating the contaminated sites by using various treatment technologies.

Hazardous waste site remediation has been an area of major environmental concern. As the above examples demonstrate, remediating hazardous waste sites has historically come to be one of the most lengthy and costly programs of U.S. environmental policy. The disturbing trend is that despite efforts to expedite remediation by the U.S. Environmental Protection Agency (EPA), cleanup times for ongoing projects are longer than those already completed because of the greater complexity and different characteristics of these ongoing projects.[11]

In terms of spending of the Superfund budget, the U.S. General Accounting Office (GAO) found that since the 1986 passage of the Superfund Amendments and Reauthorization Act (SARA), EPA has shifted the remediation budget away from the remedial study phase to the construction phase. From 1987 to 1993, EPA's budget was $2.6 billion, of which 40% was used to construct remedies. The down side, however, is that a small number of expensive projects accounted for a large portion of the budget. The total $1.1 billion construction expenditure was for just 13 of the more than 200 sites that received such funding during the period.

Under the Superfund program, the parties responsible for hazardous waste sites that are contaminating the environment are liable for the costs of remediating the sites. A recent survey of the Fortune 500 Industrial and Fortune 500 Service corporations indicates that the corporations that had been involved at Superfund sites reported having spent a median of $1.5 million since January 1, 1987.[12] Approximately two thirds of this total, or $1 million, was used for site cleanups and the remaining one third, or $500,000, was used for legal expenses.

Another observation on the national trend of cleanup costs reveals that if the U.S. maintains its course, direct remediation costs in 1990 dollars is estimated to be $750 billion over the next three decades.[13] The direct costs include the costs of physical activities that result from decisions to clean up sites, including the costs of site characterization. The total will be contributed by the Department of Energy (32%), the RCRA corrective action program (31%), the Superfund program (20%), and a combination of underground storage tank, Department of Defense, and state–private programs (17%).

The cost of site remediation is unquestionably determined by the type of technology used in remediating a site. Results from a University of North Dakota study indicate that compliance with government regulations represents the most important consideration in selecting site remediation technology, followed by economics, public acceptance, and the engineering aspect of the technology.[14] This means that in selecting an appropriate remediation technology the decision maker must combine technical judgment, cost, liability, and statutory criteria to arrive at a decision.[15]

As indicated by the preceding citations, there are certain reasons why the site remediation program has become the most lengthy and costly program in U.S. history. Some of the reasons are associated with legal jurisdiction regarding the remediation process, uncertainty of science, how clean a remediated site must be before it is considered clean enough to be an acceptable level of risk, and liability and equity issues. The following sections examine these outstanding issues in greater detail. In addition, issues related to abandoned sites are also reviewed due to a new trend occurring in the U.S. remediation program aimed at making future use of waste-contaminated sites.

1. DIVERSE REGULATORY UMBRELLA

The variability in the characteristics of sites makes site remediation a challenging undertaking. In addition, different laws may impose different remediation requirements on the contaminated sites. Some of these statutes include CERCLA and RCRA. A number of states also have laws that might be more stringent than federal laws.

Although RCRA and CERCLA apply to hazardous waste sites, the two statutes are quite different. CERCLA was enacted to clean up abandoned and uncontrolled sites through a specific cost recovery mechanism, i.e., by identifying the parties that are liable for the costs of cleaning up contaminated sites and forcing them to perform the remedial work. Liable parties include:

1. current owners and operators of a site
2. former owners and operators of a site who were involved with the site during the time any hazardous substance was disposed at the site
3. the generators of the hazardous substances
4. the transporters of the hazardous substances

EPA has always used its enforcement authority to recover cleanup costs and to seek judicial orders requiring responsible parties to conduct remedial activities. Before the agency can initiate an enforcement action, EPA must identify those parties responsible for a site's cleanup. Many CERCLA sites are the result of disposal activities by hundreds of companies; therefore, the responsible parties for such a site may reach a significant number that often times end up in court in a lengthy litigation process. Because CERCLA does little more than generally identify the categories of liable parties, it has been left to the courts to address in detail how a party may fit within each category.[16]

Unlike CERCLA, RCRA was established to prevent more abandoned hazardous waste sites from being created. Under RCRA, EPA and authorized states can bring some types of enforcement actions including the issuance of administrative orders and the filing of civil and criminal penalties. An administrative order requires the violator to comply immediately or within a specified time period, or seeks injunctive relief against the alleged violator through a civil action filed in court. A RCRA violator is liable for a civil penalty of up to $25,000 for each day of violation, regardless of whether the violator has been served with a compliance order. RCRA also imposes criminal penalties of up to $50,000, two years imprisonment, or both, for any person who knowingly commits RCRA violations.

In many cases, a site owner will have no choice as to which law governs the site; however, in a large number of instances, the owner may have the opportunity to negotiate with the agency in choosing which statute can be appropriately applied to the site. An example is where a site which has had a hazardous waste release into the soil which can be delineated spatially from a CERCLA plume (mobile columns of contaminants that are dispersed in soils and groundwater), and the time frame of the release has been determined: both RCRA and CERCLA provisions may govern the remediation of the site.

In choosing which law is advantageous for performing a cost-effective cleanup of a contaminated site, it is crucial to understand that a significant difference between CERCLA and RCRA is that only CERCLA allows for natural resource damage claims. CERCLA also exercises cost recovery claims against responsible parties. One of the advantages of cleaning up a site under

RCRA is that EPA's RCRA staff is not preconditioned to file cost recovery claims.[17] Hope reveals that in theory, EPA could use CERCLA's liability provisions to recover costs incurred in a RCRA cleanup action, but it seldom occurs.

Under RCRA, EPA is authorized to delegate the RCRA program to state environmental agencies, including the authority to enforce RCRA. In CERCLA, EPA maintains control over the Superfund program and handles CERCLA enforcement itself. Therefore, when a state is authorized to administer the RCRA program, Hope argues that the choice between CERCLA and RCRA is really a choice between federal or state enforcement. Site owners may prefer to work with state agencies because the agencies are more sensitive to local concerns and to political influence than EPA, or the owners may have a poor relationship with state regulators and prefer to work with EPA on a problem instead.

The cleanup standard under CERCLA permanently and significantly reduces the volume, toxicity, or mobility of the contaminants. One of the most difficult problems in selecting cleanup standards at CERCLA sites is determining the appropriate standards. On the other hand, CERCLA provides some level of flexibility in selecting cleanup standards.

The standards for managing RCRA sites are usually considered relevant and appropriate at CERCLA sites. This means that CERCLA sites can be remediated using RCRA standards. The standard appropriate at RCRA sites, however, can be difficult to achieve. In some cases, RCRA staff interprets the RCRA provisions for releases of hazardous wastes to include remediating the contaminants down to no-detect or background levels. It would certainly be ideal if the cleanup standards were the same for all environmental regulations. If that is the case, then what level of cleanup at a contaminated site provides an acceptable level of risk? This question will be answered in the next section.

2. HOW CLEAN IS CLEAN?

If hazardous wastes are introduced in any environmental media due to a spill or other mismanagement, there may be both short- and long-term effects on human health and the ecological system. Any chemical can cause severe health impairment or death to humans if taken in sufficiently large amounts. There are also those that can cause adverse health effects even in small doses. The effects include cancers (carcinogenesis), genetic defects, reproductive abnormalities, alterations of immunobiological homeostatis, central nervous system disorders, and congenital anomalies.[18]

Although these adverse health effects are recognized, establishing appropriate standards for remediating contaminated property represents one of the most difficult tasks in site remediation projects. One may advocate stringent standards which permanently remove contaminants and restore the sites to their pristine conditions. The drawback of this "conservative approach" is that the ability to finance this program is very limited, resulting in a slow pace of actual remediation activities. As Wildavsky[19] observes, remedial programs have been too slow, too expensive, and provide negligible health benefits because the actual amount of potentially harmful contaminants is frequently too minute and too far from people to do much harm.

As the site remediation program has matured, it is acknowledged that we do not have the financial resources necessary to implement site remediation programs using the conservative approach, nor is it necessary that all properties need to be remediated to the same standard. The recognition of these tenets has caused site remediation planners to rethink their approaches in setting remediation standards and develop appropriate alternatives. One of the options is the use of safe and less stringent standards that consider land-use factors and the real risks introduced by site contaminants.

Milloy[20] estimates that current remediation costs may be reduced by 60% if science-based risk assessment is employed in site remediation. He suggests that site remediation should be prioritized on the basis of current risk and future risk as determined by reasonably foreseeable land use. Stults[21] adds that if the potential for future exposure to existing harmful contaminants at a site is relatively low, site remediation need not be as extensive and costly as for property that has high public exposure. With this approach, site-remediation managers would no longer be in the position of spending huge amounts of money on site remediation, unless such spending was justified by appropriate scientific principles.

The adoption of less stringent remediation standards by utilizing a risk-assessment approach has recently become an appealing option in site remediation. Maritato et al.[22] observe that although it would be a stretch of the imagination to suggest that states openly embrace a risk-based approach to hazardous waste site remediation, the momentum in favor of doing so is building. This observation was also made by a study conducted by the National Governors' Association: in evaluating site remediation practices, it was discovered that states are increasingly using either risk-based standards or site assessment for setting site remediation criteria.[23]

To utilize risk assessment effectively, site remediation planners should factor in the risk-assessment equation, the current and future risks of a site as determined by its land use, i.e., whether the site is intended to be used as residential or nonresidential property. The integration of land-use factors in the risk-based remedial decision includes institutional controls such as land-use restrictions, which are established to inform future buyers of the cleanliness level of the property. Land-use restrictions are legal mechanisms to limit or prohibit certain property uses that may result in human exposure to contaminants or interfere with the integrity of a remedial action.[24] Because courts generally disfavor restraints on the use of real property, Nightingale[25] suggests that clear and unequivocal statutory authority for land-use restrictions should be provided in relevant legislation.

Massachusetts state law provides three options for defining remediation criteria.[26] Method 1 uses numeric standards for more than 100 chemicals in soil and groundwater, method 2 allows for adjustments of those standards to consider site-specific conditions, and method 3 requires the use of a site-specific risk assessment to determine the remediation level. The laws authorize the state to utilize institutional controls as part of its remediation program.

Texas' remediation program establishes three risk-reduction standards.[27] Standard 1 requires remediation to background levels; standard 2 requires remediation to health-based levels. Standard 2 uses numeric standards for soil and groundwater for more than 150 contaminants at both residential and nonresidential sites. Deed certification is required to provide notice to future property owners of the residual levels of contaminants at the site. Standard 3 uses site-specific risk assessment. Deed certification is required and, generally, engineering controls to the property are also necessary, depending upon the degree of long-term effectiveness achieved by the remedial activities.

In Michigan, state statutes specify that contaminated sites may be remediated either to type A, B, or C criteria.[28] Type A criteria require remediation to either background levels or to the method detection limits for the contaminants of concern. Type B standards provide for contaminant concentrations that do not pose an unacceptable risk based on standardized exposure assumptions. Type C criteria utilize site-specific risk assessments to establish remediation levels that do not pose an unacceptable risk. This risk assessment, among other things, must consider land use and requires the arrangement of institutional controls to the property.

A review of states' site remediation programs reveals that in establishing remediation criteria, the states generally provide flexibility in their standards; however, the three remediation standards that are common to most states are background levels, numeric health-based standards, and remediation levels derived from site-specific risk assessment. In site remediation terminology, risk assessment is viewed as a process used to determine whether contaminants at a site may pose a risk to human health and the environment, and if so, to what extent. As far as human health is concerned, the two principal components of risk assessment are toxicity assessment and exposure assessment.[29]

The toxicity assessment is aimed at determining what health effects, stated in cancer and noncancer effects, may occur as a result of exposure to the contaminants present at the site. Based on a set of assumptions, the toxicity assessment establishes a toxicity or dose level (in mg/kg/day) above which the contaminant of concern may affect human health and below which the contaminant is considered no threat to human health.

The exposure assessment is conducted to examine the magnitude of contact of human organs (skin, lung, brain) with a chemical or physical agent. The assessment identifies who may, currently and in the future, potentially be exposed to the contaminants. It identifies the pathways of exposure (e.g., inhaling air, ingesting soil) and projects the exposure frequency and intensity. The result is expressed in exposure terms such as daily intake (in mg/kg/day).

The conclusions of the exposure assessment, combined with the results of the toxicity assessment are used to calculate the health risk of the site. This process is widely known as risk characterization. The calculated risk is then compared to the target risk level which can be 1×10^{-6}, 1×10^{-5}, or any other risk level perceived by the citizens of the state as a safe level for cancer risk. For a noncancer risk, the target risk or hazard index is 1.0. A hazard index of less than 1.0 means that, considering all routes of exposure, the amount of daily intake is less than the allowable dose.

It has been shown from the above that risk assessment is basically a site-specific methodology. Instead of utilizing this process, which treats each site uniquely, the most recent national survey reveals that more and more states favor the use of generic risk-based standards, also called health-based standards, which can be applied uniformly across all sites.[30] The survey indicates that the generic health-based standards are based on a specific target risk, usually 10^{-6}.

If for example, a health-based standard for a contaminant of concern in soil is 58 mg/kg, and the sampling and laboratory results indicate that the level of that particular contaminant at a site is 40 mg/kg, it can instantly be concluded that the site is legally considered clean. The availability of such a generic standard will certainly accelerate site remediation and encourage voluntary remediation because all parties involved in the process will have a clear idea of the specific remediation goals that have to be accomplished.

In a number of instances, the generic health-based standard of a particular contaminant of concern may be higher than background levels for that contaminant. Requiring a site to be remediated to a level below background will certainly create an equity issue. For this reason, background levels are normally provided as alternative remediation standards. Depending on the remediation strategy of a state, background levels can be decided for all contaminants or only for inorganic compounds, only for natural background or for anthropogenic background where background concentrations of a contaminant have been affected by human activities such as agriculture and mining.

Background is generally defined as any area within close proximity to a site which has not been affected by a release of contaminants. When background is considered as an option, the most significant issue regarding this remediation standard is that it is often difficult to obtain the concentration level that truly represents background. The use of a generic risk-based standard is advantageous simply because of this difficulty.

While gaining widespread support, the risk assessment method has also come under tight public scrutiny. When GAO conducted a review on the consistency of approaches used in risk assessment in the Superfund program, the agency found that different assumptions are often used in estimating exposure, particularly in judging how sites might be used in the future and how much contamination would remain there.[31] Different assumptions also occurred in determining how people absorb contaminants through the skin. The agency noted that when such inconsistencies occurred, risks were estimated differently even though the sites had similar characteristics.

Criticisms of risk assessment are also directed to the assumptions used in its statistical procedure, in estimating the toxicity of contaminants, the cancer risk focus of assessment, and the public expectation of zero risk. In this regard, Carpenter[32] writes that the use of the 95% upper bound of statistical distributions and worst-case scenarios are subject to challenge as being unwarranted and philosophically pessimistic. He also reminds us that the transferability of animal response data to humans continues to be a matter of scientific controversy and argues that risk assessment could be a hard sell when the public expects and demands a zero risk along with the benefits of a highly technological civilization and a burgeoning economy.

Focusing on health risk, Locke[33] observes that the current risk assessment field is dominated by cancer risk methodology, resulting in a cancer-risk evaluation that is highly mechanistic and data intensive. In contrast, evaluation of other adverse health effects is much less sophisticated. From a scientific perspective, he argues, emphasis on cancer means that other health effects may be downplayed or perhaps ignored.

Realizing the limitations of risk assessment methodology, it is imperative that when site remediation planners present the findings of their risk assessment, their presentation should include a statement of confidence in the assessment that identifies all major uncertainties, along with discussion concerning their influence on the assessment.[34] The presentation should also discuss key scientific concepts, data, and methods used in the assessment.

3. UNCERTAINTY OF SCIENCE

Science plays a significant role in site remediation planning in at least three different project stages: during site characterization, during risk assessment, and during technology selection. Site characterization is an integral part of the remedial process that attempts to characterize a site in terms of the horizontal and vertical levels of contamination, and to evaluate the potential of contaminants to migrate from their original locations. Risk assessment examines the danger to human health and the environment created when leaving a contaminant on-site, and is used to make risk management decisions that provide a cleanup level with a resultant contamination level that is an acceptable risk. Technology selection provides a framework for choosing a remedial technology that is technically feasible, socially acceptable, and cost effective.

Site characterization begins with the generation of massive planning documents for the establishment of monitoring points, and for the acquisition and analysis of environmental samples. The parameters (such as media geometry, ambient flux) which apply the primary control on contaminant transport are estimated during the sampling or monitoring activities. Frequently, the results of initial sampling activities showed that the contamination has migrated beyond the initial limits of the study area, or the concentration gradients could not be interpreted sufficiently within the constraints of the existing database. The typical response to this situation has been to establish additional sampling points and collect and analyze more environmental samples.[35] Sara observes that investigations that follow this track rapidly sink into "plume chases" wherein the goal gradually shifts away from defining and quantifying exposure pathways to locating the leading edge of the plume.

In remediating a site using a pump-and-treat cleanup scheme, one of the important parameters to estimate is the sorption characteristics of the contamination to the soil. Data from 19 ongoing and completed pump-and-treat remediation plans analyzed by Haley et al.[36] indicate that if sorption properties of the contaminant are not considered during site characterization, gross underestimation of contaminant mass and remediation time is likely to occur.

Complete characterization of the subsurface conditions of a site is almost impossible, with the exception of comprehensive excavation. However, excavation is not feasible for all sites. The greater the degree of compositional and grain size heterogeneity of the subsurface soils, the more uncertain the characterization becomes.[37] Vance adds that the other important issue is recognizing that subsurface conditions are not steady during the remediation process. The operating conditions optimal for remediation during the first three months are unlikely to be optimal two years later. The reason is that contaminant concentrations change, subsurface geochemistry is modified by remedial processes, and the dominant mechanisms affecting mass transport efficiencies change.

The above examples illustrate issues that can be developed from the scientific uncertainty in the area of site characterization. The uncertainty factor, however, is not an unknown phenomenon in science. We can always expect an uncertainty level in site characterization particularly when we realize that we have to bind together various branches of science such as chemistry, geology, hydrogeology, and geotechnical and soil engineering in characterizing a site.

Although scientific uncertainty is common, people do not like to deal with uncertainty in any aspect of life, let alone those aspects having to do with possible endangerment to life. At the same time the scientific uncertainties regarding the health risks posed by contaminants in the environment are undeniable. Criticism of risk assessment has already been discussed in the previous section. Some of those arguments involve some levels of uncertainty. Essentially, data limitations and uncertainties are inherent in the risk assessment process. As Heath[38] observes, risk assessments are often compromised by the frequent necessity to extrapolate high-dose animal toxicologic data to low-dose human exposures and to use subclinical observations as a theoretical basis for predicting human disease.

The following are some of the factors that explain why uncertainties are unavoidable in risk assessment:[39]

- Dose extrapolation. The toxicity values of chemicals are derived by extrapolating the occurrence of adverse health effects under certain conditions of exposure, based on knowledge of adverse health effects that occur under a different set of conditions.

- Exposure evaluation to mixtures of chemicals. Because there is not yet an approved quantitative methodology for this situation, most assessments assume an additive effect of multiple chemical exposures.
- Human exposure. In the absence of actual data, human health risk assessments are based on animal exposure assumptions.
- Mathematical models. Risk assessments rely on mathematical modeling to estimate the nature and magnitude of human exposure. The models vary significantly in complexity.
- Analytical methodology. The exposure assessment is often based on laboratory results from field samples. The specificity of the analytical methodology is crucial in determining the accuracy of the risk assessment.

Uncertainties derived from a risk assessment are obviously the result of the use assumptions in the assessment process. Some selected key assumptions, involving cancer and noncancer toxicity assessments as well as groundwater and soil exposure assessments, are identified by Milloy[40] as follows:

1. Cancer toxicity assessment
   - A chemical that has been judged to cause cancer in animals is assumed to also cause cancer in humans.
   - Where laboratory animal experiments have used different species (e.g., rats vs. mice) humans are assumed to be as susceptible to cancer as the species most susceptible to cancer.
   - Each member of a class of chemicals is assumed to be as potent a carcinogen as the most potent carcinogenic member in that group.
2. Noncancer toxicity assessment
   - Biological effects observed in laboratory experiments are assumed to be adverse health effects.
   - Humans are assumed to be at least 100 times more susceptible to noncancer health effects than laboratory animals.
3. Exposure assessment for site groundwater
   - Site groundwater is always assumed to be potable.
   - Chemical concentrations in groundwater are calculated from the highest measured site values or the upper-bound value, and are assumed to remain constant throughout the duration of exposure.
   - Any chemical detected in groundwater is assumed to occur across the site at a maximum detected concentration or at an upper-bound average.
   - Groundwater is assumed to be ingested at a rate of 0.5 gallon per day.
   - Individuals are assumed to be exposed to site groundwater every day for 30 years.

4. Residential exposure to soils
   - Chemical concentrations are calculated from the highest measured site values or the upper-bound values, and are assumed to be present over 100% of the site.
   - Chemical concentrations in soils are assumed to remain constant over the entire exposure period.
   - Children ages 1 to 6 are assumed to ingest 200 mg of soil per day.
   - Daily exposure is assumed for a period of 30 years.

Clearly, scientists have developed techniques for dealing with uncertainties, but the techniques do not make uncertainties simply disappear. As Rodricks[41] observes, one of the techniques attempts to ensure that uncertainties are treated in the same way for all chemicals, therefore, the relative risks posed by different chemicals can be roughly understood and no statement can really be made regarding the absolute risk posed by a specific chemical, particularly the risk of toxicity associated with chronic, low-dose exposure of chemicals. By the same token, there is also no site that can be categorized as a zero-risk site.

In the area of technology selection, conventional methods for remediating hazardous waste sites are expensive, offer only short-term solutions subject to increasing restrictions, and do not work well for certain types of hazardous waste. Pump-and-treat for treating contaminant plumes, for example, does not remove the source of contamination, therefore, this conventional technology fails as a permanent solution.

Innovative technologies could be cheaper, faster, safer, and more efficient. However, the fact remains that even where new technology has been successfully demonstrated, managers are reluctant to try new approaches, tending instead to choose conventional techniques to clean up their sites. Characterizing this situation as "fear of trying", Prendergast[42] asserts that this type of fear represents a tug-of-war between federal and state environmental agencies, engineering consulting firms, and clients over the use of innovative technologies for cleaning up hazardous waste sites.

There are several factors that inhibit the adoption of new technology. They include the following:[43]

1. Fear that a new technology will not be accepted by regulatory agencies and rejected by contractors who may favor particular technologies on the basis of their own experiences and investments.
2. Uncertainty about the effectiveness, reliability, and cost of a new technology.
3. Reluctance by managers to gamble on a new technology, particularly if using it may lead to missing a milestone in the event the technology fails.
4. A concern that a new technology may result in an unacceptable level of risk; the fear is about liability in light of the uncertainty.

A report on environmental remediation technology[44] describes the "fear of trying" dilemma by saying that EPA requires that proven technologies be used to remediate contaminated sites, but a technology must have been used somewhere to be considered proven. Regulatory blocks are certainly part of the problem, but lack of information is the real issue. GAO reported in 1993 that the lack of reliable information has led government officials, private contractors, and investors to avoid the possible risks associated with innovative technologies.[45]

Whoever has the responsibility to promote innovative technologies in the market will continue to have a hard time in selling as long as risk and liability remain issues. This can be avoided if the users can be provided with more cost and performance data for new technologies. There is indeed a real need to reduce uncertainty surrounding the performance and cost of innovative technologies.

4. LIABILITY AND EQUITY

One of the most controversial issues in the Superfund program is its liability framework. By statute, CERCLA places "strict, joint and several" liabilities on responsible parties associated with the contaminated site.

- Strict liability means that no showing of actual fault is required. A responsible party may claim that he was not negligent, or that his activities were consistent with standard industry practice and, there-fore, he is not responsible for the site nor liable for the contamination costs. The strict liability concept does not accept this claim.
- Joint and several liability means that whenever there is any evidence of the commingling of hazardous substances by different parties, each is individually responsible for all cleanup costs at the site. This implies that multiple contributors of hazardous waste contamination at a site may all be held equally responsible unless one or more can demonstrate that its waste can be separately identified or could not possibly have contributed to the harm.

CERCLA's strict liability has been uniformly endorsed by the courts. Under strict liability, companies are liable for the full environmental costs of their activities. While CERCLA contains no statutory mandate that liability be joint and several, courts have often found such liability.[46] Under joint and several liability, EPA can go after so-called "deep pocket" potential responsible parties to require them to finance major cleanups.

The debate over Superfund liability is essentially focused on this particular joint and several liability provision. Opponents to the stipulation argue that this type of liability discourages property sales and provides no incentive to reduce the contamination before an anticipated sale of a property. The system is also unfair, is driving up transaction costs and, therefore, is slowing down the cleanup process.

Proponents to the joint and several liability scheme underline the fact that the provision is given simply to let the polluters pay. If the liability is repealed, and financing is inadequate, the Superfund program may face a slowdown or even stop. This situation will force the public to pay for cleanups. Essentially, the liability issue is also an equity issue.

To help remove liability barriers, EPA Region 1 erased 655 sites from the Superfund list aiming at reducing liability and enticing prospective buyers. Those sites have little contamination or are being handled by the states.[47] In Michigan, legislation designed to ease the process for cleaning up and developing contaminated property was enacted in June 1995. Under the new Michigan law, parties not responsible for contamination cannot be held responsible for cleanup, and a culpability standard that imposes liability only in cases in which the party was an owner or operator of a contaminated site during the hazardous release was enacted.[48,49] The U.S. District Court for the Northern District of Illinois ruled in March 1995[50] that in Illinois, a former owner of a hazardous waste contaminated site cannot be held liable for cleanup costs under CERCLA unless there is a real and concrete threat from a release that was the fault of the former owner.

The above few examples illustrate that amid the issues of liability and equity, some initiatives have been launched to prevent these issues from becoming roadblocks to site remediation. Such initiatives are essential because a remediated site represents an untapped resource for reviving urban economy.

5. URBAN REVITALIZATION CONNECTION

Abandoned contaminated hazardous waste sites, whose owners typically are bankrupt companies or delinquent taxpayers, represent distinct urban eyesores. These sites are abandoned because of numerous legal and financial barriers. Revitalizing these sites by returning them to productive use will place the properties back on the tax rolls and the local community will benefit from an improved, healthier environment, better urban landscape, additional employment opportunities, and improved economics.

The problem in revitalizing these sites is that banking and other financial institutions are normally reluctant to lend money to develop the properties because of liability concerns. CERCLA, for example, authorizes EPA to compel potentially responsible parties (PRPs) to clean up contaminated sites for which they are held liable. If PRPs are unable to remediate the site, EPA does so using the Superfund and seeks to recover the remediation costs later. The other problem is that, given strict remediation criteria, it is much more advantageous for investors to acquire suburban greenfields (i.e., clean sites) for investment than developing those sites located in contaminated areas, labeled as brownfields.

Confronted with this dilemma, EPA and some states have initiated programs to encourage the rehabilitation of brownfields sites. Through its Brownfields Action Agenda, EPA creates incentives to clean up and redevelop industrial and commercial sites located in urban areas. As a start, the agency envisions some 50 community pilot programs. On the other hand, the states have also begun with their own brownfields programs. The most aggressive and sophisticated programs are implemented in the rust-belt areas: Illinois, Indiana, Ohio, Massachusetts, and Michigan have programs that address how clean a site must be; when liability is limited; how the state will supervise the work; who and what qualifies for the program; and what, if any, dealings with the surrounding community are necessary.[51]

Other states are active as well. Connecticut's Urban Sites Remedial Action Program (USRAP), for example, shows promise for attracting businesses back into the state's numerous unused urban industrial sites.[52] Site owners or investors interested in developing contaminated properties have three options to choose from under USRAP. Each option requires interested parties to pursue sites in poor areas. The state has appropriated $10 million for the program with the possibility of another $15 million. The first two options come with partial state funding. State-funded plans require evidence that the development benefits outweigh the costs. Under the third option, the state assumes all liability for a contaminated site.

Minnesota's Voluntary Investigation and Cleanup (VIC) program has a record of returning some 550 contaminated sites to productive use.[53] New Jersey's Hazardous Discharge Site Remediation Fund (HDSRF) was established for municipalities, private citizens, and businesses to clean up contaminated properties. The fund was capitalized through a $45 million appropriation by the legislature and a $10 million allocation from the state's economic development fund.[54] Under the HDSRF, individuals and companies can qualify for a 50% innocent party grant if they meet certain requirements. Similarly, municipalities can qualify for a 100% grant supporting site characterization if they meet certain conditions. A loan program was also established for municipalities for conducting remedial action, as well as for individual and corporate applicants failing to qualify for grants.

In Kansas, under an agreement with the Kansas Department of Health and Environment (KDHE), the City of Wichita may issue a certificate of release to any owner whose property is located within the six-square-mile contaminated area in the central downtown business district. The certificate is given to the owner who makes satisfactory application to the City regarding the environmental conditions of the property. Upon issuance of the certificate and by the terms of the release and covenant not to sue of the City–KDHE agreement, both the City and KDHE release the recipient from claims arising from groundwater contamination at the site.[55] Once a certificate is obtained, the land owner may go to a bank and the bank may make a loan against the property without fear of environmental liability.

Powell reports that by the end of 1993, the City had granted thousands of certificates, resulting in the old town area development becoming a reality. There are now many restaurants, retail businesses, and new offices located in the area. The uncertainty that threatened to paralyze the economic development in the downtown area has been removed. All of this was accomplished within one year.

With the promising results of the brownfields programs, many investors argue that the programs should allow more flexible cleanup standards to induce more companies to participate. Environmental activists fear, however, that these programs contribute to ineffective cleanups, and offer generous releases from liability.

Though conceived in a positive spirit and admirably motivated, brownfields programs still face significant obstacles.[56] Three barriers are identified:

1. State laws cannot shield investors from Superfund liability; therefore, without supportive federal legislation, the best EPA can do is to establish policies that limit investors' exposure to risk and liability.
2. States need authority to fashion flexible, site-specific cleanup standards. Brownfields programs cannot offer releases from liability without sensibly strict cleanup criteria.
3. Brownfields programs must accommodate community participation without allowing revitalization to be stopped completely.

Revitalization of contaminated sites, environmental liability related to the sites, acceptable cleanup levels, and uncertainty of science, combined with the overlapping legal jurisdictions make site remediation truly a complex program. Adding to the complexity is the fact that many of the contaminated sites are found in sensitive urban areas. As Shanoff[57] reports, the National Resource Defense Council maintains that most of the contaminated sites are located in impoverished areas, often next to residential areas that are occupied by people of color. Walking through these issues, planners have to come up with a program that can be implemented faster, cheaper, and safer. In addition, the plan must be acceptable to all stakeholders.

2 COMPONENTS OF SITE REMEDIATION PLANNING

Herbert Smith once wrote that hundreds of thousands of words are written yearly about the human exploitation of natural resources, our misuse of the land, and the increasingly complex organization of our urban communities. Again and again the same conclusion is reached: if order is to be created out of chaos, the solution will be found in the process of planning.[58] Planning is then defined as the application of intelligent foresight to the future of our community and its character, and of our environment.

In this regard, planning represents the conscious organization of human activities to serve human needs. Better planning can be accomplished by greater integration of the separate components into a broader, more coherent framework.[59] Saarinen argues that in order to be effective, planning must consider not only the physical environment but the way people perceive and utilize each segment of the environment.

The need for integrating the social component into planning has put planners at the center of the decision process, in the roles of facilitators and integrators.[60] Hart argues that these particular roles require the fostering of planners with strong backgrounds in group and organizational processes as well as policy-analytic skills.

As Hemmens[61] points out, planning consists of both a technical and a moral undertaking. The dilemma is whether planners can be both advocates for a particular solution and technicians providing knowledge in a value-free way. In reviewing available planning literature Rothman and Hugentobler[62] discovered that, for the most part, planners seem committed to both advocacy as well as the traditional planning doctrine which emphasizes rational planning and neutral policy roles for practitioners.

The emphasis on rational planning requires planners to have, at a minimum, analytic and technical skills, knowledge of planning practices, a balanced role orientation, and an interdisciplinary perspective.[63] To enhance the planning practice aspect, it is essential that planners combine the element of pragmatism with vision.[64]

If planning is the control of future action and a prescriptive, problem-solving activity,[65] then it must start with the identification of what constitutes a planning problem. This planning process continues until a desired plan is formulated and implemented. Depending on the nature of the planning, different programs have different elements in their planning process. Buskirk's[66] example of these elements is provided below. They include:

1. Identify problems and specify objectives.
2. Design, study, and compile an inventory of conditions and resources.
3. Analyze data and formulate plans and policies.
4. Evaluate alternative plans and policies and select the best alternative.
5. Implement the plan or policy and continuously monitor the implementation.

The above overview of planning literature provides a glimpse of the general mission of planning. The overview also describes the scope of planning, its approach, and its processes. In thinking of caring about the earth, it is appropriate at this juncture to quote some of Caplan's[67] thoughts on planners and rumors of Mother Nature's death.

In his reflection, Caplan compares the big cities to a petri dish. The big city communities function much like bacterial colonies as they grow outward from an "infection point" and absorb the natural resources within their borders. Unfortunately, one of the subtle messages of urban life is that groceries come from the supermarket, not from the farm; or, water comes from the faucet, not from the river or well. Urban dwellers are continuously surrounded by messages that land and resources elsewhere are somehow unrelated to urban activities.

In the petri dish, bacteria eventually drain all the sustenance from the media under them and starve. As bacterial colonies mature, they begin to poison themselves with excretions. Because the bacteria are well adapted to their environment, the poisonous effects are minimized. Humans, however, may not be so well adapted. Poisons generated by human activities cause long-term disasters, kill vegetation, irreversibly taint water, and disrupt life-support systems for generations.

Caplan argues that solutions are already being implemented for certain limited conflicts and conditions such as air pollution and groundwater contaminants. He cautions, however, that the long-term solutions for complex socio-ecological problems will depend on making some changes in how planners and decision makers view the resources in their petri dish.

A contaminated site represents an unusable resource but, with adequate site remediation planning, the site can be restored for productive use. The planning aspect of site remediation can be examined from two perspectives: from the relationship of site remediation planning with the broader area of environmental planning and management, and from the subject matter of the remedial planning. The following paragraphs describe each of these two components.

1. PLANNING CONTEXT

Site remediation planning is intended to appropriately clean up chemically contaminated sites. The cleanup activities may take the form of removing the contaminants from a site or remediating the contamination on-site. As described extensively in Chapter 1, these contaminated sites have become an environmental and societal problem. They are in existence in our environment because of accidents, mishandling or mismanagement of chemicals and hazardous waste. These sites have also become a problem because communities basically want a clean site in their backyards. As Caplan[68] argues, the history of human use of the planet teaches that communities are just as dependent on healthy ecosystems for their survival as any other group of living organisms.

Unfortunately, contaminated sites will always exist as long as we continue to use chemicals and generate hazardous waste. The use of chemicals and the generation of hazardous waste is unavoidable because they have become part of our lives. The use of chemicals is necessary for sustaining and accelerating economic development. The generation of hazardous waste represents the consequence of economic development. As the second law of thermodynamics states, in converting energy to work a certain amount of waste energy must be discharged into the environment. Applying this law to the hazardous waste realm suggests that the transformation process of raw materials into final products involves the creation of waste, some of which is hazardous.[69]

Looking at the transformation of raw material into products within the broader context of environmental management, Rowe et al.[70] consider environmental planning and management as the initiation and operation of activities to manage the acquisition, transformation, distribution, and disposal of resources in a manner capable of sustaining human activities, with a minimum disruption of physical, ecological, and social processes. In order to make an economic activity feasible, Rowe et al.[70] suggest that the activity must be conducted at minimum environmental, monetary, and social costs. In this context, the goals of environmental management planning can be formulated to prevent adverse effects of economic activities with minimum monetary and environmental costs, in a just and equitable manner.[71]

Traditionally, planning is viewed as an important part of management. McDowell[72] argues that not only is planning considered part of management, it is an integral part of the whole policy-making and implementation process. This approach necessitates that planners also provide information to decision makers and the general public, respond to regulations, and support the image or position of the organization.[73] Toward this end, planners need better technical skills and better problem-solving ability so they can connect effectively with those making planning decisions.[74]

Depending on the specific planning objectives, hazardous waste planning creates plans, policies, or programs that can be classified into four major types of planning.[75] They are

1. Hazardous waste management planning.
2. Site correction and remediation.
3. Emergency response and hazard management.
4. Citizen participation planning.

Within this contextual scheme, site remediation planning represents an integral part of hazardous waste planning.

2. THE CONTENTS OF REMEDIAL PLANNING

The content of site remediation planning is conceptualized in Figure 2.1. The figure shows that upon the discovery of a contaminated site, a string of planning questions can be posed to address the remediation aspects of the site. These questions are

1. Under which statute should the remediation be conducted?
2. How widespread is the contamination? Once it is identified and evaluated, what is the likelihood that the contaminants will migrate?
3. If remediation is necessary, what cleanup level is required to restore the site to a clean site?
4. In order to clean up the site, what remedial options are currently available? Of them, which options are the preferred remedies for the site?

Figure 2.1 clearly indicates that the fundamental question of any site remediation planning relates to the regulatory factor. As Toner[76] argues, the heart and soul of planning for most planners is in the regulations that implement their plans. In their profession, planners must continue to develop their planning skills and regulatory knowledge to keep pace with the intricate institutional setting and the rapidly expanding rules and regulations.

This means that site remediation planners should have a good understanding of the legal framework of site remediation. This significant aspect of planning is described in detail in Chapters 3–5. The legal background is discussed in terms of the laws, rules, and procedures related to the planning and management of contaminated sites.

Chapter 3 specifically addresses the legal provisions established under the Comprehensive Environmental Response, Compensation, and Liability Act (CERCLA), which is popularly known as Superfund. The Superfund law provides EPA with the authority to respond to a release of contaminants which may present a danger to human health or the environment and is intended to help

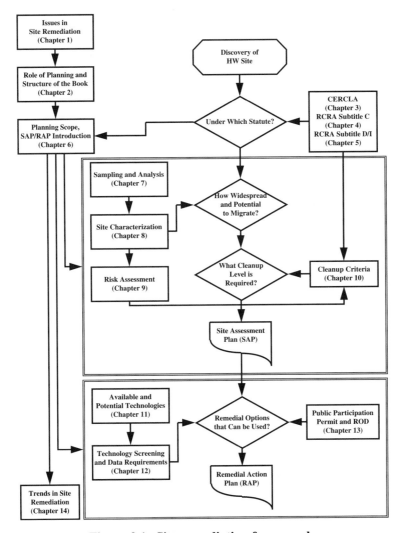

Figure 2.1. Site remediation framework.

clean up chemically contaminated sites by removing contaminated soils from the sites and/or remediating them using appropriate remedies on- or off-site.

Chapter 4 discusses the Resource Conservation and Recovery Act (RCRA), Subtitle C, which outlines the national hazardous waste management program. Under RCRA Subtitle C, the site remediation program is examined by looking at the hazardous waste handler that contributes to the contamination of a site. Separate regulatory provisions are applied to different categories of handlers in that RCRA mandates different response actions for hazardous waste generators, transporters, and treatment, storage, and disposal (TSD) facilities.

Chapter 5 reviews RCRA Subtitles D and I, which, respectively, describe the management of solid waste and underground storage tanks (USTs), particularly the description of its remedial action component. In particular, the discussion is focused on petroleum-contaminated soils (PCS) and the leaking UST (LUST) program.

Obviously, the scope of site remediation planning in CERCLA or RCRA is determined by the respective statutory provisions described in Chapters 3–5. The two major components that underline site remediation planning, as required by those statutes, are site assessment planning (SAP) and remedial action planning (RAP). The introduction to the SAP/RAP discussion is provided in Chapter 6. The detailed review of the SAP process is incorporated in Chapters 7–10, while the RAP process is delineated in Chapters 11–13.

The investigation of the vertical and horizontal extent, and the nature of the contamination involves both sampling and analysis, and site characterization. The sampling and analysis discussion is organized in Chapter 7 and the site characterization process is addressed in Chapter 8. Sampling, analysis, and data management are aimed at collecting information from representative samples of a site. Samples that are not representative of the contaminant of interest are of little use in remediation activities and are not defensible in court.

Site characterization describes whether a contaminant release has occurred at a site, at what concentration, and to what extent. This means that site characterization identifies the types of contaminants that are present in the site's soils and groundwater. It also assesses the concentrations of the contaminants at various distances to examine if they have migrated from the source of contamination.

Showing how to use the site characterization information, Chapter 9 examines the potential of harm to human health or the environment by a contaminated site in the form of a risk assessment discussion. The scope of the risk assessment discussion in this chapter covers both the human-health risk assessment as well as the ecological risk assessment. A human-health risk assessment is useful in examining the health risks associated with potential exposures of a human population to site contaminants.

The risk assessment findings can be used to identify the level of risk associated with a site as well as to determine the level of cleanup necessary to achieve the remediation goals of the site. The default cleanup levels for various regulated contaminants are also established by their respective statutes. This issue of cleanup criteria is the subject of the discussion of Chapter 10.

After the site assessment planning is completed, the subsequent planning activity is focused on reviewing existing potential remediation technologies. This review is discussed in Chapter 11. Planners seeking appropriate remedial technologies will face a variety of options and generally project goals will dictate the scoping and screening of available technologies. Considered technologies are those that permanently destroy or immobilize toxic contaminants resulting in reductions of their volume, toxicity, or mobility.

Once identified, these technologies are properly and systematically screened and evaluated. The preferred remedies are those technologies that are selected for remediating the site. The characteristics of the preferred remedies are that they are technically suitable for the site, publicly acceptable, and will achieve project goals in the most cost-effective way. The remedy selection process is addressed in Chapter 12. The chapter describes the factors used to evaluate the performance, technical applicability, cost, development status, and institutional issues of a potential technology.

As part of the planning process, and because of the importance of community involvement in site remediation, the public participation aspect of site remediation is included as a separate discussion in Chapter 13. Public participation is not only important but also required by various laws. In most instances, the success of conducting site remediation is attributed to the successful planning and implementation of a public participation plan.

Chapter 14, the last chapter, looks at the trend in site remediation. It addresses the technical, legal, and managerial aspects of the program as well as the societal impact that will influence the future of site remediation planning. Among the issues of particular importance are speeding up the remediation of sites, voluntary cleanup, and reusing the contaminated sites for urban development.

Structurally, the chapters of the book can be grouped into five major parts. Part I, which consists of Chapters 1 and 2, serves as the introductory part of the book describing the issues and problems in site remediation and outlining the planning process of site remediation once a contaminated site is discovered.

Part II combines the legal framework surrounding site remediation programs. This legal background discussion consists of Chapters 3–5, which respectively address regulatory review of the Superfund law, hazardous waste, UST and PCS regulations. The purpose of this part is to find similarities and discrepancies among various regulatory frameworks as they relate to site remediation. The two major components of site remediation that are required by the existing laws are site assessment planning and remedial action planning.

Chapters related to site assessment planning are grouped in Part III. The components of site assessment planning are environmental sampling and analysis (Chapter 7), site characterization (Chapter 8), risk assessment (Chapter 9), and cleanup standard determination (Chapter 10). The introductory portion of site assessment planning is provided in Chapter 6.

Chapters related to remedial action planning are grouped in Part IV. The topics of discussion include available remedial technology (Chapter 11), the screening and selection of remedial technology (Chapter 12), and public participation (Chapter 13).

Chapter 14 is the concluding chapter of the book and the only chapter of Part V. The chapter looks at the trends and the future of site remediation.

PART I
REFERENCES

1. Probst, Katherine N. et al., *Footing the Bill for Superfund Cleanups — Who Pays and How?*, The Brookings Institution, Washington, D.C., 1995.
2. Brown, Michael, *Laying Waste: The Poisoning of America by Toxic Chemicals*, Pantheon Books, New York, 1980.
3. Silverman, Gerald, Love Canal: A Retrospective, *Environment Reporter*, September 15, 1989, 835–850.
4. Reilly Endorses Love Canal Study Finding Two-thirds of Area Habitable, *Environment Reporter*, May 18, 1990, 191–192.
5. Silverman, Gerald, Resettlement of Love Canal Begins, But Bank Sour on Mortgage, *Environment Reporter*, December 21, 1990, 1590–1592.
6. Vig, Norman and Kraft, Michael, *Environmental Policy in the 1980s: Reagan's New Agenda*, Congressional Quarterly Inc., Washington, D.C., 1983.
7. Times Beach, Cleanup Begins in Abandoned Town, *Superfund Report*, May 8, 1991, 20.
8. DiPasquale, Suzanne, Times Beach, Judge Approves Plan to Incinerate Dioxin-tainted Soil, *Superfund Report*, January 16, 1991, 22.
9. Settlement Reached in Cleanup of Times Beach: Syntex Agribusiness Agrees to Pay $10 Million, *Environment Reporter*, July 27, 1990, 531–532.
10. U.S. Environmental Protection Agency, *Environmental Progress and Challenges: EPA's Updates*, Washington, D.C., 1988.
11. U.S. General Accounting Office, *Superfund: Status, Cost and Timeliness of Hazardous Waste Site Cleanups*, Washington, D.C., 1994.
12. U.S. General Accounting Office, *Superfund: Legal Expenses for Cleanup-Related Activities of Major U.S. Corporations*, Washington, D.C., 1995.
13. Russell, M., Colglazier, E.W., and English, M.R., *Hazardous Waste Remediation: The Task Ahead*, Waste Management Research and Education Institute, University of Tennessee, Knoxville, 1991.
14. Grasso, Domenic, *Hazardous Waste Site Remediation — Source Control*, Lewis Publishers, Boca Raton, FL, 1993.
15. Sarno, D.J., Making Cleanup Decisions at Hazardous Waste Sites: the Cleanup Sites Approach, *Journal of the Air and Water Management Association*, Vol. 41, No. 9, 1991, 1172–1175.

16. Lee, Robert T., Comprehensive Environmental Response, Compensation, and Liability Act, in J. Gordon Arbuckle et al., eds., *Environmental Law Handbook — Twelfth Edition*, Government Institutes Inc., Rockville, MD, 1993, 267–320.

17. Hope, Michael R., RCRA/CERCLA Interaction, in Kenneth M. Kastner and Brian Cave, eds., *Current Developments in RCRA*, Government Institutes Inc., Rockville, MD, 1993, 213–223.

18. Grisham, J.W., ed., *Health Aspects of the Disposal of Waste Chemicals*, Pergamon Press, Oxford, England, 1986.

19. Wildavsky, Aaron, *But Is It True? A Citizen's Guide to Environmental Health and Safety Issues*, Harvard University Press, Cambridge, MA, 1995.

20. Milloy, Steven J., *Science-Based Risk Assessment: A Piece of the Superfund Puzzle*, National Environmental Policy Institute, Washington, D.C., 1995.

21. Stults, Russel G., How Clean is Clean Enough? Risk-Based Corrective Action Employs Common-Sense Approach to Environmental Restoration, *Environmental Solution*, Vol. 8, No. 7, 1995, 24–29.

22. Maritato, Mark C. et al., Risk-Based Cleanups Form Powerful Approach to Prioritizing, Restoring Hazardous Waste Sites, *Environmental Solution*, Vol. 8, No. 1, 1995, 51–56.

23. National Governors' Association, *Risk in Environmental Decision Making, A State Perspective — Working Paper*, Washington, D.C., 1994.

24. Nightingale, Paul C., Land Use Restrictions and Waste Site Cleanups: Lessons From the States, *Toxics Law Reporter*, April 5, 1995, 1215–1221.

25. Nightingale, Paul C., Suggested Approaches to Land Use Restrictions in State Cleanup Programs, *Toxics Law Reporter*, April 12, 1995, 1243–1249.

26. Massachusetts Department of Environmental Protection, *The Massachusetts Contingency Plan: Risk Characterization and Evaluation — How Clean is Clean Enough?*, Boston, MA, 1994.

27. Texas Natural Resource Conservation Commission, *Summary of the TNRCC's Risk Reduction Rules*, Austin, TX, 1994.

28. Nightingale, Paul C., Suggested Approaches to Land Use Restrictions in State Cleanup Programs, *Toxics Law Reporter*, April 12, 1994, 1243–1249.

29. U.S. Environmental Protection Agency, *Risk Assessment Guidance for Superfund, Volume 1 — Human Health Evaluation Manual (Part A), Interim Final*, Washington, D.C., 1989.

30. National Governors' Association, *Risk in Environmental Decision Making, A State Perspective — Working Paper*, Washington, D.C., 1994.

31. U.S. General Accounting Office, *Superfund — Improved Reviews and Guidance Could Reduce Inconsistencies in Risk Assessments*, Washington, D.C., 1994.

32. Carpenter, Richard A., Risk Assessment, *Impact Assessment*, Vol. 13, No. 2, 1995, 153–187.

33. Locke, Paul A., *Reorienting Risk Assessment*, Environmental Law Institute, Washington, D.C., 1994.

34. U.S. Environmental Protection Agency, *Policy for Risk Characterization at the U.S. Environmental Protection Agency*, Washington, D.C., 1995.

35. Sara, Martin N., *Standard Handbook for Solid and Hazardous Waste Facility Assessments*, Lewis Publishers, Boca Raton, FL, 1994.

36. Haley, J.L., Hanson, B., Enfield, C., and Glass, J., Evaluating the Effectiveness of Groundwater Extraction Systems, *Groundwater Monitor Review*, Vol. 11, No. 1, 119–124.

37. Vance, David B., Site Assessment and Remediation: Focus on the Possible, *The National Environmental Journal*, Vol. 5, Issue 3, 1995, 22–23.

38. Heath, Clark W., Jr., Health Effects of Hazardous Wastes, in Harry M. Freeman, ed., *Standard Handbook of Hazardous Waste Treatment and Disposal*, McGraw-Hill, New York, 1989, 3.17–3.27.

39. Grasso, Domenic, *Hazardous Waste Site Remediation — Source Control*, Lewis Publishers, Boca Raton, FL, 1993.

40. Milloy, Steven J., *Science-Based Risk Assessment — A Piece of the Superfund Puzzle*, National Environmental Policy Institute, Washington, D.C., 1995.

41. Rodricks, Joseph V., *Calculated Risks — The Toxicity and Human Health Risks of Chemicals in Our Environment*, The Press Syndicate of the University of Cambridge, New York, 1994.

42. Prendergast, John, Fear of Trying, *Civil Engineering*, Vol. 61, No. 4, 1991, 52–55.

43. U.S. General Accounting Office, *Management Changes Needed to Expand Use of Innovative Cleanup Technologies*, Washington, D.C., 1994.

44. Business Publishers Inc., *Demonstrating Overseas: A BPI Special Report*, Silver Spring, MD, August 1994.

45. U.S. General Accounting Office, *Superfund: EPA Needs to Better Focus Cleanup Technology Development*, Washington, D.C., 1993.

46. Lee, Robert T., Comprehensive Environmental Response, Compensation, and Liability Act, in J. Gordon Arbuckle et al., eds., *Environmental Law Handbook — Twelfth Edition*, Government Institutes Inc., Rockville, MD, 1993.

47. EPA Region 1 Promises Reforms will Prompt Faster Cleanups, *Hazardous Waste News*, Vol. 17, No. 10, 1995, 76.

48. Governor Signs Amendments to Cleanup Law to Ease Development of Contaminated Sites, *Environment Reporter*, June 16, 1995, 359.

49. Michigan Official Urges Congress to Use New State Law as Model for CERCLA Rewrite, *Environment Reporter*, June 16, 1995, 346.

50. 'Real and Concrete' Threat Needed for Former Owner to be Liable, *Hazardous Waste News*, Vol. 17, No. 17, 1995, 136.

51. Shanoff, Barry, Federal and State Programs Create Cleanup Incentives, *World Wastes*, Vol. 38, No. 7, 1995, 26–28.

52. Remedy Selection May Be Roadblock to Progress in Voluntary Cleanups, *Hazardous Waste News*, Vol. 17, No. 16, 1995, 124.

53. Federal Authorities Eye Ohio Program as Superfund Brownfields Blueprint, *Hazardous Waste News*, Vol. 17, No. 19, 1995, 148.

54. Salinas, Miguel L., McCann, George G., and Jiorle, Neil P., NJ Sets Example for Cleanup Funding, *Environmental Solutions*, Vol. 8, No. 8, 1995, 42–44.

55. Powell, Thomas R., Lessening the Economic Impacts of Superfund Cleanup — the Wichita Model, *The National Environmental Journal*, Vol. 4, Issue 6, 1994, 37–41.

56. Shanoff, Barry, Federal and State Programs Create Cleanup Incentives, *World Wastes*, Vol. 38, No. 7, 1995, 26–28.

57. Shanoff, Barry, Developers Hope to Benefit from Polluted Sites, *World Wastes*, Vol. 37, No. 11, 1995, 22–23.

58. Smith, Herbert H., *The Citizen's Guide to Planning*, American Planning Association, Chicago, 1979.

59. Saarinen, Thomas F., *Environmental Planning: Perception and Behavior*, Houghton Mifflin, Boston, 1976.

60. Hart, Stuart L., Steering the Path Between Ambiguity and Overload: Planning as Strategic Social Process, in Milan J. Dluhy and Kan Chen, eds., *Interdisciplinary Planning: A Perspective for the Future*, Rutgers University, New Brunswick, NJ, 1986, 107–123.

61. Hemmens, G., Introduction, *Journal of the American Planning Association*, 46, 1980, 259–260.

62. Rothman, Jack and Hugentobler, Margrit, Planning Theory and Planning Practice: Roles and Attitudes of Planners, in Milan J. Dluhy and Kan Chen, eds., *Interdisciplinary Planning: A Perspective for the Future*, Rutgers University, New Brunswick, NJ, 1986, 3–26.

63. Dluhy, Milan J., Introduction: Planning Perspectives, in Milan J. Dluhy and Kan Chen, eds., *Interdisciplinary Planning: A Perspective for the Future*, Rutgers University, New Brunswick, NJ, 1986, xiii–xvii.

64. Forester, J., Critical Theory and Planning Practice, *Journal of the American Planning Association*, 46, 1980, 275–286.

65. LeBreton, Preston and Henning, Dale, *Planning Theory*, Prentice Hall, Englewood Cliffs, NJ, 1961.

66. Buskirk, Drannon, Environmental Impact Analysis, in Frank S. So et al., eds., *The Practice of State and Regional Planning*, American Planning Association, Washington, D.C., 1986, 238–254.

67. Caplan, James A., Some Thoughts on Bacteria, Planners, and Rumors of Mother Nature's Death, *Environmental Planning Quarterly*, Vol. 8, No. 3, 1991, 4–6, 16–17.

68. Caplan, James A., Some Thoughts on Bacteria, Planners, and Rumors of Mother Nature's Death, *Environmental Planning Quarterly*, Vol. 8, No. 3, 1991, 4–6, 16–17.

69. Thibodeaux, Louis, Hazardous Material Management in the Future, *Environmental Science and Technology*, Vol. 24, No. 4, 1990, 456–459.

70. Rowe, P. et al., *Principles for Local Environmental Management*, Ballinger, Cambridge, MA, 1976.

71. Baldwin, John, *Environmental Planning and Management*, Westview Press, Boulder, CO, 1985.

72. McDowell, Bruce D., Approaches to Planning, in Frank S. So et al., eds. *The Practice of State and Regional Planning*, American Planning Association, Washington, D.C., 1986, 3–22.

73. DeBettencourt, J.S. et al., Making Planning More Responsive to its Users: The Concept of Metaplanning, *Environment and Planning*, Volume A, Number 14, 1982, 311–322.

74. Chen, Kan, Shifting Demands on Interdisciplinary Planning: An Educational Response, in Milan J. Dluhy and Kan Chen, eds., *Interdisciplinary Planning: A Perspective for the Future*, Rutgers University, New Brunswick, NJ, 1986, 183–191.

75. Soesilo, J. Andy and Wilson, Stephanie R., *Hazardous Waste Planning*, Lewis Publishers, Boca Raton, FL, 1995.

76. Toner, William, Environmental Land Use Planning, in Frank S. So and Judith Getzels, eds., *The Practice of Local Government Planning*, International City Management Association, Washington, DC., 1988, 117–138.

PART II
LEGAL BACKGROUND

The three chapters of this part describe the legal aspects of site remediation. The chapters provide an overview of the laws, rules, regulations, policies, and procedures associated with the management and remediation of contaminated sites. The two major statutes that significantly impact the site remediation program are the Comprehensive Environmental Response, Compensation, and Liability Act (CERCLA), the Resource Conservation and Recovery Act (RCRA) Subtitles C, D, and I.

CERCLA or the Superfund law is discussed in Chapter 3. This chapter examines the U.S. Environmental Protection Agency's (EPA) authority under CERCLA, the objectives of Superfund, and the major features of the law. Characteristics of the statute are examined in terms of its liability provisions, requirements to enforce the cleanup, and stipulations regarding settlement. The chapter also focuses on the implementation process of the Superfund program, including the discussion of the types of remedial responses, the assessment of the contaminated sites, the selection of sites to be remediated and the technology appropriate for site remediation. Issues regarding the implementation of Superfund are briefly addressed.

Chapter 4 describes applicable laws and regulations for the hazardous waste program. It focuses on Subtitle C of RCRA and its implementation regulations. The discussion covers topics related to site remediation and includes the goals of RCRA, the structure of the law, the hazardous waste determination process, and the requirements for various hazardous waste handlers. The cleanup provisions in Subtitle C of RCRA are described in terms of EPA authority and RCRA process requirements for site remediation.

The underground storage tank (UST) program is regulated under Subtitle I of RCRA; petroleum contaminated soil (PCS) is under the domain of Subtitle D of RCRA. These two programs are briefly described in Chapter 5. The focus of this chapter is on the discussion of the objectives of the programs, the general responsibilities of owners and operators and, particularly, specific provisions related to site remediation, and the site remediation processes required for each of the programs.

3   SUPERFUND PROGRAM

Public environmental awareness in the United States was significantly intensified in the early 1960s with the publication of Rachel Carson's book *Silent Spring* in 1962.[1] The book is widely credited with popularizing public fears about potential health effects of synthetic chemicals and, as a result, there can be no doubt that the book's message advanced the burgeoning environmental movement and contributed to the banning of DDT in the U.S. 10 years later.[2]

Beginning with the enactment of the National Environmental Policy Act (NEPA) in 1969, and the creation of the U.S. Environmental Protection Agency (EPA) through an executive order by President Richard M. Nixon in 1970, the federal government established a comprehensive legal system, known as the Resource Conservation and Recovery Act (RCRA), to manage hazardous waste from the time it is generated to its final disposal. In the 1970s the American public also started to realize that the hazardous waste problems caused by past management practices were not addressed by RCRA or other existing environmental statutes. As described in Chapter 1, during the 1970s national attention was focused on the discovery of a large number of abandoned, leaking hazardous waste dumps that were contaminating the environment and threatening human health.

When the Love Canal incident gained national attention, the EPA had few tools to deal with a disaster of this type and magnitude. Although RCRA allowed EPA to require proper management of hazardous waste at active and properly closed facilities and to compel persons to abate endangerments to human health and environment, RCRA provided limited authority for governmental response to leaks or threatened leaks of hazardous waste at abandoned or inactive sites. It quickly became apparent to the president, Congress, and the public that some type of national legislation was needed to fill this void. This legislation took the form of the Comprehensive Environmental Response, Compensation, and Liability Act (CERCLA) which was passed by Congress in 1980 and signed into law by President Jimmy Carter.

1. PROGRAM OBJECTIVES AND OPERATING MECHANISMS

CERCLA authorized a tax to be levied on crude oil, certain petroleum products, and 42 chemical feedstocks. Title II of the Act created a Hazardous Substance Response Trust Fund financed by the CERCLA tax and appropriations from general revenues. Because of the large magnitude of the CERCLA trust fund, it is popularly known as Superfund. The 1986 Superfund Amendments and Reauthorization Act (SARA), for example, authorized an appropriation of $8.5 billion for the period of 1986–1991. In 1990, Congress reauthorized the Superfund program until September 1994 at a funding level of $5.1 billion.

Section 111 of CERCLA specified that the fund could be used to pay EPA's cleanup and enforcement costs, certain natural resource damages, and certain claims of private parties. An example of the latter is a request for reimbursement when a consultant who has performed a cleanup has been unable to obtain payment from the site owner.

CERCLA is designed to address environmental problems associated with hazardous substances. In Section 101 of CERCLA, a hazardous substance is defined as consisting of substances that are designated under other environmental laws, including hazardous wastes under RCRA; hazardous substances and toxic pollutants under the Clean Water Act (CWA); hazardous air pollutants under the Clean Air Act (CAA); and hazardous chemical substances under the Toxic Substances Control Act (TSCA).

CERCLA also provides EPA with authority to respond to a release (or substantial threat of release) of any pollutant or contaminant that may present a danger to human health or the environment. Within this provision, the term pollutant or contaminant refers to just about anything. EPA authority regarding pollutant releases is limited to response action. The agency is not authorized to recover its cleanup costs from private parties or to order the parties to conduct a cleanup when the substance involved is a pollutant, not a hazardous substance.

The Superfund program helped clean up abandoned or inactive hazardous waste sites by removing contaminated soils from the sites, and/or remediating the contaminated sites by using various treatment technologies. Structurally, the program contains three objectives:

1. to take actions against releases or threatened releases of hazardous chemicals that may present danger to the public and the environment
2. to identify individual(s) or firm(s) responsible for a contaminated site before taking any response action; these individual(s) and firm(s) are defined as responsible parties (RPs); once the RPs are identified, EPA requires them to take appropriate response action

    to clean up the site; the action may include the removal of contaminated soil and/or site remediation

3. to clean up a contaminated site where the RPs are not found or where the RPs are unable to take the responsibility; EPA can perform the necessary remediation by spending the CERCLA trust fund

Operationally, the Superfund program requires EPA to identify the parties responsible for the contamination of a site. The potentially responsible parties include past and present owners or operators of the site, generators of hazardous waste found at the site, and transporters of hazardous waste to the site. Once the parties are identified, EPA can negotiate a legal agreement or unilaterally order them to take remedial action responsibility of the site.

Parallel to the CERCLA goal of requiring the RPs to take appropriate response action to clean up a contaminated site, the Act places "strict, joint and several" liabilities on RPs associated with the site. This liability scheme is explained in Chapter 1. To pursue the recovery of cleanup costs and to seek judicial orders requiring RPs to perform cleanups, EPA has used its enforcement authority. Potential liabilities from CERCLA provisions are clearly a major motivation for hazardous waste generators to comply with all RCRA regulations.[3] Although full compliance with RCRA regulations is no defense against liability under CERCLA,[4] compliance is still the best defense against RCRA civil and criminal sanctions.[5]

The implementation of CERCLA's liability provision is illustrated below. In 1989, a group of 533 RPs in Michigan signed a consent decree in federal court for the cleanup of the contaminated Liquid Disposal Inc. site. It was the largest group of RPs to sign a consent decree in the history of Superfund during the 1980s.[6] The consent decree contained the RP's agreement to pay EPA $24.4 million for the cleanup of the site. In return, the court dismissed the case.

In the same year, the U.S. District Court approved a settlement valued at more than $66 million with over 100 RPs for the Operating Industries Inc. site in Monterey Park, CA. The monetary settlement is one of the largest awarded in the Superfund program in the 1980s.[7]

Yet, another consent decree signed in 1990 required a former steel company to pay at least $22 million for cleaning up contaminated sites south of Salt Lake City, UT. This case was the nation's largest settlement involving a bankrupt party held liable for a hazardous waste-contaminated site.[8]

These examples illustrate various Superfund enforcement actions by the federal government. The intensity of the Superfund program is reflected in the following EPA figures: from 1984–1989, the Agency issued a total of 1,015 CERCLA administrative orders, an average of 170 orders a year.[9] These figures do not include cleanup actions initiated by states. State actions to clean up abandoned sites are triggered when the federal government is slow in responding to local needs. State Superfund programs are established under state laws.

2. SUPERFUND PRIMARY ATTRIBUTES

The scope of the Superfund program is reflected by the contents of Title I of CERCLA: Hazardous Substances Releases, Liability, Compensation. The sections that make up CERCLA Title I are

Sec. 101 — Definitions
Sec. 102 — Reportable Quantities and Additional Designations
Sec. 103 — Notices, Penalties
Sec. 104 — Response Authorities
Sec. 105 — National Contingency Plan
Sec. 106 — Abatement Action
Sec. 107 — Liability
Sec. 108 — Financial Responsibility
Sec. 109 — Civil Penalties and Awards
Sec. 110 — Employee Protection
Sec. 111 — Uses of Funds
Sec. 112 — Claims Procedure
Sec. 113 — Litigation, Jurisdiction, and Venue
Sec. 114 — Relationship to Other Law
Sec. 115 — Authority to Delegate, Issue Regulations
Sec. 116 — Schedules
Sec. 117 — Public Participation
Sec. 118 — High Priority for Drinking Water Supplies
Sec. 119 — Response Action Contractors
Sec. 120 — Federal Facilities
Sec. 121 — Cleanup Standards
Sec. 122 — Settlements
Sec. 123 — Reimbursements to Local Governments
Sec. 124 — Methane Recovery
Sec. 125 — Section 3001(b)(3)(A)(i) Waste
Sec. 126 — Native American Tribes

There are some provisions from the above list that, because of their outstanding features, can be used to characterize the CERCLA program. These major attributes of CERCLA include the authority given to EPA to clean up Superfund sites, to enforce against responsible parties to perform cleanup, to provide procedural guidance on settlements with EPA, to detail reporting requirements, and to establish the procedures to be followed for the cleanup of federal installations. Each of these notable features is discussed in the following paragraphs.

2.1. Cleanup Authority

CERCLA provides EPA with the authority to clean up contaminated sites and establishes the procedures and standards which must be followed in remediating the sites. The National Contingency Plan (NCP) represents the primary document for CERCLA response action. The term response action covers removals and remedial actions and related enforcement activities. The NCP is required by Section 105 of CERCLA. It sets forth the procedures which must be followed by EPA and private parties in performing CERCLA response actions.

Although it is required by CERCLA, the NCP is also applicable to certain activities conducted pursuant to the CWA, such as removal actions pursuant to Section 311. The structure and scope of the NCP can be examined in Title 40 of the Code of Federal Regulations (CFR), Part 300. Title 40 of the CFR contains a codification of new regulations pertaining specifically to the protection of the environment. As a common practice, in addition to the promulgation of regulations, EPA also issues guidance and policy documents. Federal regulations are the legal mechanisms that spell out how the broad policy directives of a federal law are to be implemented. The guidance documents are issued by the EPA primarily to elaborate and provide direction on the implementation of the regulations. They are also used to provide the EPA's interpretation of the law and their regulations. Policy statements specify the EPA's operating procedures that must be followed.[10]

The procedure for promulgating a new regulation starts with the publication of a proposed regulation in the Federal Register (FR). The FR is an official federal government publication that specifically deals with regulation development, and is published year round, every working day. It notifies the public of proposed and final regulations promulgated by all federal agencies. Environmental regulations are proposed by EPA. In addition to the wording of a proposed regulation, a preamble provides a discussion of EPA's rationale for proposing the regulation.

The FR uses the date and the first page number of the proposed or final regulation in the register as the reference code. The commonly used notation for any reference in the FR is "(volume) FR (page)". The regulation on the listing of spent solvent, for example, can be found in 50 FR 53315; this means that this specific regulation is published in volume 50, starting on page 53315 of the FR. Volume 50 refers to the year 1985, volume 51 the year 1986, volume 56 the year 1991, and so on. The Federal Register is available in most public libraries.

Once published in the register, EPA normally allows the public 60 days to comment on the proposed regulations. It is important for industry, citizen groups, and the general public to utilize this public comment period to express their concerns to EPA. Following the comment period, EPA will revise the proposed regulation to accommodate public comments and finalize the regulation through publication in the FR. EPA may also drop the proposed regulations, based on comments received.

Annually on the first day of July, the newly promulgated regulations are added to the existing regulations in the CFR in a highly structured format. This process is called codification. Normally CFR documents are available in most public libraries. Sales of CFR documents are handled exclusively by the Superintendent of Documents, U.S. Government Printing Office, Washington, D.C. 20402, telephone 202-783-3238.

Once published, the new regulations are allocated systematically to appropriate agencies. Superfund regulations are compiled in Title 40 of the CFR in Parts 300–373, hazardous waste regulations are in Title 40 of the CFR in Parts 260–272. These regulations are cited as 40 CFR, with the part and section number listed afterward. A citation that reads 40 CFR Part 302.7 means volume 40, Part 302, Section 7 of the CFR.

The NCP has been present in various forms since it was first promulgated in 1973 pursuant to the Federal Water Pollution Control Act to address the removal of oil and other hazardous substances. With the enactment of CERCLA in 1980, EPA was required to expand the NCP to prepare for and respond to discharges of oil and releases of hazardous substances, pollutants, and contaminants.

The NCP as codified in 40 CFR Part 300 consists of 11 subparts. Each subpart contains a number of sections. The list of subparts that make up 40 CFR Part 300 is provided below:

Subpart A — Introduction
Subpart B — Responsibility and Organization for Response
Subpart C — Planning and Preparedness
Subpart D — Operational Response Phases for Oil Removal
Subpart E — Hazardous Substance Response
Subpart F — State Involvement in Hazardous Substance Response
Subpart G — Trustees for Natural Resources
Subpart H — Participation by Other Persons
Subpart I — Administrative Record for Selection of Response Action
Subpart J — Use of Dispersant and Other Chemicals
Subpart K — Federal Facilities

As indicated in the above 40 CFR Part 300, the NCP sets forth the responsibilities and coordination of various organizations which take part in responses to releases, provides standards, methods, and procedures for appropriate response actions including the preparation of administrative records to support response actions. A more detailed discussion on the NCP is provided in the next section of this chapter.

2.2. Enforcing Clean Up

The second major attribute of CERCLA is its enforcement provisions which can be used to identify the classes of parties liable under the statute. CERCLA defines the parties responsible for the contamination of a site to include past and present owners or operators of the site, generators of hazardous waste found at the site, and transporters of hazardous waste to the site. Once the RPs are identified, EPA can negotiate a legal agreement or unilaterally order them to take remedial action responsibility of the site. Parallel to the CERCLA goal of requiring the RPs to take appropriate response action to clean up a contaminated site, CERCLA places "strict, joint and several" liabilities on RPs associated with the site.

Liability provisions under CERCLA include:[11]

1. allowing EPA and private parties to recover their cleanup costs
2. permitting EPA to seek a judicial order to require a liable party to abate an endangerment to human health, welfare, or the environment
3. permitting EPA to take administrative actions to compel private parties to undertake actions to protect public health, welfare, or the environment
4. allowing private parties to bring citizen suits to enforce CERCLA's provisions
5. providing authority for the natural resource trustee to bring actions for damages of natural resources

Allowing EPA and private parties to recover cleanup costs is common in the Superfund program. This represents the majority of Superfund litigation and is brought pursuant to CERCLA Section 107. This section allows EPA, federal agencies, individual states, and private parties to bring an action to recover costs they have incurred in conducting a Superfund response action. Procedures for the arbitration of EPA cost-recovery claims arising under Section 107 of CERCLA are established under 40 CFR Part 304. The procedure governs the arbitration of EPA claims for recovery of response costs not to exceed $500,000, excluding interest (40 CFR 304.11).

In order to recover costs, the claimant must justify that the costs are associated with Superfund response activities which consist of removals and remedial action. Before any costs are recovered, it must be shown that the release or threatened release caused the incurrence of the costs. Also, in any cost-recovery action (from either EPA or a private party), the party seeking reimbursement of such costs must have complied with the provisions of the NCP in incurring such costs. Examples of recoverable costs include those costs associated with sampling, services of environmental consultants and attorneys, and the identification and disposal of hazardous substances; examples of unrecoverable costs are medical monitoring costs, lost profits, and general damages.

Another major Superfund litigation is pursued under CERCLA Section 106. This section permits EPA to seek judicial relief (i.e., an injunctive relief obtained through court order) requiring a RP to abate imminent and substantial endangerment to human health or the environment because of a release of a hazardous substance from a facility. Such an action is available only to EPA and not to private parties. While Section 107 is aimed at obtaining costs associated with performing a Superfund response action, Section 106 of CER-CLA is designed to abate an endangerment.

In addition to authorizing the injunctive relief mechanism, Section 106 of CERCLA also allows EPA to issue a unilateral administrative order to require a private party to conduct a response action. EPA has always used its enforcement authority to seek court orders to recover the cleanup costs, and to issue its own administrative orders requiring RPs to perform remediation. In practice, when a site requires remediation and RPs are successfully identified, EPA compels the RPs to remediate the site using the authority of Section 106, rather than conduct the cleanup with Superfund funds. As stated in 40 CFR 300.400(c)(3), EPA must conserve the fund by requiring private parties to initiate response.

The Section 106 order is normally issued to the RPs which are the largest contributors of contamination and most financially viable. Failure to comply with the order may result in a penalty of up to $25,000 per day of violation and punitive damages of up to three times the amount of costs incurred as a result of the RP's failure to take the necessary action described in the order.

Superfund requirements for a citizen suit can be found in CERCLA Section 310, under Title III — Miscellaneous Provisions. The provision allows any person to file a civil action in two instances:

- against any other person (including the United States) for violations of any CERCLA requirements
- against any officer of the United States for failure to perform a nondiscretionary act under CERCLA

A person initiating a citizen suit must provide 60 days notice of intended action to EPA or the alleged violator.

CERCLA's natural resource damages provision (Section 107) is intended to recover costs associated with the loss of a contaminated area's natural resources. Monies recovered for damages are to be used for restoration of the resources or for acquisition of an equivalent resource.

Natural resources are broadly defined to include wildlife, fish, biota, land, air, water, drinking water supplies, etc. However, the provision is applied only to those resources which are owned, held in trust, or controlled by the federal government, a state, or an Native American nation. Damages to private property are not recoverable.

2.3. Cleanup Settlements

The third major characteristic of CERCLA details the provisions of cost recovery, and provides guidance on cleanup settlements with EPA. CERCLA's settlement provisions are covered in Section 122 of the statute. The section provides specific instructions for when and how various settlement options may be used in the Superfund program.

Section 122(b) provides the options for partial funding, or mixed funding agreements, whereby EPA agrees to use the Superfund funds to reimburse settlers a portion of the response action costs. The partial funding is a function of the cumulative volume of waste contributed by the settler at the contaminated site. The use of this provision is advantageous for a multi-party CERCLA site when some responsible parties wish to settle with EPA and others do not.

Section 122(f) details the circumstances in which EPA can provide a covenant not to sue in a settlement agreement. This option is used if it would expedite a response action and is in the interest of the public. The use of this option is not appropriate in situations where future additional information may reveal that the chosen remedy is no longer protective of human health or the environment. Generally, there is a re-opener clause to handle the additional information issue.

Section 122(g) addresses multi-party Superfund sites where there are a large number of companies which have disposed of small quantities of hazardous substances. These companies, called *de minimis* parties, may enter *de minimis* settlements with EPA. The *de minimis* party also includes an owner of a property where a facility is located who did not conduct or allow the generation, handling, or disposal of hazardous substances at the facility. EPA will usually provide *de minimis* parties with a covenant not to sue. The covenant not to sue in this case is revocable only if subsequent information reveals that the party's waste contribution to the site contamination did not qualify for a *de minimis* designation.

Settlements with EPA are established in a consent decree, which is filed and signed in a federal court, or in a consent agreement drafted and signed by EPA and the responsible party. Settlements are often preferable because they may speed up the process and minimize costs; however, settlement negotiations can also be argumentative, lengthy, and expensive. This is particularly true where RPs must negotiate with EPA and also with each other.

In any type of settlement, there is always a provision for stipulated penalties in the event that a settler fails to meet the conditions specified in the settlement agreement. The applicability and dollar amount of the penalties are subject to negotiation between the settling parties. EPA policy on this is to tie stipulated penalties to the party's compliance with the remediation schedule and in meeting the performance standards and reporting requirements.[12]

2.4. Reportable Quantities

In the fourth provision, CERCLA specifies when releases of hazardous substances must be reported. The statute does not specify the amount of a hazardous substance that must be present before a response action can be taken, or if a party is found liable for the release. However, CERCLA's reporting requirements mandate reporting a release of hazardous substances only when a specified reportable quantity is released. Mandated by Section 102 of CERCLA, this reporting requirement is specified in 40 CFR Part 302. Failure to report the release of a reportable quantity of hazardous substance may result in a civil penalty of up to $25,000 per day of violation or a criminal penalty with a maximum of three years in prison for the first conviction and five years in prison for subsequent convictions. Table 3.1 provides a list of selected hazardous substances and their reportable quantities.

The numbers under the column headed "CASRN" are the chemical abstracts service registry numbers for each substance. Other names by which each hazardous substance is known by other statutes are provided in the "Regulatory Synonyms" column. The "Statutory RQ" refers the reportable quantities under CERCLA Section 102. The "RCRA Waste Number" column shows whether a substance is regulated under RCRA and in what RCRA waste-code classification. The "Final RQ" in "Pounds/Kg" column provides the reportable quantity regulatory adjustment for each hazardous substance in pounds and kilograms. The "Final RQ" is the reportable quantity that must be reported to the National Response Center (toll-free 1-800-424-8802).

CERCLA provides authority for EPA to respond to a release or a substantial threat of a release of any contaminant which may present an imminent and substantial danger to human health or welfare. Under CERCLA, the term contaminant or pollutant encompasses just about anything. While EPA can respond to and clean up a site polluted by a contaminant, CERCLA does not authorize EPA to recover its cleanup costs from private parties or to issue an order compelling the parties to conduct a cleanup when the contaminant is not a hazardous substance.

Another related reporting requirement is required by Title III of SARA. Also known as the Emergency Planning and Community Right-to-Know Act (EPCRA), this requirement is intended to bolster emergency planning efforts at the state and local level, and to provide citizens and governmental agencies with information concerning potential chemical hazards in their communities through "community right-to-know" reporting mechanisms.

There are provisions in 40 CFR that mandate the owner or operator of a facility to notify the Local Emergency Planning Committees (LEPC) and State Emergency Response Commission (SERC) likely to be affected if there is a release into the environment of a listed extremely hazardous substance that exceeds the reportable quantity for that substance. EPCRA regulations are codified in 40 CFR Parts 355, 370, 372, and 373. The first part (40 CFR Part 355) contains provisions regarding the emergency planning component of EPCRA, and the three remaining parts (40 CFR Parts 370, 372, and 373) provide specifics to the community right-to-know component of EPCRA.

2.5. Remediating Federal Sites

In the fifth provision, CERCLA establishes procedures to be followed for the cleanup of federal facilities. The greatest environmental concerns in

Table 3.1 Selected Hazardous Substances and Reportable Quantities

Hazardous substance	CASRN	Regulatory synonyms	Statutory RQ	Code†	RCRA Waste Number	Final RQ Category	Pounds (Kg)
Acenapthene	83329		1*	2		B	100 (45.4)
Acenaphthylene	208968		1*	2		D	5000 (2270)
Acetaldehyde	75070	Ethanal	1000	1,4	U001	C	1000 (454)
Acetaldehyde, chloro-	107200	Chloroacetaldehyde	1*	4	P023	C	1000 (454)
Acetaldehyde, trichloro-	75876	Chloral	1*	4	U034	D	5000 (2270)
Acetamide, N-(aminothioxomethyl)-	591082	1-Acetyl-2-thiourea	1*	4	P002	C	1000 (454)
Acetamide, N-(4-ethoxyphenyl)-	62442	Phenacetin	1*	4	U187	B	100 (45.4)
Acetamide, 2-fluoro-	640197	Fluoroacetamide	1*	4	P057	B	100 (45.4)
Acetamide, N-9H-fluoren-2-yl-	53963	2-Acetylaminofluorene	1*	4	U005	X	1 (0.454)
Acetic acid	64197		1000	1		D	5000 (2270)
Acetic acid (2,4-dichlorophenoxy)-	94757	2,4-D Acid; 2,4-D salts and esters	100	1,4	U240	B	100 (45.4)
Acetic acid, lead(2+) salt	301042	Lead acetate	5000	1,4	U144		#
Acetic acid, thallium(1+) salt	563688	Thallium(I) acetate	1*	4	U214	B	100 (45.4)
Acetic acid, (2,4,5-tricholorophynoxy)	93765	2,4,5-T; 2,4,5-T acid	100	1,4	U232	C	100 (454)
Acetic acid, ethyl ester	141786	Ethyl acetate	1*	4	U112	D	5000 (2270)
Acetic acidfluoro-, sodium salt	62748	Fluoroacetic acid, sodium salt	1*	4	P058	A	10 (4.54)
Acetic anhydride	108247		1000	1		D	5000 (2270)
Acetone	67641	2-Propanone	1*	4	U002	D	5000 (2270)
Acetone cyanohydrin	75865	Propanenitrile, 2-hydroxy-2-methyl-2-; Methyllactonitrile	10	1,4	P069	A	10 (5.54)
Acetonitrile	75058		1*	4	U003	D	5000 (2270)
Acetophenone	98862	Ethanone, 1-phenyl-	1*	4	U004	D	5000 (2270)
2-Acetylaminofluorene	53963	Acetamide, N-9H-fluoren-2-yl-	1*	4	U005	X	1 (0.454)
Acetyl bromide	506967		5000	1		D	5000 (2270)
Acetyl chloride	75365		5000	1,4	U006	D	5000 (2270)
1-Acetyl-2-thiourea	591082	Acetamide, N-(aminothioxomethyl)-	1*	4	P002	C	1000 (454)
Acrolein	107028	2-Propenal	1	1,2,4	P003	X	1 (0.454)
Acrylamide	79061	2-Propenamide	1*	4	U007	D	5000 (2270)
Acrylic acid	79107	2-Propenoic acid	1*	4	U008	D	5000 (2270)

Table 3.1 Selected Hazardous Substances and Reportable Quantities (Continued)

Hazardous substance	CASRN	Regulatory synonyms	Statutory RQ	Statutory Code†	RCRA Waste Number	Final RQ Category	Final RQ Pounds (Kg)
Acrylonitrile	107131	2-Propenenitrile	100	1,2,4	U009	B	100 (45.4)
Adipic acid	124049		5000	1		D	5000 (2270)
Aldicarb	116063	Propanal, 2-methyl-2-(methylthio)-, O-(methylamino)carbonyl]oxime	1*	4	P070	X	1 (0.454)
Aldrin	309002	1,4,5,8-Dimethanonaphthalene, 1,2,3,4,10,10-hexachloro-1,4,4a,5,8,8a-hexahydro-, (1α,4α,4aβ,5α,8α,8β)-	1	1,2,4	P004	X	1 (0.454)
Allyl alcohol	107186	2-Propen-1-ol	100	1,4	P005	B	100 (45.4)
Allyl chloride	107051		1000	1		C	1000 (454)
Methacrylonitrile	126987	2-Propenenitrile, 2-methyl-	1*	4	U152	C	1000 (454)
Methanamine, N-methyl-	124403	Dimethylamine	1000	1,4	U092	C	1000 (454)
Methanamine, N-methyl-N-nitroso-	62759	N-Nitrosodimethylamine	1*	2,4	P082	A	10 (4.54)
Methane, bromo-	74839	Methyl bromide	1*	2,4	U029	C	1000 (454)
Methane, chloro-	74873	Methylchloride	1*	2,4	U045	B	100 (45.4)
Methane, chloromethoxy-	107302	Chloromethyl methyl ether	1*	4	U046	A	10 (4.54)
Methane, dibromo-	74953	Methylene bromide	1*	4	U068	C	1000 (454)
Methane, dichloro-	75092	Methylene chloride	1*	2,4	U080	C	1000 (454)
Methane, dichlorodifluoro-	75718	Dichlorodifluoromethane	1*	4	U075	D	5000 (2270)
Methane, iodo-	74884	Methyl iodide	1*	4	U138	B	100 (45.4)
Methane, isocyanato-	624839	Methyl isocyanate	1*	4	P064		##
Methane, oxybis(chloro-	542881	Dichloromethyl ether	1*	4	P016	A	10 (4.54)
Methanesulfenyl chloride, trichloro-	594423	Trichloromethanesulfenyl chloride	1*	4	P118	B	100 (45.4)
Methanesulfonic acid, ethyl ester	62500	Ethyl methanesulfonate	1*	4	U119	X	1 (0.454)
Methane, tetrachloro-	56235	Carbon tetrachloride	5000	1,2,4	U211	A	10 (4.54)
Methane, tetranitro-	509148	Tetranitromethane	1*	4	P112	A	10 (4.54)
Methane, tribromo-	75252	Bromoform	1*	4	U225	B	100 (45.4)
Methane, trichloro-	67683	Chloroform	5000	1,2,4	U044	A	10 (4.54)
Methane, trichlorofluoro-	75694	Trichloromonofluoromethane	1*	4	U121	D	5000 (2270)
Methanethiol	74931	Methylmercaptan, Thiomethanol	100	1,4	U153	B	100 (45.4)

Chemical Name	CAS Number	Regulatory Synonyms	Statutory Code	Statutory Category	RCRA Waste No.	RQ Code	Final RQ Pounds (Kg)
6,9-Methano-2,4,3-benzodioxathiepin, 6,7,8,9,10,1 0-hexachloro-1,5,5a,6,9,9a-hexahydro-, 3-oxide	115297	Endosulfan	1	1,2,4	P050	X	1 (0.454)
1,3,4-Metheno-2H-cyclobuta[cd]pentalen-2-one,1,1a,3,3a,4,5,5,5a,5b,6-decachlorooctahydro-	143500	Kepone	1	1,4	U142	X	1 (0.454)
4,7-Metheno-1H-indene, 1,4,5,6,7,8,8-heptachloro-3a,4,7,7a-tetrahydro-	76448	Heptachlor	1	1,2,4	P059	X	1 (0.454)
4,7-Metheno-1H-indene, 1,2,4,5,6,7,8,8-octachloro-2,3,3a,4,7,7a-hexahydro-	57749	Chlordane; Chlordane, α and γ isomers; Chlordane, technical	1	1,2,4	U036	X	1 (0.454)
Methanol	67581	Methyl alcohol	1*	4	U154	D	5000 (2270)
Methapyrilene	91805	1,2-Ethanediamine, N,N-dimethyl-N'-2-pyridinyl-N'-(2-thienylmethyl)-	1*	4	U155	D	5000 (2270)
Methomyl	16752775	Ethanimidothioic acid, N-[[(methyl-amino)carbonyl]oxy]-, methyl ester	1*	4	P066	B	100 (45.4)
Methoxychlor	72435	Benzene, 1,1'-(2,2,2-trichloroethylidene) bis[4-meth-oxy-	1	1,4	U247	X	1 (0.454)
Methyl alcohol	675561	Methanol	1*	4	U154	D	5000 (2270)
Methyl bromide	74839	Methane, bromo-	1*	2,4	U029	C	1000 (454)
1-Methylbutadiene	504609	1,3-Pentadiene	1*	4	U186	B	100 (45.4)
1-Propanol, 2-methyl-	78831	Isobutyl alcohol	1*	4	U140	D	5000 (2270)
2-Propanone	67641	Acetone	1*	4	U002	D	5000 (2270)
2-Propanone, 1-bromo-	598312	Bromoacetone	1*	4	P017	C	1000 (454)
Propargite	2312358		10	1		A	10 (4.54)
Propargyl alcohol	107197	2-Propyn-1-ol	1*	4	P102	C	1000 (454)
2-Propenal	107028	Acrolein	1	1,2,4	P003	X	1 (0.454)
2-Propenamide	79061	Acrylamide	1*	4	U007	D	5000 (2270)
1-Propene, 1,1,2,3,3,3-hexachloro-	1888717	Hexachloropropene	1*	4	U243	C	1000 (454)
1-Propene, 1,3-dichloro-	542756	1,3-Dichloropropene	5000	1,2,4	U084	B	100 (45.4)
2-Propenenitrile	107131	Acrylonitrile	100	1,2,4	U009	B	100 (45.4)
2-Propenenitrile, 2-methyl-	126987	Methacrylonitrile	1*	4	U152	C	1000 (454)
2-Propenoic acid	79107	Acrylic acid	1*	4	U008	D	5000 (2270)
2-Propenoic acid, ethyl ester	140885	Ethyl acrylate	1*	4	U113	C	1000 (454)
2-Propenoic acid, 2-methyl-, ethyl ester	97632	Ethyl methacrylate	1*	4	U118	C	1000 (454)
2-Propenoic acid, 2-methyl-, methyl ester	80626	Methyl methacrylate	5000	1,4	U162	C	1000 (454)
2-Propen-1-ol	107188	Allyl alcohol	100	1,4	P005	B	100 (45.4)
Propionic acid	79094		5000	1		D	5000 (2270)
Propionic acid, 2-(2,4,5-trichlorophenoxy)-; 2,4,5-TP acid	93721	Silver (2,4,5-TP)	100	1,4	U233	B	100 (45.4)

Table 3.1 Selected Hazardous Substances and Reportable Quantities (Continued)

Hazardous substance	CASRN	Regulatory synonyms	Statutory			Final RQ	
			RQ	Code†	RCRA Waste Number	Category	Pounds (Kg)
Propionic anhydride	123626		5000	1		D	5000 (2270)
n-Propylamine	107108	1-Propanamine	1*	4	U194	D	5000 (2270)
Propylene dichloride	78875	Propane, 1,2-dichloro- 1,2-Dichloropropane	5000	1,2,4	U083	C	1000 (454)
Propylene oxide	75569		5000	1		B	100 (45.4)
1,2-Propylenimine	75558	Aziridine, 2-methyl-	1*	4	P067	X	1 (0.454)
2-Propyn-1-ol	107197	Propargyl alcohol	1*	4	P102	C	1000 (454)
Pyrene	129000		1*	2		D	5000 (2270)
Pyrethrins	121299 121211 8003347		1000	1		X	1 (0.454)
3,6-Pyridazinedione, 1,2-dihydro-	123331	Maleic hydrazide	1*	4	U148	D	5000 (2270)
4-Pyridinamine	504245	4-Aminopyridine	1*	4	P008	C	1000 (454)
Pyridine	110861		1*	4	U196	C	1000 (454)
Pyridine, 2-methyl-	109068	2-Picoline	1*	4	U191	D	5000 (2270)
Pyridine, 3-(1-methyl-2-pyrrolidinyl)-, (S)-	54115	Nicotine and salts	1*	4	P075	B	100 (45.4)
2,4-(1H,3H)-Pyrimidinedione, 5-[bis(2-chloroethyl)amino]-	66751	Uracil mustard	1*	4	U237	A	10 (4.54)
4(1H)-Pyrimidinone, 2,3-dihydro-6-methyl-2-thioxo-	56042	Methylthiouracil	1*	4	U164	A	10 (4.54)
Pyrrolidine, 1-nitroso-	930552	N-Nitrosopyrrolidine	1*	4	U180	X	1 (0.454)
Quinoline	91225		1000	1		D	5000 (2270)
Wastewater from the reaction vent gas scrubber in the production of ethylene bromide via bromination of ethene.							
K118 Spent absorbent solids from purification of ethylene dibromide in the production of ethylene dibromide.			1*	4	K118	X	1 (0.454)
K123 Process wastewater (including supermates, filtrates, and washwaters) from the production of ethylenebisdithiocarbamic acid and its salts.			1*	4	K123	A	10 (4.54)

K124 Reactor vent scrubber water from the production of ethylenebisdithiocarbamic acid and its salts.	1*	4	K124	A	10 (4.54)
K125 Filtration, evaporation, and centrifugation solids from the production of ethylenebisdithiocarbamic acid and its salts.	1*	4	K125	A	10 (4.54)
K126 Baghouse dust and floor sweepings in milling and packaging operations from the production or formulation of ethylenebisdithiocarbamic acid and its salts.	1*	4	K126	A	10 (4.54)
K131 Wastewater from the reactor and spent sulfuric acid from the acid dryer in the production of methyl bromide.	100	4	K131	X	100 (45.4)
K132 Spent absorbent and wastewater solids from the production of methyl bromide.	1000	4	K132	X	1000 (454)
K136 Still bottoms from the purification of ethylene dibromide in the production of ethylene dibromide via bromination of ethene.	1*	4	K136	X	1 (0.454)

† Indicates the statutory source as defined by 1, 2, 3, and 4 below.

†† No reporting of releases of this hazardous substance is required if the diameter of the pieces of the solid metal released is equal to or exceeds 100 micrometers (0.004 inches).

1 —indicates that the statutory source for designation of this hazardous substance under CERCLA is CWA Section 311(b)(4).

2 —indicates that the statutory source for designation of this hazardous substance under CERCLA is CWA Section 307(a).

3 —indicates that the statutory source for designation of this hazardous substance under CERCLA is CAA Section 112.

4 —indicates that the statutory source for designation of this hazardous substance under CERCLA is RCRA Section 3001.

1* —indicates that the 1-pound RQ is a CERCLA statutory RQ.

# Indicates that the RQ is subject to change when the assessment of potential carcinogenicity is completed.

The agency may adjust the statutory RQ for this hazardous substance in a future rulemaking; until then the statutory RQ applies.

§ The adjusted RQs for radionuclides may be found in Appendix B to this table.

**—indicates that no RQ is being assigned to the generic or broad class.

this case are for those Department of Defense (DOD) and Department of Energy (DOE) sites that were constructed decades ago. The contamination at these facilities was caused by the pre-RCRA disposal practices that included the use of unlined pits, discharge to the ground, and the on-site burning of wastes.

CERCLA waives sovereign immunity for federal facilities. This allows individuals and states to litigate against federal facilities on site remediation matters. Under Section 120, federal facilities are required to comply to the same extent as private entities such as requirements to have their qualified sites listed on the National Priorities List (NPL). Section 120 also establishes requirements that are unique to federal facilities such as the creation of a Federal Agency Hazardous Waste Compliance Docket that lists federal facilities which manage hazardous waste or have potential hazardous waste problems. The list is used to determine whether a federal facility site should be placed on the NPL.

3. IMPLEMENTING SUPERFUND

As discussed earlier in this chapter, the NCP sets forth the procedures which must be followed by EPA and private parties in conducting CERCLA response actions. It also clarifies that in terms of implementation, CERCLA sets forth two categories of response actions: removals and remedial actions. These activities are described and codified in 40 CFR Part 300.

Subpart E of 40 CFR Part 300 establishes methods and criteria for determining the appropriate extent of response authorized by CERCLA due to the release of a hazardous substance or contaminant into the environment that may present an imminent and substantial danger to human health or welfare. Superfund response action performed on-site does not require federal, state, or local permits (40 CFR 300.400(e)). The term on-site generally refers to the area within a facility's boundaries; however, in 40 CFR 300.400(e) on-site means the areal extent of contamination and all suitable areas in very close proximity to the contamination necessary for implementation of the response action. Permits, if required, shall be obtained for all response activities conducted off-site.

A release may be discovered through reports submitted in accordance with reportable quantity requirements, required by a permit, or other similar mandatory reports (40 CFR 300.405(a)). A release may also be discovered through inspections, referrals from other agencies, citizen complaints, or petitions to EPA requesting that a preliminary assessment be conducted on a potentially contaminated site. 40 CFR 300.410(a) defines a removal site evaluation as a removal preliminary assessment and, if more information is needed, a removal site inspection. This activity must be undertaken by the lead agency as promptly as possible.

A removal preliminary assessment can be done based on readily available information and personal interviews and may include (40 CFR 300.410(c)):

- identification of the nature and source of the release
- assessment of the threat to public health performed by the Agency for Toxic Substances and Disease Registry (ATSDR) or the state public health agency
- examination of the magnitude of the threat
- evaluation of factors necessary to make a decision if removal is necessary
- determination of whether a nonfederal party is undertaking proper response

A removal site evaluation shall be terminated when it is determined that there is no release, the source is not regulated under CERCLA, the release involves neither a hazardous substance nor contaminants that may present an imminent and substantial danger to public health or welfare, the amount or concentration of the release does not warrant a response, or the removal has been completed (40 CFR 300.410(e)). The scope of the removal action is based on the information produced from the removal site evaluation and the current site conditions.

Regardless of whether the site is listed on the NPL, whenever a release threatens public health, welfare, or the environment, the lead agency may take any appropriate removal action to abate, prevent, minimize, stabilize, mitigate, or eliminate the release or the threat of release as soon as possible (40 CFR 300.415(b)). The factors that shall be considered include:

- determining the appropriateness of a removal action
- actual/potential exposure to nearby populations, animals, or the food chain from hazardous substances or contaminants
- actual/potential contamination of drinking water supplies or sensitive ecosystems
- hazardous substances or pollutants found in bulk storage containers (e.g., drums, tanks, barrels) that may pose a threat of release
- threat of fire or explosion
- potential of high levels of hazardous substances or contaminants in soils at the surface to migrate
- weather conditions that may cause the migration

Removals can occur as part of the initial action to an extensive remediation. In this circumstance, to the extent practicable, the removal action must be planned to contribute to the efficient performance of any anticipated long-term remedial action at the site.

Removal actions are conducted promptly in response to environmental emergencies to eliminate or minimize the threat of a chemically contaminated site. Examples of removal actions include the following (40 CFR 300.415(d)):

- the immediate cleanup of hazardous waste spilled from a container using appropriate chemicals and other materials that retard the spread of the release or mitigate its effects
- the installation of a fence around a chemically contaminated site, warning signs, and other security precautions where humans or animals have access to the site
- run-off diversion and other drainage controls to avoid migration of hazardous substances or contaminants off-site
- capping of contaminated soils to avoid migration through the soil to surface water, groundwater, or air
- stabilization of berms, dikes, and similar structures to maintain the integrity of the structures
- excavation, consolidation, and removal of contaminated soils
- removal of drums, tanks, and containers that contain chemicals likely to leak, catch fire, or explode
- the provision of alternate water supplies for a local community

If necessary, temporary community relocation may be initiated through coordination with the Federal Emergency Management Agency (FEMA). A spokesperson shall be designated by the lead agency to inform the community of any immediate removal action performed (40 CFR 300.415(m)). The person is responsible for responding to inquiries and providing information concerning the release and the response action. A community relation plan is necessary for removal actions conducted beyond 120 days from the initiation of the activities. Public notice is required for a removal action that will be conducted at least six months after the initiation. If the removal action will be performed after a period of six months, the lead agency is required to have an engineering evaluation/cost analysis (EE/CA) to examine the removal alternatives for the site. If necessary, sampling activities may be required.

Removal actions can occur as part of the initial action of a more extensive remedial action at a NPL site; they can also occur at a site not listed on the NPL. Any removal action must be completed within one year and at a cost of less than $2 million, unless there are reasons for extending the time and the cost, e.g., the action is required as part of a larger approved remedial action, or that there is an immediate risk to public health or the environment.

Although Superfund is a federal program, EPA encourages states to enter into State–EPA Superfund Memorandum of Agreements (SMOA) to increase state involvement and strengthen the State–EPA partnership. This can be done in two ways (40 CFR 300.500(b)):

- The state assumes the lead through a cooperative agreement.
- The state assumes the support role in EPA-lead remedial actions.

The lead agency provides oversight for actions taken by RPs to ensure that a response is conducted consistent with response requirements. EPA will provide oversight when the response action results from an EPA order or consent decree (40 CFR 300.400(h)). The requirements for state involvement in EPA-lead actions and comparable requirements for EPA involvement in state-lead remedial and enforcement responses are set forth in 40 CFR 300.515. 40 CFR 300.520 specifies provisions for involvement in enforcement negotiations and 40 CFR 300.525 sets forth removal action requirements.

The SMOA establishes the nature and extent of EPA and state interaction during EPA- and state-lead responses. It describes the roles and responsibilities of each; the general requirements for EPA oversight; the general nature of interaction regarding the review of key documents; the decision points in removal, remedial, and enforcement response; and the procedures for modification of the SMOA (40 CFR 300.505(a)). Specific requirements for EPA oversight are detailed in cooperative agreements.

Section 104(d)(1) of CERCLA authorizes EPA to enter into cooperative agreements or contracts with states, political subdivisions, or Native American tribes to carry out Superfund response actions when EPA determines that they have the capability to undertake such actions (40 CFR 300.515(a)). The capability measures include an assurance for the operation and maintenance of implemented remedial actions for the expected life of such actions; the availability of storage, treatment, and disposal facilities for the management of contaminants; and the availability of institutional controls restricting the use of the property (40 CFR 300.510).

Annually, EPA and the state determine priorities, exchange information, discuss site-specific response actions, and make lead and support agency designations for the following fiscal year. A person primarily responsible for directing and coordinating response efforts at a site is called the on-scene coordinator (OSC), or remedial project manager (RPM). An OSC primarily deals with removals and an RPM deals with remedial actions.

If natural resources are affected by the release of chemicals, the lead agency shall ensure that state and federal trustees of the impacted resources are promptly notified (40 CFR 300.410(g)). Natural resources include those over which the United States has sovereign rights, and those resources belonging to, managed by, appertaining to, or held in trust by the United States (40 CFR 300.600(a)).

If the site evaluation indicates that removal is not necessary but remedial action is required, then a remedial site evaluation must be initiated (40 CFR 300.410(h)). Remedial actions are long-term, permanent cleanups to eliminate any threat that a site may pose. Only sites listed on the NPL qualify for remedial actions. Examples of remedial actions are excavations, installation of clay covers, and the neutralization or destruction of hazardous substances. Because remedial action is a complex, lengthy, and costly process, detailed and comprehensive requirements are established by CERCLA. Major components of this process are discussed in the following section.

3.1. Choosing Candidate Sites

The NPL, which is part of the NCP and is defined in 40 CFR 300.5, prioritizes CERCLA cleanup sites based on the levels of risk to public health, welfare, or the environment. The list, called the CERCLA Information System (CERCLIS) must be updated annually. The priority listing of CERCLA sites takes into consideration a variety of factors including the extent of population at risk, the hazard potential, the potential for drinking water supply contamination, and the threat to ambient air because of the existence of a site.

This process starts with listing the sites with releases or threatened releases in the CERCLIS. This database represents the official national inventory of CERCLA sites. The list is compiled by EPA from various information sources such as CERCLA reporting, submissions from states, and citizen complaints. Once compiled, EPA then performs a remedial preliminary assessment (PA) and site inspection (SI). The description of the methods, procedures, and criteria for implementing a PA/SI is contained in 40 CFR 300.420.

A PA must be performed by the lead agency on all sites listed in CERCLIS. The objectives of a PA are

1. to eliminate sites that pose no threat to human health or the environment from further consideration
2. to determine if there is a need for removal action
3. to collect available data for facilitating the hazard ranking system (HRS) evaluation process
4. to set priorities for remedial site inspection

A remedial PA consists of a review of available information on the pathways of exposure of the released chemical, exposure targets (human population and the environment), sources of exposure, and the nature of the release. On-site and off-site reconnaissance may be conducted as appropriate. After the completion of the PA, the findings must be reported. The PA report form is available from EPA offices and consists of a description of the extent and nature of the release, and a recommendation of appropriate future action.

If the PA indicates that a site presents a threat and will probably score high enough to be listed on the NPL, EPA will undertake additional site inspections to obtain more information about the hazardous substances at the site, the movement (or migration) of the contaminants and the potential exposure to humans and the environment. A remedial SI is conducted by the lead agency with the following objectives:

1. to eliminate sites that pose no significant threat to human health or the environment from further consideration
2. to determine the need for removal action, if any
3. to collect additional data for facilitating the HRS evaluation process
4. to collect data for better characterization of the site in the subsequent remedial investigation (RI) and feasibility study (FS) process

As part of the SI, on-site and off-site field investigatory efforts may be performed, including sampling activities. If sampling is to be done, a sampling and analysis plan that includes the information listed below must be prepared (40 CFR 300.420(c)(4)):

- a field sampling plan which describes the number, type, and location of samples and the type of laboratory analyses required
- a quality assurance project plan (QAPP) which describes the policy, organization, and functional activities, and the data quality objectives (DQO) and measures necessary to achieve adequate data for use in site evaluation and HRS activities

Once completed, the SI findings must be reported. The report consists of a description of the waste-handling history, the known contaminants, the pathways of migration of the contaminants, the human and environmental targets, and the recommendations for action. All of the information collected in the PA/SI is used to score the site in accordance with the NPL's HRS.

The NPL is the list of priorities for remedial action; in other words, only the sites included in the NPL will be considered eligible for Superfund remedial action. Removal action and remedial planning activities are not limited to NPL sites (40 CFR 300.425(b)). Inclusion of a site on the NPL does not imply that Superfund funds will be expended on the site nor does the rank of a site in the NPL establish the priorities for Superfund funds allocation.

A site can be included in the NPL if it meets one of the following criteria (40 CFR 300.425(c)):

1. The site scores significantly high in the HRS. The HRS procedure is elaborated in Appendix A of 40 CFR Part 300.
2. A state has designated the site as its highest priority. Only one such designation is allowed from a state. This one-time designation was completed by the states in the early 1980s.
3. The site satisfies all of these requirements:
   - The ATSDR has issued a health advisory that recommends dissociation of individuals from the site.
   - EPA determines that the site poses a significant threat to public health.
   - EPA decides to utilize remedial action instead of removal action.

When a site is qualified to be placed on the NPL, the lead agency submits the candidate site to EPA anytime throughout the year. EPA then reviews the submission and makes necessary revisions. To facilitate public involvement, when a site is going to be added to the NPL, EPA will publish the proposed decision in the FR and solicit comments on the action (40 CFR 300.425(d)). The update of the NPL is conducted annually.

Once a site has been remediated and no further response is necessary, the site will be deleted from the NPL. EPA shall consult with the respective state on the proposed deletion prior to establishing a notice of intent to delete in the FR. Whenever there is a significant release from a site deleted from the NPL, the site will be restored to the NPL without requiring the HRS evaluation process (40 CFR 300.425(e)).

The HRS utilizes a variety of scoring factors such as the toxicity of the substances; the location of the population at risk; the exposure pathways; and the threats to the human food chain, ambient air, surface water, and groundwater. The scoring system runs from 1 to 100 with a 100 score representing the worst site. A site will be listed on the NPL if it scores above 28.5.

3.2. Selecting Appropriate Remedial Technology

The selection of remedial technologies is determined by the findings of the RI/FS, which assesses the conditions of a contaminated site and examines various alternatives to remedy the site. The overall process is intended to select remedies that eliminate, reduce, or control risks to human health and the environment, maintain protection over time, and minimize untreated waste.

In reviewing alternative remedies, the following preferences must be taken into consideration (40 CFR 300.430(a)(1)):

- to use treatment technology to address the principal threats posed by a site; the principal threats include liquids, highly mobile materials, and areas contaminated with high concentrations of toxic compounds
- to use engineering controls (e.g., containment) for contaminants that pose a relatively low long-term threat, or where treatment is impractical
- to use a combination of treatment and engineering controls as appropriate
- to use institutional controls (e.g., water use restriction, deed restriction) to supplement engineering controls as appropriate
- to use innovative technology when such technology offers the potential for comparable or superior treatment performance or implementability
- to return usable groundwaters to their beneficial uses wherever practical

The overall RI/FS process includes project scoping, data collection, risk assessment, treatability studies, and analysis of remediation alternatives (40 CFR 300.430(a)(2)). Described in 40 CFR 300.430(b), scoping a project refers to the following activities:

- Assemble and evaluate existing data on the site from the PA/SI findings, NPL listing process, and any removal actions data.
- Develop a conceptual understanding of the site based on the evaluation of the existing data.
- Identify likely response scenarios including potential technologies and operable unit designations.
- Identify the need for treatability studies.
- Identify the type, quality, and quantity of the data that will be collected during the RI/FS.
- Prepare health and safety plans that specify employee training and protective equipment, medical surveillance requirements, standard operating procedures, and a contingency plan.
- Develop a sampling and analysis plan.
- Initiate the identification of applicable or relevant and appropriate requirements (ARARs).

CERCLA defines applicable requirements as cleanup criteria and requirements under environmental laws that specifically address a hazardous substance, pollutant, remedial action, or other circumstance found at a CERCLA site. Relevant and appropriate requirements refer to cleanup criteria and requirements under environmental laws that, while not applicable to a hazardous substance, pollutant, remedial action, or other circumstance found at a CERCLA, address problems or situations similar to those encountered at the CERCLA site and their use is well suited to the particular site.

Prior to commencing field work for the RI, community relation activities must be thoroughly considered. These activities, as specified in 40 CFR 300.430(c), are

1. to conduct interviews with local officials, community residents, and interested parties in order to solicit their concerns and information needs
2. to prepare a formal Community Relation Plan (CRP)
3. to establish a local information repository that houses copies of documents for the public to review
4. to inform the community of the availability of a technical assistance grant.

Once all the preparatory steps are completed, the RI can be initiated. The purpose of the RI is to evaluate the nature and extent of releases of hazardous substances and to determine those areas of a site where releases have created damage or the threat of damage to public health or the environment. In essence, the RI collects data necessary to adequately characterize a site for the purpose of establishing effective remedial alternatives (40 CFR 300.430(d)). This is done by conducting field investigations, treatability studies and a baseline risk assessment.

The treatability studies provide additional data for the detailed analysis and support engineering design of remedial alternatives. The baseline risk assessment characterizes the threats of the site to human health or the environment from contaminants migrating to groundwater, leaching through soil, releasing to air, or bioaccumulating in the food chain, etc. The result of the baseline risk assessment will help establish acceptable exposure levels for use in developing remedial alternatives in the FS.

Depending on the complexity of a site, site characterization may be conducted in phases. Because estimates of exposures and associated impacts on human and environmental receptors can be refined throughout the phases of RI as new information is obtained, site characterization activities are generally integrated with the evaluation of alternatives in the FS. Factors to be evaluated in characterizing a site are (40 CFR 300.430(d)92)):

1. physical characteristics of the site (e.g., surface features, soils, geology, hydrogeology, meteorology)
2. classifications of groundwater, surface water, and air
3. characteristics of the waste (e.g., quantities, toxicity, concentration, persistence, mobility, propensity to bioaccumulate)
4. extent to which the source can be adequately identified and characterized
5. exposure pathways through environmental media
6. exposure routes (e.g., ingestion, inhalation)
7. other factors (e.g., the existence of sensitive population subgroups in the area)

The RI requires extensive soil and groundwater sampling and voluminous reports describing the detailed remedial investigations. These investigations examine the nature of the site's geology and hydrogeology; identify the sources, types and mobility of contamination; and determine the nature of the threat to public health or the environment.

The conclusions of the RI provide valuable documentation and information on the sources of contamination, the nature and extent of contamination, and the actual and potential exposure pathways. The RI finding provides the lead agency with the basis for designing an FS of the site, the purpose of which is to develop a range of remedial action alternatives to address the contaminated site. Of the options evaluated, the lead agency will select the best alternative as the remedy of the site.

The entire RI/FS process can take years to complete. It is time consuming and very costly. For efficiency, the RI/FS recommendations may include phasing the cleanup effort by dividing the site into operable units. The two major reasons for using operable units are

- when early actions are necessary to achieve risk reduction quickly
- given the size or complexity of a site, when phased analysis and response is necessary to expedite the completion of overall site cleanup

An operable unit means a discrete action that comprises an incremental step toward comprehensively addressing site problems (40 CFR 300.5). When designed, operable units should be consistent with the expected final remedy. Operable units may address geographic portions of a site, specific site problems, initial phases of an action, a set of actions performed over time, or a set of actions that are concurrent but located in different parts of a site. For example, a site may consist of an operable unit that will address the remediation of groundwater at the site and another operable unit that is designed to clean up and isolate the sources of contamination. By using this approach it is possible to remediate certain portions of the site, while other portions are undergoing further study and evaluation.

The FS is intended to ensure that appropriate remedial alternatives are evaluated and presented to the decision maker so that an appropriate remedy can be selected (40 CFR 300.430(e)). An FS can be developed to address a specific site problem or the entire site. The number and type of alternatives to be analyzed is determined at each site based on the scope, characteristics, and complexity of the site problem.

Steps in developing and screening remedial alternatives are as follows:

- establish remedial action objectives by specifying contaminants and media of concern, exposure pathways, and remediation goals; remediation goals, expressed in terms of acceptable exposure levels, should be modified as necessary as more information becomes available during the RI/FS process
- identify and evaluate potentially suitable technologies
- assemble suitable technologies into alternative remedial actions

Remediation goals must be protective of human health and the environment. They must be developed by considering the following factors (40 CFR 300.430(e)(2)):

1. ARARs, if available. In addition, consider uncertainties, technical limitations, and other limiting factors. For systemic toxicants, the goals must represent concentration levels to which the human population may be exposed without adverse effects during a lifetime or part of a lifetime. For carcinogens, the goals are concentrations that, in statistical terms, represent an excess upper-bound lifetime cancer risk to individuals of between 10^{-4} and 10^{-6}.
2. For groundwaters or surface waters that are sources of drinking water, the maximum contaminant level goals (MCLGs) established

under the Safe Drinking Water Act shall be attained as appropriate. If the MCLG is determined not to be relevant and appropriate, the corresponding maximum contaminant level (MCL) shall be attained.

3. Where the MCLG for a contaminant has been set at a level of zero, the MCL shall be attained.
4. In cases involving multiple contaminants or pathways, the 10^{-6} risk level shall be used as the point of departure for determining remedial goals.
5. Where relevant and appropriate, the water quality criteria established under Sections 303 and 304 of the Clean Water Act shall be attained.
6. Where appropriate, an alternative concentration limit (ACL) may be established.
7. For a site with potential threats to the environment, especially sensitive habitats, environmental evaluations shall be performed to identify suitable technologies for alternative remedial actions.

To the extent sufficient information is available, the screening of remedial alternatives should be based on effectiveness, implementability, and cost. Effectiveness refers to the degree to which the technology reduces toxicity, mobility, or volume. Implementability means technical feasibility and availability. Cost factor covers construction costs and any long-term costs to operate and maintain the technology.

Once the screening of alternatives is completed, a detailed analysis is conducted on a limited number of alternatives that represent viable options to remediation. The detailed analysis consists of evaluating the individual alternatives against each of the evaluation criteria. The nine criteria, specified in 40 CFR 300.430(e)(9), are

1. overall protection of human health and the environment
2. compliance with ARARs
3. long-term effectiveness and permanence
4. reduction of toxicity, mobility, or volume through treatment
5. short-term effectiveness
6. implementability
7. cost
8. State acceptance
9. community acceptance

The output of the RI/FS process is a list of remedies that are suitable for a site. Generally, one of the most critical issues in remedying any CERCLA site is the level of cleanup that must be achieved before the site can be considered clean. The preceding paragraphs have described a set of remedial goals that must be attained for various circumstances. The following section addresses the procedure in the selection of remedies that will ensure the achievement of the remedial goals.

3.3. Designating the Preferred Technology

The significant issues regarding the cleanup standard in the CERCLA program are the required level of groundwater remediation and the level of residual soil contamination that is allowed to remain at the site. The decision on these two issues determines the type of remedial technology that can be used, and the extent of required excavation at the contaminated site. Consequently, this decision will affect the remedial cost; therefore, all parties with a stake in the CERCLA site have serious concerns about how cleanup standards are determined.

CERCLA provides a solid preference for remedies that are permanent and involve the treatment of hazardous substances to reduce their volume, toxicity, and mobility. Remedies that are designed to contain contamination within a site, as well as the off-site transport and disposal of untreated hazardous substances, are not favored.

The statute requires that a selected remedy achieves all ARARs. The application of ARARs at a CERCLA site implies that the selected remedies must achieve the highest cleanup level, which in practice is extremely conservative and costly. CERCLA does not specify what ARARs apply to a specific site. The selection of ARARs represents the most important part of the remedial investigation and feasibility study process.

The NCP provides more detailed criteria for selecting remedies and applying ARARs (40 CFR 300.430(f)). The first set of criteria is the threshold criteria which requires the remedy to provide both overall protection of human health and the environment, and to comply with ARARs. If a proposed remedy fails to meet both criteria, it will not be selected.

If the proposed remedy meets the threshold criteria, it will be further screened by the primary balancing criteria:

1. long-term effectiveness and permanence
2. reduction (through treatment) of volume, toxicity, and mobility
3. short-term effectiveness
4. implementability
5. cost

Balancing the first three criteria with the fifth measure provides the planner with a mechanism to check if a proposed remedy is in fact cost effective.

The last screening tool is the modifying criteria, which refer to State and community acceptance. Normally if a proposed remedy passes the threshold and primary balancing criteria, it also meets the modifying criteria.

In evaluating relevance and appropriateness, the NCP lists eight comparisons that must be examined to determine whether a remedy addresses problems or situations sufficiently similar to the circumstances of the release or proposed remedial action, and whether the requirement is well suited to the site and, therefore, is both relevant and appropriate. The comparison factors, specified in 40 CFR 300.400(g)(2), are

1. the purpose of the requirement and the purpose of the CERCLA action
2. the medium affected by the requirement and the medium contaminated at the CERCLA site
3. the substances regulated by the standard and the substances found at the site
4. the activities regulated by the requirement and the activities contemplated at the CERCLA site
5. any variances, waivers, or exemptions of the requirement and their applicability to the site
6. the type of place regulated and the type of place affected by the release or by the CERCLA action
7. the type and size of the facility regulated and the type and size of the facility affected by the release or contemplated by the CERCLA action
8. any consideration of the use of the affected resources in the requirement and the potential use of the affected resources at the CERCLA site

In the remedy selection process, the lead agency shall identify the preferred alternative and present that alternative to the public in a proposed plan (40 CFR 300.430.(f)(2)). The plan provides a brief description of the alternatives evaluated in the detailed analysis, a discussion of the rationale that supports the preferred alternative, a summary of the comments received from the support agency, and a summary of any proposed waiver from an ARAR. The proposed plan will supplement the RI/FS and provide the public with an opportunity to comment on the preferred alternative for remedial action.

After preparation of the proposed plan, the lead agency will publish a notice of availability of the proposed plan in a major local newspaper, make the plan available to the public, provide an opportunity for submission of comments on the plan, provide the opportunity for a public meeting, and prepare a written summary of comments. The responsiveness summary shall be made available with the record of decision.

Once the selection of ARARs is completed and the RI/FS is finalized, EPA issues a record of decision (ROD), which contains all of the facts, analyses, and policy determinations (40 CFR 300.430.(f)(5)). It explains how the selected remedy is protective of human health and the environment, how the ARARs will be achieved, why the remedy is cost effective and represents a permanent solution. It also describes whether hazardous substances or contaminants will remain at the site.

After the signing of a ROD, the lead agency shall publish a public notice in a major local newspaper and make the ROD available for public inspection. The ROD, therefore, must be placed in the site's administrative record. This record is critical because it will be used in any judicial proceedings.

The issuance of the ROD is followed by the completion of a remedial design (RD). Through RD, the selected remedy which is recorded in the ROD is translated into a detailed design for construction and operation of the selected remediation technique. The construction and operation phase is called the remedial action (RA) of a CERCLA site. All RD and RA activities must conform with the remedy selected and set forth in the ROD.

3.4. Issues in Implementation

Site remediation is currently being confronted with the intricate relationship between cost, timing, and risk. While the goal of site remediation is to expedite cleanup without ignoring the risk to public health and the environment, the elements of cost and time to implement remediation represent two major barriers to site remediation. Stated differently, Travis and Doty[13] assert that the two fundamental elements of the decision-making process are selecting cost-effective remedial alternatives that provide for effective and permanent cleanup, and deciding which sites to remediate.

In the early 1990s the average cost of remediation for a CERCLA site was estimated to be $25 million per site, which was up from $8.5 million two years before. The time required to clean up a CERCLA site can be as much as 10 to 15 years. Costs are rising rapidly, and sites with expenditures of $50 to $70 million are increasingly common.[14]

Statistically, more than 33,000 CERCLA sites have been identified as possible candidates for listing on the NPL; the candidacy of approximately 19,000 of these sites has been dropped after preliminary assessment.[15] Since 1980 more than 1,200 sites have been listed in the NPL and remedial action has been started at over 600 of these sites. Additionally, more than 2,600 remedial actions have been performed.[16]

The timing and cost problems are not necessarily particular to Superfund sites. Similar issues also apply to the remediation program under RCRA. The reasons for this type of problem are that each contaminated site is unique and many have an unusual combination of technical, regulatory, and politically sensitive issues.

Burke[17] argues that much of the delay in the remediation process can be attributed to the difficulties of decision making. The decision making represents an interface of science and values of those affected by remediation decisions. According to Burke, the interface is most evident in the areas of priority setting, establishing cleanup standards, assessing health risks, and communicating risks to policy makers and the general public.

Many analysts believe that more attention should be given to speeding basic cleanups and relatively less attention to fine-tuning the remedial alternatives or to achieving perfection in site restoration.[18] In this regard, efforts to streamline remedial planning and implementation are numerous.

Examples of such effort include the establishment of state Superfund programs which address CERCLA contaminated sites that fail to be included in the federal Superfund program and the introduction of the Superfund Accelerated Cleanup Model (SACM). SACM is an EPA pilot project developed in early 1990 to make the federal Superfund process simpler, more flexible, faster, cheaper, and quicker in reducing acute risks. The main SACM features are streamlining the site assessment process, blending the removal and remedial processes, one-stop sampling, using regional decision-making teams, and public involvement throughout the process.

As identified in Chapter 1, one of the major regulatory issues confronting site remediation is the fact that for some contaminated sites, both RCRA and CERCLA provisions can potentially be applied to the site. Specific discussion of the RCRA program is provided in Chapter 4. The following sections, however, provide a comparison of the Superfund and RCRA programs, particularly as they relate to site remediation.

4. RELATIONSHIP WITH HAZARDOUS WASTE PROGRAMS

RCRA and CERCLA are separate but interrelated programs. For example, when wastes from a CERCLA site are taken off-site for TSD, the receiving facility must be an RCRA TSD facility and the waste must be handled in accordance with applicable RCRA requirements, including the hazardous waste generator and transporter regulations.

The difference between the two programs is that RCRA handles hazardous waste which is narrowly and precisely defined in 40 CFR Part 261, while CERCLA deals with remediation caused by releases of hazardous substance. As discussed previously, the definition of a hazardous substance is extremely broad and covers most chemicals.

When first introduced, the focus of CERCLA was to clean up contamination at inactive or abandoned sites. Prior to the 1984 amendments of RCRA, known as the Hazardous and Solid Waste Amendments (HSWA), no other environmental laws were in place to address the cleanup of active hazardous waste sites. EPA's only authority for requiring RCRA facilities to conduct site cleanups in case of contamination was under CERCLA. With the signing of HSWA, EPA was given corrective action authority which requires RCRA facilities to perform cleanup activities ranging from quick-fix measures to full cleanups of their contaminated sites.

Under CERCLA, EPA has discretion in deciding the level of cleanup as long as the cleanups permanently and significantly reduce the volume, toxicity, or mobility of the hazardous substances. However, permanent cleanups are extraordinarily expensive and difficult to accomplish.[19] Regarding active RCRA sites, EPA's philosophy is that these sites are different from CERCLA sites in that they represent a controlled situation. This controlled situation designation implies that a facility can perform prompt cleanup to levels consistent with current use, but can defer final cleanup as long as the facility remains under a RCRA permit.[20]

Besides the fact that RCRA focuses its attention on current and future activities and CERCLA is principally retrospective, the major differences between RCRA and CERCLA are the role of regulations in implementing each statute and the time periods on which each statute focuses its primary attention.[21] First, CERCLA creates a program that establishes liabilities and obligations, and does not necessarily require promulgation of regulations for implementation. RCRA, on the other hand, is not self-implementing. Congress directed EPA to promulgate regulations that would control company activities. The primary exceptions are certain obligations created by the 1984 HSWA which do not require implementing regulations.

In a number of cases, a site may be subject to both CERCLA and RCRA requirements, e.g., when a RCRA site is contaminated with both an RCRA hazardous waste and a CERCLA hazardous substance. A Superfund site in Cincinnati, OH, for example, was remediated as both a CERCLA removal and a RCRA remediation (i.e., closure) activity. Walsh[22] found that complications in this case arose regarding the disposition of clean debris from the site, the treatment of contaminated media, the duration of cleanup activities, and the cleanup certification requirements. These factors are regulated differently by RCRA and CERCLA.

As an example Walsh[22] argues that removal actions under CERCLA can be completed in time frames ranging from days to years while the RCRA closure requirement is usually 180 days after the approval of the closure plan with the possibility of an extension. Cleanup certification requirements represent another example of regulatory differences. Under RCRA, the certification must be provided by a professional engineer (PE) stating that closure activities have been performed in accordance with an approved closure plan. Under CERCLA, removal and remedial action reports serve a similar purpose as the RCRA certification report but the CERCLA reports do not require PE approval.

In any situation where a site may be subject to CERCLA and RCRA requirements, the owner must focus his attention on the RCRA/CERCLA interface very early in the business deliberations and assessment of environmental responsibility, liability, and costs. As Robertson[23] observes, the regulations implementing these laws are still evolving and progressive corporate planners, risk managers, loss control specialists, and environmental engineers need to identify not only the health and safety aspects but also the administrative and societal requirements and expectations. Robertson cautions that even the most comprehensive program of risk management cannot ensure that all future exposures to liability will be eliminated or even identified.

4   SITE REMEDIATION UNDER RCRA

As described briefly in Chapter 1, the U.S. Congress in 1965 passed the Solid Waste Disposal Act (SWDA) which provided technical and financial assistance to states, interstate agencies, and local governments for the development of solid waste management plans. The Act also provided training grants for the design, operation, and maintenance of solid waste disposal systems. The enactment of the SWDA was the result of a pressing public concern over our inability to manage garbage, refuse, sludge, and other discarded materials sufficiently. Although the 1965 Act expressed concerns over hazardous waste management practices, it was passed with the primary purpose of improving solid waste disposal practices.

The SWDA describes hazardous waste as a solid waste or combination of solid wastes, which, because of its amount, concentration, physical, chemical, or infectious characteristics, may contribute to adverse health effects; and when improperly treated, stored, transported, disposed of, or otherwise managed, it may pose hazards to human health and the environment. Physically, solid waste may be either a solid, a liquid, or a contained gas.

Amendments to the SWDA raised solid waste management to a higher level of sophistication. The amendments are the Resource Recovery Act (RRA) of 1970 and the Resource Conservation and Recovery Act (RCRA) of 1976. The latter amendment greatly expanded the provisions pertaining to the management of hazardous waste. Although the term RCRA is the acronym for the Resource Conservation and Recovery Act, it is often used to refer to the overall program resulting from the Act and its amendments.

RCRA has been amended twice since 1976: in 1980 and 1984. The 1984 amendments, called the Hazardous and Solid Waste Amendments (HSWA), expand both the scope and detail of the requirements of the RCRA. As Case[24] observes, the HSWA has added many new requirements which represent a challenge to the aspiration and imagination of the regulated community. These will include the use of innovative and emerging technologies to improve the treatment of hazardous waste and the changing of raw materials or modifications to production processes to prevent or reduce the generation of hazardous waste.

RCRA describes the kind of waste management program that Congress wants to establish. The goals set by RCRA are straightforward. They are

1. the protection of human health and the environment
2. the reduction of waste and the conservation of energy and natural resources
3. the reduction or elimination of the generation of hazardous waste as expeditiously as possible

Structurally, RCRA has ten subtitles, labeled A through J (Table 4.1). Of these ten, Subtitles C, D, and I outline the framework for the three major programs that make up RCRA:

*   Subtitle C of RCRA establishes a management system of hazardous waste from the time it is generated until its ultimate disposal. This management scheme is known as the cradle-to-grave approach.
*   Subtitle D encourages states to develop comprehensive plans for the management of solid waste.
*   Subtitle I describes the management program for underground storage tanks (USTs).

Table 4.1 List of RCRA Subtitles

Subtitle A	General provisions
Subtitle B	Office of Solid Waste; authorities of the administrator
Subtitle C	Hazardous waste management
Subtitle D	State or regional solid waste plans
Subtitle E	Duties of the Secretary of Commerce in resource and recovery
Subtitle F	Federal responsibilities
Subtitle G	Miscellaneous provisions
Subtitle H	Research, development, demonstration, and information
Subtitle I	Regulation of underground storage tanks
Subtitle J	Demonstration program for medical waste tracking

Although hazardous waste is by definition part of solid waste, Subtitle D does not regulate hazardous waste. It specifically refers only to the nonhazardous portion of solid waste. Subtitle I was created by Congress under the HSWA to control and prevent releases from USTs storing petroleum and other hazardous substances, but it does not regulate tanks storing hazardous waste.[25] Such tanks are regulated under Subtitle C.

When RCRA is mentioned, most people think of Subtitle C. In other words, RCRA has always been associated with hazardous waste even though the association is not entirely accurate. When people mention solid waste, they allude to Subtitle D and, similarly, UST refers to Subtitle I of RCRA. Site remediation under Subtitles D and I is addressed exclusively in Chapter 5, while this chapter focuses on Subtitle C site remediation.

Structurally, this chapter consists of two parts. The first part outlines the provisions of the RCRA program that are critical to hazardous waste site remediation. The second part describes the two cleanup programs that fall under the RCRA program. Included in the second part is a brief discussion of land disposal restrictions (LDR) as well as the linkage of the LDR provisions to the recent regulatory developments in RCRA site remediation.

1. BASIC RCRA PROVISIONS

The objective of the RCRA program is to assure that hazardous waste is handled in a manner that protects human health and the environment. The program has resulted in perhaps the most comprehensive regulations EPA has ever developed.[26] RCRA regulations specify wastes that are hazardous and establish administrative requirements for the three categories of hazardous waste handlers: generators, transporters, and owners or operators of treatment, storage, and disposal (TSD) facilities.

These three groups have specific requirements in the handling and management of their hazardous waste. The groups also have distinctive responsibilities in addressing site contamination caused by the release of hazardous waste. This particular aspect will be examined later in this chapter. The next section, however, is specifically devoted to the discussion of the process for determining whether a waste is a hazardous waste. If a waste which contaminates a site is not a hazardous waste, the contaminated site is not subject to RCRA. This process is essential because the waste determination finding will affect the legal status of the contaminated site along with all of its ramifications.

1.1. Determination of Hazardous Waste

To be in the RCRA program, a waste must first meet the definition of a solid waste. If it is not a solid waste, it is also not a hazardous waste. Solid waste is defined in 40 CFR 261.2a(1) as any discarded material that is not qualified for either solid waste exclusions or for recycling exclusions. To clarify the solid waste definition, RCRA regulations define a discarded material in 40 CFR 261.2a(2) as any material that is abandoned (such as material being disposed of), recycled (such as material burned for energy recovery), or considered inherently waste-like (such as recycled reactant wastes of hexachlorobenzene under alkaline conditions).

RCRA specifies certain materials that are excluded from the definition of solid waste. The following wastes are not solid waste due to the exclusion from the definition of solid waste rule as found in 40 CFR 261.4(a):

1. domestic sewage and mixtures of domestic sewage and other wastes passing through a sewer system to a publicly owned treatment works (POTW)
2. industrial point source wastewater discharge requiring a National Pollutant Discharge Elimination System (NPDES) permit
3. irrigation return flow
4. radioactive materials defined under the Nuclear Regulatory Commission (NRC)
5. *in situ* mining waste not removed from the ground
6. pulping liquors reclaimed in a pulping liquor recovery furnace and then reused in the pulping process
7. spent sulfuric acid used to produce virgin sulfuric acid
8. secondary materials reclaimed and returned to the original process(es) in which they are generated under certain provisions (specified in 40 CFR 261.4(a))

A material not considered a solid waste cannot be a hazardous waste. If a material is a solid waste, it might be a hazardous waste. On the other hand, a hazardous waste might be considered nonhazardous based on the regulatory exclusion. 40 CFR 261.4(b) provides such exclusion. Under this provision, the following wastes are excluded from the definition of hazardous waste:

1. household hazardous wastes
2. solid wastes from agricultural crops and animals returned to the soil as fertilizer
3. mining overburden returned to the mine site
4. fly ash waste, bottom ash waste, slag waste, and flue gas emission control waste, generated primarily from the combustion of coal or other fossil fuels. This exclusion does not apply to a residue derived from the burning or processing of hazardous waste in a boiler or industrial furnace unless certain specifications are met
5. drilling fluids, produced waters, and other wastes associated with the exploration, development, or production of crude oil, natural gas, or geothermal energy
6. wastes containing chromium in the trivalent state and failing the toxicity test solely because of chromium (examples are given in 40 CFR 261.4(b)(6)(ii))
7. solid wastes from the extraction, beneficiation, and processing of ores and minerals (with examples and exceptions given in 40 CFR 261.4(b)(7))
8. cement kiln dust waste
9. discarded wood or wood products generated by users of arsenical-treated wood or wood products

Also excluded in the definition of solid waste are certain recycled materials that are specified in 40 CFR 260.30 and 260.31. The recycling of hazardous waste presents a unique situation concerning regulatory requirements. Historically, provided that a generator was going to legitimately recycle a hazardous waste, few regulatory controls were instituted. Because of recent widespread sham recycling, EPA decided to institute controls to prevent sham recycling while trying to encourage legitimate recycling. EPA concluded that by redefining solid waste, these goals could be met. The result has been one of the most complicated and confusing sets of regulations ever developed by EPA.[27]

Under 40 CFR 261.2, used or spent materials, sludges, and by-products uncovered in a facility are considered recyclable (secondary) materials. These recyclable materials are further classified as follows:

1. spent materials
2. sludges listed in 40 CFR 261.31 and 261.32
3. sludges exhibiting characteristics of hazardous waste
4. by-products listed in 40 CFR 261.31 and 261.32
5. by-products exhibiting characteristics of hazardous waste
6. commercial chemical products listed in 40 CFR 261.33
7. scrap metal

Sludges are residues from treating air, wastewater, or other residues from pollution control. By-products are residual materials resulting from industrial, commercial, mining, and agricultural operations that are not primary products, are not produced intentionally, and are not fit for a desired end use without further processing. Characteristics of hazardous waste are determined in terms of ignitability, corrosivity, reactivity, and toxicity.

By knowing the type of recyclable material and the recycling process to be used, 40 CFR 261.2 provides a matrix (Table 4.2) showing the relationship between the two elements. A recycling process or recycling activity can be identified by one of the following components:

1. Used in a manner constituting disposal. This practice involves placing wastes onto the land that contains hazardous wastes as an ingredient. A classic example is that certain wastes are incorporated into asphalt, then used in road construction.
2. Burning for energy recovery. This includes using wastes to produce a fuel.
3. Reclaimed. This refers to the recovery of usable products or regeneration of materials from the wastes.
4. Accumulated speculatively. Speculative accumulation occurs because certain wastes are potentially recyclable, but no feasible recycling market exists for them. It also involves recycling less than 75% of accumulated wastes during a one-year period.

Table 4.2 Classification of Secondary Materials

Secondary material	Use constituting disposal	Reclamation	Speculative accumulation	Burned as fuel
Spent materials	Y	Y	Y	Y
Sludges, listed	Y	Y	Y	Y
Sludges, characteristics	Y	No	Y	Y
By-products, listed	Y	Y	Y	Y
By-products, characteristics	Y	No	Y	Y
Commercial chemical products	Y	No	No	Y
Scrap metal	Y	Y	Y	Y

Note: Y means the material is a solid waste; No means the material is not a solid waste.

Of the 28 combinations available in the matrix, only four combinations (the "No" cells in Table 4.2) represent recycling exemptions. They are:

1. reclamation of sludges exhibiting characteristics of hazardous waste
2. reclamation of by-products exhibiting characteristics of hazardous waste
3. reclamation of commercial chemical products
4. accumulation of commercial chemical products. Failure to recycle at least 75% of these materials within one year may result in the revocation of this exemption

Materials described in the above exemptions are not considered solid waste and, therefore, they are not regulated as hazardous waste. Once a material is found to be a solid waste, the next question is whether it is a hazardous waste. If a solid waste does not qualify for an exemption, it will be deemed a hazardous waste when the waste either is listed by EPA in 40 CFR 261.30 through 261.33, or exhibits any of the four hazardous waste characteristics described in 40 CFR 261.20 through 261.24.

This determination process results in the designation of a hazardous waste as a listed or characteristic hazardous waste. A listed hazardous waste consists of the following groups:

1. Nonspecific source wastes (40 CFR 261.31) are commonly generated by manufacturing and industrial processes. An example from this list is spent halogenated solvents used in degreasing. EPA has coded this hazardous waste as F001. Another example is wastewater

treatment sludge from electroplating processes, which is coded as F006. EPA codes for nonspecific source wastes always start with "F"; therefore, nonspecific source wastes are commonly known as the F-wastes.

2. Specific source wastes (40 CFR 261.32) are generated from specifically identified industries such as wood preserving, petroleum refining, and organic chemical manufacturing. The wastes include sludges, still bottoms, wastewaters, spent catalysts, and residues. They are known as K-wastes. Examples are bottom-sediment sludge from the treatment of wastewaters from wood-preserving processes (K001) and oven residue from the production of chrome oxide green pigments (K008).

3. Discarded commercial chemical products (40 CFR 261.33(e) and 261.33(f)) consist of acute hazardous waste (P-wastes) and toxic hazardous waste (U-wastes). Examples are arsenic acid (P010) and xylene (U239).

Characteristic hazardous wastes are commonly called D-wastes. These wastes are determined either by the generator through general knowledge based on daily operation, or through laboratory testing. D001 represents ignitable waste (40 CFR 261.21). A solid waste that exhibits any of the following properties is considered a hazardous waste due to its ignitability:

- a liquid, other than aqueous solutions containing less than 24% alcohol, that has a flashpoint less than 60°C or 140°F
- a nonliquid that, under standard temperature and pressure, is capable of spontaneous and sustained combustion
- an ignitable compressed gas, specified by the Department of Transportation (DOT) under 49 CFR 173.300
- an oxidizer per DOT regulations in 49 CFR 173.51

Under 40 CFR 261.22, EPA specifies the criteria for a corrosive waste. D002 is used for coding such waste. A solid waste that exhibits any of the following properties is considered a hazardous waste due to its corrosivity:

- an aqueous material with a pH of less than or equal to 2, or greater than or equal to 12.5
- a liquid that corrodes steel at a rate greater than 0.25 in or 6.35 mm per year at a temperature of 55°C or 130°F

D003 is a reactive waste under 40 CFR 261.23. A solid waste that exhibits any of the following properties is considered a hazardous waste due to its reactivity:

- normally unstable and readily undergoes violent change without detonating
- reacts violently with water
- forms an explosive mixture with water
- generates toxic gases, vapors, or fumes when mixed with water
- contains cyanide or sulfide and generates toxic gases, vapors, or fumes at a pH between 2 and 12.5
- capable of detonation if heated under confinement or if subjected to a strong initiating source
- capable of detonation at standard temperature and pressure
- listed by the DOT as explosive in 49 CFR 51, 53, and 88

A solid waste that exhibits the characteristics of toxicity is coded by 40 CFR 261 as D004 through D043. The waste may have metal constituents (D004 through D011), insecticides (D012 through D015), herbicides (D016, D017), or organics (D018 through D043). The toxicity characteristics are tested using the Toxicity Characteristics Leaching Procedure (TCLP) test, which is designed to identify wastes likely to leach hazardous concentrations of particular toxic constituents into surrounding soils or groundwater as a result of improper management of hazardous waste.

The procedure extracts constituents from the tested waste to simulate the leaching actions that occur in landfills. The extract is then analyzed to determine if it possesses any of the TCLP contaminants specified in 40 CFR 261.24(b). If the concentrations of the toxic constituents exceed the levels listed in 40 CFR 262.24(b), the waste is classified as hazardous. The list of EPA waste codes, TCLP contaminants, and the maximum concentration of contaminants (in milligram/liter) of each contaminant is provided in Table 4.3.

In addition to the listed wastes and characteristics wastes, EPA also defines certain other wastes based on what are called the mixture rule and derived-from rule. Under the mixture rule, a mixture of a listed waste and a solid waste must be considered a hazardous waste unless the mixture qualifies for an exemption (40 CFR 261.3(a)(2)(iv)). If a characteristic waste is mixed with a solid waste, the resulting mixture is only hazardous if it continues to exhibit the characteristic (40 CFR 261.3(b)(3)). Under the derived-from rule, a waste that is generated from the treatment, storage, or disposal of a listed hazardous waste is also a hazardous waste (40 CFR 261.3(c)(2)(i)); and if it is generated from a characteristic waste, the waste is hazardous if it exhibits a characteristic (40 CFR 261.3(d)(i)).

With regard to P- and U-wastes, RCRA regulations state that spill residues of these chemicals are hazardous waste (40 CFR 261.33(d)). Under certain circumstances, soil contaminated with pesticides (grouped in P- or U-wastes) would not be a RCRA listed waste. If the pesticide was applied to land as a function of its normal use, it is not considered to be a P- or U-waste. If the pesticide was spilled as an unused commercial chemical product, then the soil would be classified as hazardous waste.[28]

**Table 4.3 Maximum Concentration of Contaminants
for the Toxicity Characteristics (40 CFR 261.24)**

Waste code	Contaminant	Regulatory level (mg/l)
D004	Arsenic	5.0
D005	Barium	100.0
D018	Benzene	0.5
D006	Cadmium	1.0
D019	Carbon tetrachloride	0.5
D020	Chlordane	0.03
D021	Chlorobenzene	100.0
D022	Chloroform	6.0
D007	Chromium	5.0
D023	*o*-Cresol	200.0[a]
D024	*m*-Cresol	200.0[a]
D025	*p*-Cresol	200.0[a]
D026	Cresol	200.0[a]
D016	2,4-D	10.0
D027	1,4-Dichlorobenzene	7.5
D028	1,2-Dichloroethane	0.5
D029	1,1-Dichloroethylene	0.7
D030	2,4-Dinitrotoluene	0.13[b]
D012	Endrin	0.02
D031	Heptachlor (and its epoxide)	0.008
D032	Hexachlorobenzene	0.13[b]
D033	Hexachlorobutadiene	0.5
D034	Hexachloroethane	3.0
D008	Lead	5.0
D013	Lindane	0.4
D009	Mercury	0.2
D014	Methoxychlor	10.0
D035	Methyl ethyl ketone	200.0
D036	Nitrobenzene	2.0
D037	Pentachlorophenol	100.0
D038	Pyridine	5.0[b]
D010	Selenium	1.0
D011	Silver	5.0
D039	Tetrachloroethylene	0.7
D015	Toxaphene	0.5
D040	Trichloroethylene	0.5
D041	2,4,5-Trichlorophenol	400.0
D042	2,4,6-Trichlorophenol	2.0
D017	2,4,5-TP (Silvex)	1.0
D043	Vinyl chloride	0.2

[a] If *o*-, *m*-, and *p*-cresol concentrations cannot be differentiated, the total cresol (D026) concentration is used. The regulatory level of total cresol is 200 mg/l.

[b] The quantification limit is greater than the calculated regulatory level. The quantification limit therefore becomes the regulatory level.

Soil contaminated with hazardous waste poses a substantial problem to regulators because it does not fit easily into the RCRA regulatory scheme; it also causes problems for hazardous waste handlers because it is difficult to manage and dispose of properly.[29] Under RCRA, the initial step in managing contaminated soil is to determine whether the soil is considered a hazardous waste. The waste determination process that EPA developed, however, did not work very well for soil. It became evident that the mixture rule and the derived-from rule were not applicable to contaminated media such as soil, groundwater, surface water, and debris.

According to RCRA regulations, if leachate from a hazardous waste landfill contaminates underlying soil and groundwater, the soil and ground-water have not been discarded and, therefore, are not solid wastes; conse-quently, the mixture rule and the derived-from rule cannot apply. To deal with this situation, EPA argues that because contaminated media may contain listed wastes, they must be managed as hazardous wastes until they no longer contain the listed waste. This contained-in policy states that contaminated media remain hazardous wastes until the listed constituents are removed by treatment.

Under the RCRA program, if a generator determines that the waste gen-erated is hazardous, the company must comply with RCRA regulations. As an initial step, the generator must file a notification with EPA using Form 8700-12. Upon receipt of the form, EPA will assign an identification number to the generator. The notification requirement is applicable to hazardous waste generators, as well as to transporters and the owners or operators of TSD facilities.

Having an EPA ID number is, in fact, mandatory for all hazardous waste handlers; without it, the handler is barred from storing, transporting, or offering for transportation, treating, and disposing of any hazardous waste. Also, a generator is forbidden from offering hazardous waste to any transporter or TSD facility that does not have an EPA ID number. Notification is not, how-ever, the only requirement for hazardous waste handlers; they still are subject to other RCRA regulations. These other requirements, as they relate to site remediation, are described below.

1.2.   Requirements for Hazardous Waste Handlers

The RCRA program creates the cradle-to-grave chain of hazardous waste management. Generators of hazardous waste are the first link in the chain, followed by the transporters; TSDs are the last link. In general, the generator has the responsibility of ensuring the proper storage and handling of hazardous waste. A generator is defined by 40 CFR 260.10 as any person, by site, whose acts or processes produces hazardous waste identified or listed in 40 CFR Part 261, or whose acts first cause a hazardous waste to become subject to regu-lation.

A generator may accumulate hazardous waste on-site without a hazardous waste storage permit for 90 days or less. The 90-day period provision allows a generator to collect enough waste for one big shipment to make transportation more cost effective. In addition to proper storage practices, the generator must ensure that facility personnel have adequate training in the proper handling of hazardous waste, and that the generator has developed a written contingency plan addressing emergency response situations.

To transport hazardous waste off-site, the generator must ensure proper packaging and proper labeling, marking, and placarding of the packaged waste. The purpose of proper packaging is to prevent leakage of hazardous waste during the normal course of transportation and during unexpected situations such as a traffic accident or a waste-handling mishap (e.g., a hazardous waste drum dropped from a truck bed). Labeling, marking, and placarding of the packaged waste is done to inform the shipping crew, firefighters, and emergency responders about the characteristics and dangers associated with the waste being transported. This information is vital in an emergency situation.

The generator must prepare a shipping document called the Uniform Hazardous Waste Manifest (EPA Form 8700-22), which accompanies the hazardous waste at all times. The generator specifies on the manifest the name and EPA ID numbers of each transporter, the designated TSD facility that will receive the waste, the type of waste, the amount, certification that it is properly packaged and labeled, and his handwritten signature on the multi-copy manifest certification. Sufficient copies of the manifest must be prepared so that all handlers listed on it will be provided a copy. It is the generator's responsibility to trace the whereabouts of the shipment by tracking the flow of the manifest.

Generators are subject to extensive record keeping and reporting requirements. By March 1 of even-numbered years for the preceding year, a generator must file biennial reports with EPA or a state environmental agency using EPA Form 8700-13A. Some states may require annual reports instead of biennial reports. The report identifies each transporter used and each TSD facility to which wastes were shipped, and it describes the type and amount of each type of waste handled by the generator.

The above requirements apply to generators producing at least 2,200 lb of hazardous waste, or 2.2 lb of acutely hazardous waste, in any month of a calendar year. This type of generator is known as a large quantity generator (LQG). A small quantity generator (SQG) generates between 220 and 2,200 lb of hazardous waste, or less than 2.2 lb of acutely hazardous waste, in any calendar month. An SQG is not required to comply with the biennial reporting requirement and may store hazardous waste on-site for up to 180 days. A generator that produces less than 220 lb of hazardous waste in any month is called a conditionally exempt small quantity generator (CESQG) and is only subject to certain minimum standards.

Anyone who moves a hazardous waste off-site from where it was generated is subject to the hazardous waste transporter requirements. This means that the provision is not applicable to generators who engage in on-site transportation of their hazardous waste. A TSD facility may become a generator if the facility also produces hazardous waste as a result of a hazardous waste treatment process within the facility. This TSD is subject to the generator requirements in addition to applicable TSD standards. It is not uncommon to find a TSD facility that is a generator and a transporter of hazardous waste.

A transporter can only accept hazardous waste that is accompanied by a manifest signed by the generator. The transporter is responsible for the shipment until the cargo arrives at the TSD facility. If the waste cannot be delivered in accordance with the manifest, the transporter must contact the generator for further instruction and revise the manifest according to the generator's instructions.

The transporter may hold a hazardous waste in containers for up to 10 days at a hazardous waste transfer facility without being required to obtain a RCRA storage permit. Transfer facility is defined by 40 CFR 260.10 as including loading docks, parking areas, and storage areas where shipments are held during the normal course of transportation. If an accidental spill or leakage occurs during transportation, the transporter must take immediate response action to protect human health and the environment, such as containment of the spill and notification of local police and fire departments.

Because of the potential liabilities from accidental spills, leaks, improper handling, and illegal disposal during transportation, generators prefer to maintain control of their wastes by transporting them in company-owned trucks. Alternatively, Wentz observes, generators feel more confident that their wastes will reach the designated TSD if they are picked up and transported by the TSD's trucking operation.

As far as a TSD facility is concerned, the RCRA program sets both the technical standards for the design and safe operation of TSD facilities, and the requirements for issuing the permits that each TSD facility is required to have. The technical standards and the permitting process are interrelated in that the permitting process enforces the application of technical standards to TSD facilities.

40 CFR 260.10 describes a TSD facility as follows:

- A facility is regulated as a treatment facility if it utilizes any method or process designed to change the physical, chemical, or biological character or composition of hazardous waste so as to neutralize such waste, to recover material or energy from the waste, or to render the waste nonhazardous or less hazardous, safer to handle, or reduced in volume.
- A storage facility is a facility that engages in the holding of hazardous waste for a temporary period; at the end of that period, the waste is treated, disposed of, or stored elsewhere.
- A disposal facility is a facility where hazardous waste is intentionally placed into water or on land, where waste will remain after closure of the facility.

Separate requirements have been issued for permitted TSD facilities (40 CFR Part 264) and interim status TSD (40 CFR Part 265). Both Parts 264 and 265 include specific provisions for hazardous waste management units (HWMUs) and general requirements; examples of hazardous waste management units are a landfill, surface impoundment, incinerator, tank, and container.

- An interim status facility is allowed to continue its operation without permit. The facility must have been in existence prior to November 19, 1980, notified EPA of its operation, and filed a permit application.
- Permitted facilities include new TSD facilities and existing facilities that did not qualify for interim status, both of which must first obtain RCRA permits to begin or continue operations.

General requirements are similar for both permitted and interim status TSDs. For example, a TSD must conduct a detailed chemical and physical analysis of a representative sample of a hazardous waste before the waste is managed, install a security system, implement an employee training plan, have specific emergency preparedness equipment, have an adequate contingency plan, and be covered by liability insurance. Similar to the generator and transporter provisions, TSD facilities are also subject to manifest and biennial reporting requirements.

All TSD facilities are subject to the closure requirements specified in 40 CFR 264.110 through 264.116 and 265.110 through 265.116. These requirements are intended to cover a period when a TSD facility must cease treatment, storage, or disposal operations. The facility must address this situation by having a detailed written closure plan, an implementation schedule, and a cost estimate for closure.

Specifically for land disposal facilities (i.e., landfills, surface impoundments, waste piles, and land treatment facilities), they also are subject to the post-closure requirements specified in 40 CFR 264.117 through 264.120 and 265.117 through 265.120. Post-closure is the 30-year period after closure when facilities must perform certain monitoring and maintenance activities including groundwater monitoring, maintenance of the monitoring system, and maintenance of the waste-containment systems. Financial responsibility requirements have been established by RCRA regulations to ensure that any TSD facility has adequate and available funds for closure and post-closure care.

In addition to those general requirements, EPA has also promulgated specific design, construction, and operating standards for each hazardous waste management unit of a TSD facility. The specific standards for each unit are coded under the different subparts of 40 CFR Parts 264 and 265. Subpart I, for example, contains standards for containers, Subpart J for tanks, Subpart K for surface impoundments, Subpart L for waste piles, etc.

RCRA requires every TSD facility to obtain a permit. The permit application is divided into two parts, A and B.

- Part A is a shorter form that collects general information about a facility, such as name, location, nature of business, regulated activities, and a topographic map of the facility.
- Part B is much more extensive than Part A in that it requires the facility to supply detailed and highly technical information based on standards specified in 40 CFR Parts 264 and 270.

A new facility must submit Part A and Part B simultaneously. Interim status facilities submitted Part A when they applied for their interim status and must submit Part B applications in accordance with deadlines established by EPA or the state. Failure to meet a deadline may result in the facility losing its interim status and, therefore, the facility must close.

The requirements for a permitted TSD and an interim status TSD facility differ in a number of ways. A set of stricter standards to prevent the release of hazardous waste are applicable to a permitted TSD. When groundwater contamination is detected, or when releases of hazardous waste are found from a solid waste management unit (defined as a unit where any solid waste was placed), corrective actions must be taken.

In this case, corrective action requirements for a permitted facility (40 CFR 264.100) are broader than those for an interim status facility (40 CFR 265.93). However, EPA still maintains legal authority to order interim status facilities to undertake broader corrective actions outside the permitting status.[30] As specified in Section 3004(u) of RCRA, the corrective action provision applies similarly to permitted facilities as well as those facilities operating under interim status.[31]

The RCRA cleanup requirements are not only directed to the TSD facilities. All categories of hazardous waste handlers are essentially subject to certain cleanup provisions. A generator, transporter, or owner or operator of a TSD facility who releases hazardous waste into the environment is required to perform necessary action to clean up the site. The requirements are specified in the federal hazardous waste law and regulations, and reviewed in the following paragraphs.

2. RCRA CLEANUP PROGRAM

The cleanup program under RCRA can be examined by looking at the party that contributes to the contamination of a site. As stated in the preceding paragraphs, the transporter is responsible for the shipment of hazardous waste until the cargo arrives at the TSD facility. Within this responsibility, 40 CFR 263.30 states that if an accidental spill or leakage occurs during shipment, the transporter is responsible for its cleanup. The transporter must take immediate response action to protect human health and the environment by containing the spill and notifying local police and fire departments.

The generators are responsible for ensuring the proper storage and handling of their hazardous waste, including the responsibility for cleaning up the contamination, if it occurs. Section 3005 of RCRA indicates that the treatment, storage, and disposal of hazardous waste is prohibited, except in a permitted facility. Within this context, site contamination can be interpreted as an act of disposal because Section 1004 of RCRA defines the term disposal as including the discharge, deposit, injection, dumping, spilling, leaking, or placing of any solid waste or hazardous waste into or on any land or water so that such solid waste or hazardous waste or any constituent thereof may enter the environment or be emitted into the air or discharged into any waters, including groundwaters.

As described in the previous discussion, when a TSD facility must cease treatment, storage, and disposal operations, it must have a closure plan and implement it. Owners and operators of selected TSDs also are responsible for post-closure care. Closure, post-closure, and corrective action are important RCRA cleanup requirements that warrant further discussion. These topics are addressed in the next section. After the discussion of the TSD cleanup provisions, the hazardous waste generator cleanup program will subsequently be discussed in order to get a better understanding of the spectrum of site remediation under the RCRA program.

2.1. CORRECTIVE ACTION, CLOSURE, AND POST-CLOSURE

As originally written, the RCRA program was intended primarily to regulate current hazardous waste storage, treatment, and disposal practices. It did not provide EPA with enough authority to require TSD facilities to remediate contamination resulting from the TSD's pre-RCRA operations. EPA had to rely on its enforcement authority under Section 7003 of RCRA which requires TSD facilities to clean up hazardous waste releases resulting from facility operations. However, this authority could be used only when EPA could demonstrate that the release presented a threat of imminent and substantial endangerment to human health or the environment. Additionally, under Section 3013 of RCRA, EPA could require facilities to conduct investigations when the releases or the presence of hazardous waste might present a substantial hazard to human health or the environment.

When RCRA was amended in 1984 with the HSWA, the statute not only gave EPA authority, but under Section 3004(u), HSWA required the agency to force TSD facilities to undertake corrective action for contamination from solid waste management units (SWMU) at the facility regardless of when the wastes were disposed. SWMU is any area at a TSD facility at which hazardous wastes have been routinely or systematically released to the environment. If an area at a TSD facility has released or has the potential to release hazardous wastes to the environment on a nonroutine basis, such as sporadic spills, it is called an area of concern (AOC).

As previously indicated, the corrective action requirement applies to facilities seeking closure or post-closure permits, as well as those seeking ordinary operating permits. For permitted facilities, Section 3004(u) requires that the

TSD permit shall contain schedules of compliance for corrective action where such action cannot be completed prior to issuance of the permit and assurances of financial responsibility for completing such corrective action.

The geographic extent of RCRA corrective action is described in Section 3004(v). Under Section 3004(v), TSD facilities have to undertake corrective action beyond the facility's boundaries when necessary to protect human health or the environment, unless the owner or operator of the facility can demonstrate that despite his best efforts, he was unable to obtain permission from other property owners for conducting corrective action. The facility, however, is still required to take necessary on-site measures to prevent off-site releases.

The regulations implementing the corrective action program for permitted TSDs are codified in 40 CFR Part 264 Subpart F. An interim status TSD may be compelled to perform a corrective action in response to an EPA order known as the 3008(h) order. Section 3008(h) of RCRA authorizes EPA to require owners and operators of interim status facilities to perform corrective action when there is or has been a release of hazardous waste from the TSD when such action is necessary to protect human health or the environment.

RCRA corrective action is a three-step program consisting of the following:[32]

1. Site characterization that identifies the contamination level of the site. This step consists of the RCRA Facility Assessment (RFA) and the RCRA Facility Investigation (RFI). Conducting an RFA is the responsibility of the regulatory agency, while completing an RFI is the responsibility of the TSD facility.
2. Remedy selection describes how the facility should clean up the site. This step, which is known as the Corrective Measures Studies (CMS), starts with determining the need for remedial measures and continues with selecting appropriate remedies, including the "no-action" alternative. The CMS is the responsibility of the facility.
3. Site remediation is the responsibility of the facility and is defined as the Corrective Measures Implementation (CMI). CMI includes the design, construction, operation, and maintenance of the selected remedies. This step requires active regulatory agency oversight.

As part of the corrective action process, short-term measures can be conducted at any time to respond to immediate threats. These activities are known as interim measures. They represent the individual actions to stabilize site conditions. The goals are to control or abate immediate threats to human health or the environment, to prevent or minimize the spread of contaminants, and to immediately initiate cleanup actions; examples of interim measures are installing a landfill cap to reduce infiltration of surface water into a landfill, installing a gas collection system to intercept migrating landfill gas, removing contaminated surface soils, and shutting down a contaminated drinking water supply well.

The implementation scheme of the corrective action provisions described in the preceding paragraphs is governed by EPA's National RCRA Corrective Action strategy, which was announced by the agency in 1985. On July 27, 1990, EPA proposed a new comprehensive regulatory framework for implementing the corrective action program. The new framework, which will be contained in Subpart S of 40 CFR Part 264, essentially codifies many of the provisions currently found in the strategy materials. As of this date, extensive regulatory provisions in Subpart S have not been promulgated. Current Subpart S regulations cover only provisions regarding corrective action management units (CAMUs) and temporary units (TUs). This will be elaborated upon later in this chapter when LDR and its relationship with the RCRA corrective action program are discussed.

As an integral part of the RCRA corrective action program, the site characterization objective is to examine the vertical and lateral extent of site contamination. This is accomplished through the RCRA Facility Assessment and the RCRA Facility Investigation. Both processes are explained in the following paragraphs in terms of goals, expected outputs, and processes.

2.1.1. RCRA Facility Assessment

The purpose of the RFA is to identify and gather information on releases of pollutants and make preliminary determinations about the releases, the need for corrective measures, further investigations, and interim measures. The assessment also screens those units in the facility which do not pose a threat to human health and the environment. The overall process will furnish a thorough description of the facility in terms of:

- operation, permitting history, and the characteristics of the units
- waste generation and management
- past and present actual and potential releases of hazardous waste or constituents to all environmental media
- probable contaminant movement, or migration, through five environmental pathways: groundwater, surface water, air, soil, and subsurface gas (e.g., methane gas at landfill)
- likely human receptors that may be or have been exposed or affected by those actual or potential releases
- need for corrective measures, interim measures, and further investigations

Although the RFA addresses all releases of pollutants, some releases are lawful if the concentration level of the pollutants is below acceptable limits. Releases of concern are those that have a pollution level above the standard. In order to assess releases of concern, the RFA provides a prestructured evaluation of release potential through a four-step process:

1. preliminary review (PR)
2. visual site investigation (VSI)
3. sample collection (if necessary)
4. corrective action need assessment as specified in the RFA Report

The PR may be the most time-consuming aspect of the RFA process; however, a thorough review will pay major dividends in reduced time and effort required to complete the RFA.[33] In terms of time frame, depending on the complexity of the site, the typical RFA can be conducted within 16 to 21 weeks (Table 4.4).

Table 4.4 Typical RFA Schedule

Activity	Number of days
File review	5
Draft PR report	30–45
Draft review	7
Final PR report	7
Notify TSD facility, allow the facility to gather requested information	21–28
Visual site investigation	1–7
Draft RFA report	30–40
Draft review	7
Final RFA report	7
Overall RFA process	115–153

The PR provides a clear understanding of the facility, focuses the subsequent VSI and identifies information gaps early in the RFA. Information needed during the PR includes a facility map, a topographic map, a copy of the RCRA permit application, the facility hydrogeological study, closure plans, spill reports, biennial reports, inspection reports, enforcement orders, county soil survey, and aerial photos.

The PR report describes the facility, permit history, waste management practice, environmental setting, the facility units, and data gaps in terms of the need of additional information specific to each unit. These data gaps will be filled during the VSI. The data needed are not necessarily about the releases.

In addition to filling in data gaps identified during the PR process, the VSI entails the on-site collection of visual evidence of the facility units, current hazardous waste management practices, pollution controls, releases, and routes of exposure. The VSI includes interviews with facility personnel. During the interview, Wesling[34] suggests asking all questions necessary to fill data gaps and to talk with not only the environmental manager, but also the employees involved in the day-to-day operations.

Although the VSI is intended to fill in data gaps, the gaps may still exist after the completion of the VSI. In this instance, field data and environmental sampling may be required. As previously noted, the regulatory agency is the responsible party in conducting the RFA; however, environmental sampling as part of the RFA can be implemented by requiring the facility to do it pursuant to a permit condition.

Once environmental sampling is completed, all information obtained from the PR, VSI, and sampling are analyzed and described in the RFA report. The report consists of two major parts: nonconfidential information and confidential information. The first part describes the purpose of the RFA, facility description, permit history, waste management practice, environmental setting, the facility units, and exposure pathways. The confidential part of the report is the summary and recommendations for corrective action.[35] Based on the characterization of releases and the potential for exposure to humans and the environment, the possible next steps in the corrective action process can be (1) interim measures; (2) a referral to the appropriate program such as Superfund, air program, or NPDES; or (3) the initiation of a RFI.

2.1.2. RCRA Facility Investigation

The goal of the RFI is to identify the contamination problem in a TSD facility and examine the extent of the problem. Specific information pertinent for the RFI includes the description of the facility problems, the types of contaminants released, the sources of contaminants, the pathways, the extent of contaminant migration, and the threats to human health or the environment.

The RFI is the responsibility of the respective TSD facility. There are five elements that comprise the RFI:

- examination of existing condition
- development of the RFI workplan
- evaluation of the RFI workplan
- health and safety plan
- public involvement plan

When conducting an assessment of current conditions, the investigations must focus on visual observation of the facility, the facility's compliance history, the environmental setting, and the existing level of contamination. Once a conceptual model of contaminant migration is developed, the RFI examines the impacts of contamination based on the model, evaluates possible interim measures, and identifies the need for additional data.

The RFI workplan elaborates the overall management of the process and details the field investigation plan, sampling plan, quality assurance and quality control (QA/QC), and data management. The procedures for sample collection include a map showing sample locations, personnel and equipment requirements, equipment calibration, equipment decontamination procedures, disposal of contaminated materials, sample packaging and shipment, and sampling documentation. For a large or complex site, phasing the RFI into more than one workplan is preferred, prioritizing the worst areas first.[36]

Once the RFI workplan is completed, the TSD facility must obtain approval from the environmental agency. The agency will review the workplan with the assumption that what is written on the workplan may not necessarily happen in the field. When reviewing the plan the agency will ensure that all areas are adequately covered and that the sampling plan is designed to look for the right contaminants in the right places. Site visits and audit sampling procedures are necessary; the agency will take additional samples during site visits.

In addition to the workplan, the RFI report also contains a health and safety plan and a public involvement plan. The purpose of the health and safety plan is to ensure safety in all aspects of the project. The plan should include an assessment of hazards and preventive measures, a list of the personal protection equipment necessary for the site and how to access it, site organization, and a list of names and phone numbers of emergency contacts.

A public involvement plan is not required, but recommended. The goal of the plan is to communicate to the public the objectives of the project, methods to achieve those objectives, and the status of the project. When developing the plan, ensure that the public's concerns are identified and addressed.

The completion of the RFI report represents the end of the site characterization stage of the RCRA corrective action. Remedy selection and site remediation phases can subsequently be implemented. The selection describes how the facility should clean up the site. In the RCRA corrective action terminology, remedy selection is known as the corrective measures studies while site remediation is called the corrective measures implementation.

2.1.3. Corrective Measures Studies (CMS)

The goal of the CMS is to develop and evaluate the corrective action alternative(s) and to recommend the corrective measure(s) to be taken at the site. The owner/operator of the TSD provides the personnel, materials, and services to develop and implement the CMS.[37] EPA recommends that the study be structured into four tasks:

1. the development of corrective measure alternatives
2. the evaluation of the alternatives
3. the recommendation of the corrective measure(s)
4. the CMS reporting

Corrective measure alternatives are developed by identifying and screening technologies which are applicable at the site. At this stage, the screening will eliminate technologies that may prove infeasible to implement or do not achieve the corrective measure objective within a specified time period. Given the set of waste and site-specific conditions, the screening process focuses on eliminating those technologies which have severe limitations for those conditions. The development of alternatives relies on engineering practices to determine which of the candidate technologies appears most suitable for the site. Each alternative may consist of an individual technology or a combination of technologies. The exclusion of technologies must be documented with the reasons for excluding them.

Once the alternatives have been screened, they must be further evaluated. The evaluation of each alternative is based on technical, environmental, human health, institutional, and cost factors. The technical evaluation examines the performance, reliability, implementability, and safety aspects of a technology. The environmental assessment focuses on the site conditions and pathways of contamination actually addressed by each alternative. The human health evaluation is based on the extent to which a technology mitigates short- and long-term potential exposure to any residual contamination and protects human health during and after implementation of the corrective measures. Institutional concern refers to the effect of federal, state, and local environmental and public health regulations, standards, and community acceptance of the design, operation, and timing of each alternative. The last evaluation is the cost estimate for each technology. The estimate includes capital costs as well as operation and maintenance costs.

The recommendation of corrective measure(s) is conducted by justifying why a particular alternative is recommended based on the results of the preceding evaluation. The recommendation includes summary tables which allow the alternative(s) to be easily understood and highlights the trade-offs among technical, environmental, human health, institutional, and cost factors. Based on this justification, the regulatory agency will select the most appropriate alternative.

The CMS reporting consists of documentation and reporting requirements in the form of progress reports, and draft and final reports. The monthly or bimonthly progress report documents a description and estimate of the percentage of the CMS completed and summarizes findings, changes, and problems encountered during the reporting period. The draft report includes a description of the facility, a summary of the corrective measure(s), design and implementation precautions, cost estimates and project schedule. The final report incorporates comments received from the regulatory agency on the draft report. The completion of the final report is the end of the remedy selection process. Once the remediation alternative is chosen, the implementation of the RCRA corrective measures can be initiated.

2.1.4. Corrective Measures Implementation (CMI)

CMI refers to the design, construction, operation, maintenance, and monitoring of the performance of the corrective measure(s) selected in the CMS. The CMI program consists of:

1. program planning
2. corrective measures design
3. construction of corrective measures
4. CMS reporting

The CMI program planning includes the development and implementation of the project management plan as well as the update of the community relations plan. The project management plan documents the overall management strategy for performing the corrective action project. It describes the organization of the implementing team and the roles and qualifications of key personnel involved with project implementation. The community relations plan developed from the previous task is updated to reflect the responses to the community's concerns regarding the design and construction of the corrective action project.

The corrective measure design step refers to the preparation of the final construction plans and specifications for implementing the corrective measure(s). Included in this step is the completion of cost estimate, i.e., a revision of the cost estimate developed from the previous task. The revision is necessary to ensure that the TSD facility has the financial resources available to construct and maintain the project. The health and safety plan from the previous task is also updated to address any changes based on the scope of the project design. The project scheduling, the construction quality assurance (CQA) objectives, and the phasing of project design are developed in this phase. When the design effort is approximately 30% complete, the work product is called the preliminary design and suggests that the technical requirements of the project have been addressed and indicates that the final design will provide an operable unit and usable corrective measures. The submission of the final design implies that the design plans and specifications are 100% complete.

Following the regulatory approval of the final design, the TSD facility is required to develop a CQA plan. The purpose is to ensure that the construction of the corrective measure is performed with a reasonable degree of certainty, that the completed corrective measures meet or exceed all design criteria, plans, and specifications. The CQA plan must be submitted to the regulatory agency for approval prior to the start of construction. The facility is also required to implement the approved operation and maintenance plan.

The CMI reporting is similar to the CMS reporting. The documents that are required to be completed in the CMI process are the progress report, and draft and final reports. Essentially, at the end of the construction of the project, the facility has to submit the CMI report that contains the synopsis of the project, an explanation of any modification to the implementation plans, the results of the monitoring activities showing that the corrective measures meet or exceed the performance criteria, and the description of the operation and maintenance of the project to be taken at the facility.

The stages of RCRA corrective action described in the preceding sections are designed for remediating a TSD site in which the TSD facility is currently in operation. If a TSD facility intends to close its business, the facility is required to develop and implement a closure plan. Depending on the specific condition of the site, a TSD facility may perform closure activities without doing corrective action because not a single part of the facility is found to have a hazardous waste release. However, if a release is detected during closure, the site is subject to the corrective action provisions.

2.1.5. *Partial Closure, Final Closure, and Post-Closure*

Closing a TSD facility can mean performing partial closure or final closure. Partial closure refers to the closing of one or more hazardous waste management units within the facility. If the owner or operator of a TSD closes all of the units in the facility, it is called final closure. For some types of units, post-closure must also be conducted for both partial as well as final closures.

Closure must accomplish a number of objectives. First, it must achieve cleanup performance standards, i.e., to remediate the contaminated site and leave the contaminants at a level that is allowed by the law. This is called clean closure. If clean closure is not achievable, appropriate measures must be taken to protect human health and the environment, such as containment of the contaminated site and post-closure care activities. The second objective is that closure must provide proper disposal or decontamination of equipment, structures, and soil. The third objective is that it must be conducted in an acceptable time frame.

In order to achieve these objectives, the RCRA closure plan must incorporate the following components (40 CFR 264.112(b) and 265.112(b)):

- a description of how each HWMU achieves the cleanup performance standards
- an explanation of how final closure of the facility will be conducted including closure and post-closure care required for each HWMU
- a description of the methods used during partial and final closures, such as the methods for removing, transporting, treating, storing, and disposing
- an estimate of the maximum inventory of hazardous wastes on-site over the life of the facility
- a description of the steps needed to remove or decontaminate all hazardous waste residues and contaminated containment-system components, equipment, structures, and soils during partial and final closures; examples of the steps are to clean equipment, to remove contaminated soils, to conduct sampling
- a summary of the criteria for determining the extent of the decontamination required to satisfy the closure performance standards

- a description of other related activities such as leachate collection, groundwater monitoring, and run-off control
- a closure schedule for each HWMU as well as the closure of the facility

After completion of facility closure, post-closure care will start and continue for a 30-year period. RCRA regulations require that all post-closure care activities be conducted in accordance with the conditions provided in an approved post-closure plan.

Closure, post-closure, and corrective action regulations are designed to address contaminated TSD sites. A site contaminated with hazardous waste as a result of hazardous waste generating activities normally cannot be treated as a TSD site, but must be addressed differently. The next section addresses the remediation of a RCRA contaminated site, i.e., a contaminated site in which a hazardous waste generator is currently conducting its regular business.

2.2. Hazardous Waste Generator Cleanup

Remediating a RCRA contaminated site can be problematic in a number of ways. First, before any decision is made on remediating a RCRA site, it is imperative that the regulatory status of the contaminants is determined. For example, when a hazardous waste is found as a contaminant, the site may not necessarily be a RCRA contaminated site if the contamination occurred prior to 1980. Additionally, when a hazardous waste commingles with another hazardous substance regulated under CERCLA, the mix of contaminants under more than one statute will certainly create a policy decision issue.

There are a number of statutes addressing contaminated sites such as CERCLA and RCRA Subtitles C, D, and I. Although each statute refers to specific regulated contaminants, cleanup requirements for each medium (soils, water, and air) are not consistently provided. Basically, there is no single statute that comprehensively addresses multi-media contamination. In reality, multi-media contamination is normal and is usually found at most RCRA and CERCLA contaminated sites.

As a consequence of differing statutory requirements, the level of cleanup is also different among statutes. There are three common options regarding the cleanup level that is required by current laws: (1) no-detect level, (2) background level, and (3) risk-based or health-based level.

The no-detect option is the most stringent requirement because it does not allow the existence of any suspected contaminant at a remediated site. The background option requires the cleanup efforts to be conducted with the goal of reducing the contamination of each suspected contaminant to a level similar to the area surrounding the site. The risk-based level requires the cleanup efforts to reach a level that is acceptable based on risk assessment criteria. The criteria are contaminant and site specific. Depending on the nature of contamination, the risk-based level can be higher or lower than the background level.

In reality, all of these requirements may not be financially feasible. Remediating a site even to the highest of those three levels can be very costly. This may result in the containment of the site by capping it with a notice to the deed to inform the public about the contamination status of the site. Even worse than containment of the site is the possibility of closing the business operating on the property and leaving the site as is, until the site is included in the federal or state Superfund program.

Clearly, the combination of the above issues results in the need to address RCRA contaminated sites as comprehensively as possible by integrating the regulatory, technical, and managerial aspects into the decision-making process. The following paragraphs examine an integrated model that can be used as an example to address RCRA contaminated sites.

2.2.1. A Comprehensive Approach to RCRA Contaminated Sites

An active RCRA contaminated site means that while there is evidence of contamination in the business property, the business is still in operation to produce products and services to customers. In this regard, the active RCRA contaminated site differs from a Superfund site in that it represents a controlled use situation. This unique characteristic provides an opportunity for considering a variety of cleanup mechanisms and cleanup levels appropriate for certain RCRA contaminated sites. Soesilo and Curry[38] suggest that this opportunity be evaluated in a comprehensive decision-making process (Figure 4.1).

Figure 4.1 shows that the decision-making process starts with determining the regulatory status of the contaminants. A number of possibilities may be discovered during the waste determination process; some wastes may be RCRA wastes, some may be CERCLA wastes, UST wastes, or RCRA Subtitle D solid waste. A property contaminated with RCRA hazardous waste prior to 1980 is considered a Superfund or a CERCLA site. A site or property contaminated with RCRA hazardous waste after 1980 is considered an RCRA contaminated site. The dividing line between CERCLA and RCRA contaminants is 1980 because the respective contaminants were regulated as RCRA hazardous waste beginning in 1980.

A waste which is not listed as a RCRA hazardous waste but shows a characteristic of hazardous waste is considered a RCRA waste. A site contaminated with petroleum products because of a release from a leaking UST is regulated under the UST program. However, if the release comes from an aboveground tank, the site may be defined as a RCRA contaminated site or a Subtitle D solid waste site, depending on the hazardous constituent of the contaminants.

The above discussion shows that determining the regulatory status of the contaminants is significant because it clearly affects the remediation planning

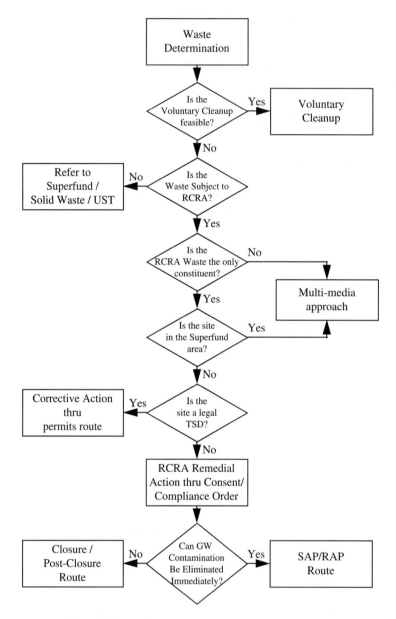

Figure 4.1. RCRA corrective/remedial actions.

for a site. After the properties of the contaminants are known, and if the magnitude of the contamination is found to be "not extensive", it is appropriate for the owner of the site to perform voluntary cleanup upon approval of the state regulatory agency, if a question exists in the mind of the owner regarding the regulatory requirements of the site.[39] In this case the owner (or the representative consultant) and the agency must mutually agree that the degree of contamination warrants voluntary cleanup.

The property owner must concur with the agency regarding whether the contamination occurred as a result of a discharge of RCRA hazardous waste, CERCLA hazardous substance, UST waste, or Subtitle D solid waste. The remediation planning is eventually performed according to the requirements specified by the respective statute or regulation.

In some states, the UST program does not "approve" voluntary cleanups. Owners or operators are expected to remediate releases expeditiously without prior approval. Power and Leifer[40] assert that the benefits of facilitating voluntary cleanups are numerous, both from an environmental protection standpoint and from the perspective of private parties. They argue that voluntary cleanup will result in a far greater number of sites being remediated, many sites will be cleaned up faster and more cost effectively, the resulting expedited action will decrease the potential for contaminants to migrate, and the cleanup will render property usable and valuable to companies and local communities.

Once it is determined that the contaminant is a RCRA hazardous waste, the next question is whether it is the only contaminant or whether it commingles, for example, with hazardous substances regulated under Superfund laws. Similarly, it is also possible that the site is located in a Superfund area. This assessment may result in a situation where the regulatory agency can address the site using RCRA regulations or Superfund provisions.

If the only contaminant is a RCRA hazardous waste, the business that is responsible for the contamination can be classified in two ways:

1. The business is a generator and is responsible for cleaning up the contamination. The generator is compelled to conduct the cleanup through the signing of an enforcement order issued by the regulatory agency.
2. The business is an illegal disposal facility because the contamination is viewed as the result of an act of disposing hazardous waste on-site without a permit.

Depending on how the business owner and the regulatory agency reach a consensus, these two classifications will definitely affect the scope of the cleanup activity. If the business is considered an illegal disposal facility, the facility is required to obtain a RCRA corrective action permit.[41] If the business is regarded as a generator, the facility may end up conducting the cleanup activities either through the Site Assessment Plan (SAP) and Remedial Action Plan (RAP) route, or through the closure and post-closure route under a formal compliance agreement.

The model illustrates that the flexibility of this approach lies in the fact that there are alternatives for the cleanup of RCRA contaminated sites. The first and most favorable alternative is the voluntary cleanup option. This option can be used if the remediation can be accomplished "immediately". However, if this is not possible, there are three other options that can also be considered: (1) the RCRA corrective action permitting option, (2) the compliance agreement with a SAP/RAP requirement, and (3) the closure and post-closure alternative.

2.2.2. Assessment of the Alternative Cleanup Routes

Among the three options, the RCRA corrective action permitting represents the alternative with the most comprehensive requirements. This option is preferred if both soil and groundwater contamination exist. The RCRA corrective action permitting process is specified in the RCRA regulations. Before conducting the cleanup, it is required that the facility develops an approvable closure plan that describes the method and standard used to clean up the contaminated site.

If clean closure is not achievable, i.e., a certain level of contamination is still left in the soil or groundwater, post-closure care must be conducted by the facility. This includes the installing, monitoring, and maintaining of groundwater wells based on an approved post-closure plan. The permitting option requires the facility to provide valid financial assurance to guarantee that a substantial amount of money is available to finance the closure and post-closure activities.

The SAP/RAP alternative is a cleanup process which uses the "comparable authority" process[42] to facilitate cleanup at a RCRA contaminated site. The SAP/RAP does not require financial assurance. The procedure is less comprehensive than the first alternative so that the cleanup activity can be conducted much faster. It is appropriate if no groundwater contamination is detected or if the groundwater contamination can be expeditiously eliminated upon the completion of the cleanup.

The SAP/RAP method and cleanup standard must be approved by the regulatory agency. In essence, the scope and content of a SAP is similar to a RFA and a RFI, while the RAP is comparable to the CMI and the CMS of the RCRA corrective action planning.

The closure and post-closure alternative is similar to the RCRA corrective action permitting option in that it requires the facility to perform closure and post-closure based on approved plans. As in RCRA corrective action permitting, the term closure means cleaning up the site. The closure and post-closure alternative is executed through an enforcement action consisting of either a consent or compliance order, not the permitting process.

The above discussion reveals that if immediate voluntary cleanup is not possible, the next best option is the SAP/RAP, followed by the closure and post-closure alternative. The RCRA corrective action permitting process represents the least desirable option. The voluntary cleanup and SAP/RAP deal with contamination where clean closure is possible and the closure/post-closure and RCRA corrective action permitting are used where clean closure is not achievable and, therefore, post-closure monitoring of the site is necessary. The applicability of this alternative scheme, however, is dependent on the provisions in state laws, policies, and regulations.

Regardless of the option used, the issue of "how clean is clean" is still prevalent. In other words, the choice of the cleanup level, whether it is a no-detect, background, or risk-based cleanup standard, is still subject to debate and is normally decided on a case-by-case basis. In addition to the choice of the cleanup level, other planning issues in remediating a CERCLA site, RCRA corrective action site, or RCRA contaminated site include the streamlining of the process, cost effectiveness, and the risk to human health and the environment.

Because RCRA is a prevention-oriented program, its focus is on preventing the generation of hazardous waste and, if a hazardous waste is generated, preventing the release of hazardous waste to minimize the risk to human health and the environment. Essentially, the RCRA program is designed to promote waste minimization and pollution prevention in lieu of or prior to land disposal.[43] Within the RCRA regulatory scheme, one of the most notable preventative provisions can be found in the LDR, also known as the land ban rules.

2.3. Land Disposal Restrictions

The goal of LDR is to discourage activities that involve placing untreated hazardous wastes in or on the land when a better treatment or immobilization alternative exists.[44] The LDR regulations established treatment standards for the majority of RCRA hazardous wastes ultimately destined for land disposal.[45] The regulations differentiate this majority into restricted and prohibited wastes. Restricted wastes are RCRA wastes for which an LDR treatment standard has been established, e.g., D004, D005, D006, D007, D008, D009, D010, and D011 wastes. The standard is expressed in concentration levels or as methods of treatment that have either substantially diminished the toxicity of wastes or reduced the likelihood that hazardous constituents from wastes will migrate from the disposal site.[46] Before the regulated wastes can be land disposed, LDR treatment standards must be met. Prohibited wastes, which constitute a subset of restricted wastes, are ineligible for land disposal; examples of prohibited wastes are solvent wastes and dioxin-containing wastes.

EPA has promulgated regulations implementing the LDR program according to a specified rule-making schedule. LDR regulations in November 1986 included prohibitions on the land disposal of dioxin and solvent containing hazardous wastes. The July 1987 regulations banned the land disposal of certain hazardous wastes which California had already banned.[47] With the LDR program fully effective since May 1992, a significant quantity of RCRA wastes have to be treated properly prior to land disposal.

The stringent treatment standards imposed by the LDR have encouraged many hazardous waste generators to minimize the generation of hazardous wastes; however, when the LDR provisions are applied to RCRA remediation, they often create a disincentive to the program. In February 1993, EPA promulgated final regulations for CAMUs and TUs. CAMU is defined as an area within a TSD facility that is approved by EPA or the authorized state to be used for the management of remediation wastes pursuant to implementing corrective action requirements at the facility. TU refers to approved temporary tanks and container storage area within a TSD facility used for the treatment or storage of remediation wastes as part of corrective action activities.

The introduction of the concept of CAMU and TU clearly represents a means of providing some relief from the burdens that LDRs impose on cleanup activities. 40 CFR 264.552 states that placement of remediation waste into an approved CAMU does not trigger LDR requirements, so that appropriate site-specific treatment can be conducted. 40 CFR 264.553 specifies that a design, operating, or closure standard applicable to TU may be replaced by alternative requirements which are still protective of human health and the environment.

In September 1993, EPA proposed the "phase II LDR regulations" which would establish a constituent-specific treatment standard for soils contaminated with hazardous waste. The proposal will impose somewhat less stringent standards for soils based on the fact that many cleanup projects generate contaminated soils that contain complex mixtures of multiple contaminants and different types of soils, with wide variations even within a single cleanup site.

In September 1995, EPA proposed the hazardous waste identification rule (HWIR) for contaminated media which would authorize EPA and authorized states to exempt certain lower-risk contaminated media from regulation as hazardous wastes. The higher-risk contaminated media would remain under the RCRA Subtitle C regulations, but would be subject to modified treatment standards. The dividing line between lower-risk and higher-risk contaminated media is called the bright line and is established through a scientific protocol known as risk assessment (see Chapter 9). With the promulgation of this proposed regulation, the CAMU rules would be withdrawn.

This most recent development indicates the direction that the RCRA site remediation program is currently heading. As previously described in this chapter, EPA has historically sought to address hazardous waste site remediation through the use of RCRA's prevention oriented standards to site remediation, while providing the states with the flexibility and tools necessary to develop site-specific remedies. The current trend shows that EPA has realized that a prevention-oriented approach is not sufficient and is inappropriate for addressing hazardous waste contaminated sites. Therefore, a response oriented approach must be pursued in order to better align the regulatory controls for contaminated sites with the risks associated with them. This urgency requires the development of common sense and a more flexible regulatory mechanism for the management of RCRA contaminated sites.

5   LEAKING UST AND PCS REMEDIATION

Underground storage tanks (USTs) are distributed widely throughout the U.S. Environmental Information Ltd. (EI)[48] estimates that approximately 1.4 million regulated operating underground storage tanks (USTs) are currently in operation, while more than 900,000 tanks have been closed since 1988. Of these total tanks, EI estimates that 463,000 tanks are considered leaking underground storage tanks (LUSTs), of which more than 100,000 LUSTs have been remediated. According to U.S Environmental Protection Agency (EPA) data, so far the majority of the cleanups have been conducted by responsible parties with state oversight.[49]

By substracting the 107,000 LUST cleanups that have already been completed, it is expected that 356,000 cleanups remain to be funded. Based on a survey of state officials and UST contractors, EI estimated the average cost of cleanup at $100,000 per site, somewhat lower than EPA's current estimate of $125,000 per site. The estimation suggests that the cost to industry and government of cleaning up current and future UST sites will be about $35.6 billion.

The figures displayed in the preceding paragraphs illustrate the massive scope of LUST remediation, both in terms of the number of contaminated sites and its remediation cost. Because of the magnitude of the problem, remediation of LUSTs is considered a major environmental issue in the U.S. Industry and environmental experts, however, have only recently recognized the important role USTs play in environmental contamination.[50]

The contamination of soil and groundwater from LUSTs was not addressed by the 1980 Comprehensive Environmental Response, Compensation, and Liability Act (CERCLA) or the 1976 Resource Conservation and Recovery Act (RCRA) provisions. In 1984, Congress responded to the environmental threats from USTs by adding Subtitle I, Regulation of Underground Storage Tanks, as part of the Hazardous and Solid Waste Amendments (HSWA) to the RCRA.

1. OBJECTIVES OF THE UST PROGRAM

Subtitle I charged EPA with the task of developing regulations for the federal program which included existing and new tanks. The final version of the initial set of UST regulations was adopted on September 23, 1988. Although EPA has continued to adopt new regulations since then, the number and volume of UST regulations are significantly less than those promulgated for hazardous waste under Subtitle C of the RCRA.

The purpose of Subtitle I which addresses both existing and future UST systems is to prevent releases from a UST system from occurring, and to remediate the environment if a release occurs.

- For tank systems that were in existence when Subtitle I went into effect, the main requirements are the identification and upgrading (or closure) of the USTs.
- For new tanks the law focused more on the prevention of leaks by mandating stringent tank design and operating standards.

Under the UST program, discovering and cleaning up existing chemical spills from leaking USTs and enforcing strict standards for new tanks present a serious challenge to the government and to private companies responsible for USTs.[51] If a tank is a UST and the contents are a petroleum-based or hazardous substance, the owner and operator of the UST is responsible for complying with UST requirements. These requirements may vary depending upon whether EPA or state or local government is the designated agency.

Like the RCRA hazardous waste program, the UST program was designed to be delegated to those states that EPA determines have UST laws and rules that meet the minimum federal UST program requirements. States and, in some instances, local governments will not receive delegation of the UST program if their requirements are less stringent than the federal requirements. However, some states and local governments are not prohibited from promulgating requirements that are stricter than the federal program.

UST is defined under 40 CFR 280.12 as any tank or combination of tanks (including underground pipes connected thereto) that is used to contain an accumulation of regulated substances, and the volume of which (including the volume of underground pipes connected thereto) is 10% or more beneath the surface of the ground. UST does not include any of the following:

1. farm or residential tanks of 1,100 gallons or less, used for storing motor fuel for noncommercial purposes
2. tanks used for storing heating oil for consumptive use on the premises where stored
3. septic tanks
4. pipeline facilities (including gathering lines) regulated under the Natural Gas Pipeline Safety Act of 1968, or the Hazardous Liquid

Pipeline Safety Act of 1979, or state laws comparable to the provisions of the law referred above

5. surface impoundments, pits, ponds, or lagoons
6. stormwater or wastewater collection systems
7. flow-through process tanks
8. liquid traps or associated gathering lines directly related to oil or gas production and gathering operations
9. storage tanks in an underground area (e.g., basement) but the tank is located upon or above the surface of the floor
10. pipes connected to any of the tanks described above
11. any UST systems holding hazardous wastes
12. any wastewater treatment tank systems that are part of a wastewater treatment facility regulated under the Clean Water Act
13. equipment or machinery that contains regulated substances for operational purposes such as hydraulic lift tanks and electrical equipment tanks
14. any UST systems whose capacities are 110 gallons or less
15. any UST systems that contain *de minimis* concentrations of regulated substances
16. any emergency spill or overfill containment UST systems that are expeditiously emptied after use

Additionally, the UST systems listed below have been deferred from regulation under Subtitle I:

1. wastewater treatment tank systems
2. any UST systems containing radioactive materials that are regulated under the Atomic Energy Act of 1954
3. any UST systems that are part of an emergency generator system at nuclear power plants regulated by the Nuclear Regulatory Commission under 10 CFR Part 50, Appendix A
4. airport hydrant fuel distribution systems
5. UST systems with field-constructed tanks

Many of the exclusions and exemptions presented above are based on the contents of the UST. Specifically, the UST regulations only apply to regulated substances which consists of:

- hazardous substances as defined in Section 101(14) of the 1980 CERCLA
- petroleum and petroleum-based substances derived from crude oil and used oils

The definition of owner and operator are extremely important in the UST program. The definition determines whether or not an individual is responsible for complying with Subtitle I and the associated liability for cleaning up any release from the UST. Although the federal definition of owner and operator are presented below, several states have modified this definition. Before purchasing property in states that have EPA-approved programs, the state definition of owner and operator should be determined to ensure that ownership of the property does not automatically include ownership of any known or unknown USTs.

Under EPA's definition (40 CFR 280.12), an owner is any person who owns a UST system used for storage, use, or dispensing of regulated substances after November 8, 1984. In the case of a UST system in use before November 8, 1984, but no longer in use on that date, owner means any person who owned such UST immmediately before the discontinuation of its use. EPA defines operator as any person in control of, or having responsibility for, the daily operation of the UST system. In both definitions, the term person means an individual, trust, firm, joint stock company, federal agency, corporation, state, municipality, commission, political subdivision of a state, or any interstate body. Person also includes a consortium, a joint venture, a commercial entity, and the U.S. government. A person can be both the owner and operator of an UST system.

2. PROGRAM REQUIREMENTS

The owner of USTs is subject to a number of Subtitle I requirements. The requirements include notification of the UST to the designated agency within 30 days of bringing the UST into use (40 CFR 280.22). In addition to providing information about the status, size, location, composition, protection mechanisms, etc., of the UST, notification forms require the owner of new USTs to certify that they are in compliance with the following requirements:

1. Installation of tanks and piping were done properly.
2. Cathodic protection was installed on steel tanks and piping.
3. Financial responsibility was demonstrated.
4. Release detection was properly installed.

The installer of a new UST must certify that the methods used to install the UST system comply with UST requirements. The UST owner is responsible for ensuring that the notification form contains the installer's certification (40 CFR 280.20(e)). Individuals who sell tanks that are to be used as USTs must advise the buyer of the UST notification requirements (40 CFR 280.22).

If the owner of a UST system, which by definition contains a regulated substance, decides to store an unregulated substance in the UST, the owner must notify the delegated agency of this change in service 30 days prior to beginning the change over or within the time period specified by the implementing agency. The owner must have all liquids and sludge removed from the UST and have a site assessment conducted prior to the change-in-service (40 CFR 280.71). If contamination is discovered through the site assessment process, the owner must take corrective action, e.g., remove the contamination, which must be completed prior to the change-in-service (40 CFR 280.72).

Owners and operators of regulated petroleum UST systems are required to demonstrate financial responsibility. Financial responsibility means that the owner and operator are able to pay for cleanups and to compensate third parties for bodily injury and property damage arising from an accidental release of petroleum from an UST.

The amount of financial responsibility is calculated on a per-occurrence basis. Owners and operators of petroleum USTs that are located at petroleum marketing facilities or that handle an average of more than 10,000 gallons of petroleum monthly based on the previous calendar year's throughput must show $1,000,000 of financial responsibility. All other owners and operators of regulated petroleum USTs must demonstrate $500,000 of financial responsibility (40 CFR 280.93(a)). If the owner or operator are two different people, only one of them is required to meet the financial responsibility requirement; however, both individuals are liable if there is noncompliance with the financial responsibility regulation (40 CFR 280.90(e)).

Most states have established funds to assist owners and operators who clean up the contamination resulting from a release from one or more of their UST systems. Some of these assurance funds provide the minimum amount of financial responsibility required by Subtitle I. This relieves the owners and operators of demonstrating financial responsibility.

In Arizona, for example, the State Assurance Fund (SAF) was established by state law in 1990 to receive money for the payment of claims through the imposition of a one-cent-per-gallon excise tax on regulated products placed into USTs. The fund will reimburse the cost only for corrective actions performed in response to releases of regulated substances from UST systems during specified time periods. The law also specifies eligible claimants and reimbursable costs.[52]

For those owners and operators who do not have a state assurance fund with a $1,000,000 or higher cap, financial responsibility can be demonstrated by any one or any combination of the following mechanisms:

- liability insurance (including risk retention group coverage)
- self-insurance
- surety bond
- letter of credit
- guarantee
- trust fund

40 CFR 280.108 adds that an alternate financial assurance mechanism can be used providing that the owner or operator satisfies the $1,000,000 or $500,000 (whichever is appropriate) requirement at all times.

Leaking USTs generally occur because tanks are made of steel and may lack corrosion protection. In addition to failed UST systems caused by corrosion or structural failure, contamination from USTs can occur from improper filling or removal practices. The performance standards for new UST systems are intended to prevent releases of regulated substances from the systems into the environment due to either structural failure, corrosion, or spills and overfills. The federal regulations provide performance standards for the following components:

- tanks
- piping
- spill and overfill protection

40 CFR 280.20 underlines that UST systems must be properly designed and constructed. Any portion of the tank and piping that is underground and routinely in contact with the regulated substance must be protected against corrosion. The corrosion protection depends upon the type of material used to construct the tank and piping. The appropriate method of corrosion protection must be developed by a nationally recognized association or independent testing laboratory. The upgrading requirements for older tanks call for lining the interior of a steel UST, or installing cathodic protection, or a combination of the two (40 CFR 280.21).

One of the cathodic protection systems, called the impressed current system, consists of:

- a rectifier that converts alternating current to direct current
- anodes which are special metal bars buried in the soil near the UST

The direct current is sent from the rectifier to the anodes through an insulated wire. The current then flows through the soil to the UST system and returns to the rectifier through an insulated wire attached to the UST. The current going to the UST system overcomes the corrosion-causing current normally flowing away from it. This is how the system protects the tank from corroding.[53]

All new UST systems must also be equipped with spill and overfill prevention equipment to prevent contaminating the environment when product is being transferred into the tank. Spill prevention equipment, such as a catchment basin, prevents the product from reaching the environment when the transfer hose is removed from the UST's fill pipe. Catchment basins are also called spill containment manholes or spill buckets. Essentially a catchment basin is a bucket sealed around the fill pipe. The larger the bucket, the more spill protection it provides.

Overfill prevention equipment must satisfy provisions established in 40 CFR 280.20(c)). The equipment must do one of the following:

- automatically shut off the flow into the tank when the tank is no more than 95% full
- alert the transfer operator by restricting the flow of product to the tank or emitting a loud alarm when the tank is no more than 90% full
- restrict the flow of product 30 min prior to overfilling, alert the operator with a loud alarm one min before overfilling, or automatically shut off the flow of product before any of the fittings on the top of the tank are exposed to the product

The spill and overfill prevention measures presented above are not necessary if no more than 25 gallons of product are transferred at any one time or the designated agency determines that alternate equipment is as protective as the previously listed methods.

3. CLOSING A TANK

An owner may discontinue using a UST, either temporarily or permanently. The requirements for permanently closing a tank are almost identical to those for a change-in-service. The differences are that if permanent closure is made in response to a release from the UST, the 30-day prior notification requirement is not necessary; and after proper cleaning, the UST must be removed from the ground or filled with an inert solid substance such as concrete (40 CFR 280.71).

Before permanent closure is complete, sampling at the points where contamination is most likely to be present must be conducted. The federal requirements for where sampling should occur are vague. They do indicate that the method of closure, the nature of the stored substance, the type of backfill, the depth to groundwater, and other appropriate factors are to be considered (40 CFR 280.72(a)). The delegated agency may have developed more specific guidelines, rules, or regulations for closure sampling.

There is no need to sample if vapor or groundwater monitoring was used as the release detection method, the system complied with the applicable requirements, and the system did not detect a release. If the sampling or any other means reveals that the regulated substance stored in the tank has contaminated the soil or groundwater or is present as a free product as either a liquid or vapor, the owner and operator must immediately begin corrective action (40 CFR 280.72(b)).

When an owner and operator temporarily closes a UST, the following requirements (40 CFR 280.70(a)) must be complied with:

- the operation and maintenance of the corrosion protection system; if the UST is not empty, the release detection system must be in place
- release reporting

- release investigation
- release confirmation
- corrective action

If the temporary closure lasts more than three months, the owner and operator must also leave the vent lines open and functional. They are also required to cap and secure all other lines, pumps, manways, and ancillary equipment (40 CFR 280.70(b)). UST systems that have been temporarily closed for more than a year must be permanently closed if the system does not meet tank performance standards for new UST systems or upgrading requirements, with the exception of spill and overfill equipment.

4. RELEASES FROM THE TANK AND CORRECTIVE ACTION REQUIREMENT

Under the federal UST regulations, the owner and operator has 24 hours to report a suspected or confirmed release and seven days to confirm a suspected release to the implementing agency. Within the first 24 hours after confirming a release, the owner and operator must also take immediate action to prevent further release of the regulated substance into the environment, and identify and mitigate any fire, explosion, and vapor hazards caused by the release (40 CFR 280.61). The 24-hour requirement may differ depending on the implementing agency.

After reporting the confirmed release, the owner and operator must begin initial abatement measures. These measures include:

- removal of as much of the regulated substance from the UST system as is necessary to prevent further release into the environment
- visual inspection of any aboveground or exposed underground releases and the prevention of further migration of the released substance into the surrounding soils and groundwater
- continuance of monitoring and mitigating any fire and safety hazards posed by vapors or free product that have migrated from the UST excavation zone and entered into subsurface structures, e.g., sewers or basements
- remediation of hazards posed by contaminated soils that are excavated or exposed as a result of release confirmation, site investigation, abatement, or corrective action activities
- sampling for the presence of a release where contamination is most likely to occur, unless the presence and source of the release have been confirmed in accordance with the UST site check or the site closure assessment protocols
- investigation to determine the possible presence of free product, and initiation of removal as soon as practicable in accordance with UST procedures; an example of a free product is released gasoline that floats on water

Within 20 days after confirming the release, the owner and operator must submit a report to the implementing agency that summarizes the initial abatement actions listed above and any resulting information. If a state is the implementing agency, the reporting period may differ from EPA's 20-day requirement (40 CFR 280.62).

Unless the implementing agency directs the owner and operator to do other activities, the next step in the process is for the owner and operator to perform an initial site characterization. This includes the information from the release confirmation and initial abatement measures, plus at a minimum:

- data on the nature and estimated quantity of the release
- data from available sources and/or site investigations concerning the surrounding populations, water quality, use and approximate locations of wells potentially affected by the release, subsurface soil conditions, locations of subsurface sewers, climatological conditions, and land use
- results of the required site check
- results of the free product investigation

Site characterization must be performed by a qualified party who is knowledgeable and experienced in the proper procedures, methods, and techniques of conducting investigations. Notification to and permits from state, county, and local government agencies are required for certain assessment and remedial activities in response to leaking USTs.

The owner and operator must submit the information found as a result of the initial site characterization to the implementing agency within 45 days after release confirmation or when required by the implementing agency (40 CFR 280.63). If there is free product and unless directed to do otherwise by the implementing agency, the owner and operator have 45 days or a time period determined by the implementing agency, to submit a free product removal report that at a minimum includes (40 CFR 280.64):

- the name of the person responsible for implementing the free product removal
- the estimated amount, type, and thickness of the free product observed in various places
- the type of free product recovery system
- whether there will be any discharge on-site or off-site and where the discharge will be located
- the type of treatment applied to and the effluent quality expected from any discharge
- the steps taken to obtain necessary permits for discharge
- the disposition of the recovered free product

The removal of the free product shall be conducted in a manner that prevents the spread of contaminants into previously uncontaminated zones.

The design objective of the free product removal system should be the abatement of free product migration. During removal, all flammable products shall be handled in a safe, competent manner and in accordance with all applicable local, state, and federal regulations. Recovered by-products shall be properly treated and discharged, or disposed of in accordance with all local, state, and federal requirements (40 CFR 280.64).

The owner and operator are responsible for determining the location and full extent of the release by conducting studies of the soil and groundwater at the release site and the surrounding area. If any of the conditions listed below exist, the owner and operator are required to perform the investigation:

- Groundwater wells have been impacted by the release.
- Free product needs to be recovered.
- Contaminated soils may be in contact with groundwater.
- The implementing agency requests an investigation based on the potential effects of the contaminated soil or groundwater on nearby surface water or groundwater.

The investigation information must be submitted as soon as practicable or within the time period set by the implementing agency (40 CFR 280.65). The implementing agency can also require the owner and operator to prepare a corrective action plan that provides information about the long-term permanent corrective action measures that will adequately protect human health and the environment. Under some circumstances, the owner and operator may voluntarily submit a corrective action plan.

Before the implementing agency will approve the plan, it must determine that implementation of the corrective action plan will adequately protect human health, safety, and the environment. To make its determination the implementing agency considers the following factors if applicable:

- the physical and chemical characteristics of the regulated substance
- the hydrogeological characteristics of the facility area
- the proximity, quality, and current and future uses of nearby water resources
- the potential effects of residual contamination on the nearby water resources
- an exposure assessment
- any information found during the course of the corrective action measures

Once the implementing agency has approved the plan, it must be implemented and the implementation must be monitored, evaluated, and reported as required by the implementing agency. Owners and operators can begin to clean up the soil and groundwater prior to receiving approval from the implementing agency as long as (40 CFR 280.66):

- The implementing agency is notified of the intention to clean up.
- The owners and operators agree to comply with any conditions from the implementing agency.
- The owners and operators agree to incorporate the cleanup measures in the corrective action plan.

The public participation provisions of the UST program are triggered when the implementing agency requires the owner and operator to submit a corrective action plan. The implementing agency must provide notice to the members of the public affected by the release and planned long-term corrective action. The implementing agency must provide site release information and decisions on the corrective action plan for the public to review. A public meeting may be held by the implementing agency prior to approving the corrective action plan if there is public interest or any other reason for the meeting (40 CFR 280.67).

A historical backlog of LUST remediation has helped prompt federal and state regulatory agencies to consider remediation strategies such as risk-based approaches that do not require cleaning up tank release sites to pristine conditions. EPA has endorsed the risk-based decision-making approach to corrective action because it could help state and local UST programs deal with UST releases more quickly and efficiently.[54]

In order for a site owner to respond to a UST release, the question arises as to what level of scientific interpretation is necessary for determining how significant the release is. After the question is answered, the next inquiry is whether comprehensive cleanup, limited cleanup, monitoring only, or no action is required to address the release. To help answer this question, a tiered approach to manage the development of risk-based cleanup goals was recently developed by the American Society for Testing and Materials.[55]

ASTM's investigative process, known as risk-based corrective action (RBCA) consists of three tiers that correspond to the different levels of complexity in the analysis based on the conditions present at a particular site. Tier I is a screening of the contamination level against cleanup standards such as Maximum Contamination Levels (MCLs). If some of the MCLs are exceeded, then some level of risk assessment is called for; otherwise, the no-action option can be used.

Tier I as a basic risk assessment may be a sufficient approach if the contamination appears to be contained within the site boundaries. The tier I risk assessment normally costs $3,000 to $5,000 to perform. The cost of a tier II risk assessment is between $6,000 and $10,000.[56]

A tier II assessment requires more subsurface evaluation of contaminant fate and transport and is intended to reduce the uncertainty in the assessment conclusions, by introducing more site-specific information. A site with free product would often be classified as at least a tier II site. Tier II sites require the use of more complicated fate and transport models. Usually they have contaminants that may not all be from the same family of chemicals. Large data sets for statistical analysis and multiple chemical exposure scenarios are typical features of a tier III assessment.

RBCA is basically an effort to standardize practical application of risk assessment practices to petroleum contamination problems. The approach allows resources to be focused on sites which pose the greatest threats to human health and the environment. RBCA is intended as a tool for determining the amount and urgency of corrective action necessary.

5. REPORTING AND RECORDKEEPING

The preceding few sections have shown that the owners and operators of USTs are responsible not only for managing USTs and LUSTs appropriately, but also for several reporting and recordkeeping activities. To summarize, the UST reporting and recordkeeping requirements include the following (40 CFR 280.34):

1. Reporting. The following information must be submitted to the implementing agency:
    *   notification for all UST systems, which includes certification of installation for new UST systems
    *   reports of all releases including suspected releases, spills and overfills, and confirmed releases
    *   corrective actions planned or taken including initial abatement measures, initial site characterization, free product removal, investigation of soil and groundwater cleanup, and corrective action plan
    *   notification before permanent closure or change-in-service
2. Recordkeeping. Owners and operators must maintain the following information:
    *   a corrosion expert's analysis of site corrosion potential if corrosion protection equipment is not used
    *   documentation of operation of corrosion protection equipment
    *   documentation of UST system repairs
    *   recent compliance with release detection requirements
    *   results of the site investigation conducted at permanent closure
3. Availability and maintenance of records. The records must be kept either at the UST site and immediately available for inspection by the implementing agency, or at a readily available alternative site and be provided for inspection to the implementing agency upon request. In the case of permanent closure, certain records that cannot be kept at the site or an alternative site can be mailed to the implementing agency.

6.   PETROLEUM-CONTAMINATED SOIL

Petroleum-contaminated soils (PCS) are soils that contain chemical constituents of benzene, toluene, ethylbenzene, xylenes (BTEX) and total petroleum hydrocarbon (TPH) in concentrations in excess of levels determined by the UST standards. These standards may vary among states, e.g., Arizona:

- benzene   = 0.13 mg/kg (ppm)
- toluene   = 200 mg/kg (ppm)
- ethylbenzene = 68 mg/kg (ppm)
- xylenes, total = 44 mg/kg (ppm)
- total TPH   = 100 mg/kg (ppm)

When concentrations of these constituents in a UST site exceed the above standards, proper cleanup must be conducted until the concentrations are reduced to below or at the standards. Gasoline, kerosene, jet fuel, and waste oils are subject to BTEX and TPH analyses. Diesel and heavy oils are subject to TPH analysis only, while solvents are required to be tested for the BTEX concentrations.

Soils contaminated with petroleum that come from non-LUST sites are not regulated by Subtitle I. This category of PCS is under the domain of solid waste. Petroleum-contaminated soils can, therefore, be classified as either UST PCS, or solid waste PCS. Check with the local state environmental agencies on the best management practices for solid waste PCS. Regulations on solid waste PCS varies from state to state.

PART II
REFERENCES

1. Carson, Rachel, *Silent Spring*, Houghton Mifflin, Cambridge, MA, 1962.
2. Wildavsky, Aaron, *But is it True? A Citizen's Guide to Environmental Health and Safety Issues*, Harvard University Press, Cambridge, MA, 1995.
3. Reinhardt, John R., Summary of Resource Conservation and Recovery Act Legislation and Regulation, in Harry M. Freeman, ed., *Standard Handbook of Hazardous Waste Treatment and Disposal*, McGraw-Hill, New York, 1989, 1.9–1.26.
4. Quarles, John, *Federal Regulations of Hazardous Wastes, A Guide to RCRA*, Environmental Law Institute, Washington, D.C., 1982.
5. Green, Carol L., Dunn, James R., and Oppenheimer, Peter H., RCRA Enforcement in the 90s: Avoiding Civil and Criminal Liability, in Kenneth M. Kastner, ed., *Current Developments in RCRA*, Government Institutes Inc., Rockville, MD, 1993, 75–110.
6. 533 PRPs Sign $22.4M Pact for Michigan Site Cleanup, *Waste Tech News*, June 5, 1989, 5.
7. U.S. Environmental Protection Agency, *Enforcement Accomplishments Report*, Washington, D.C., 1990.
8. EPA Reveals Largest Superfund Consent Decree, *Waste Tech News*, September 10, 1990, 1 and 15.
9. U.S. Environmental Protection Agency, *Enforcement Accomplishments Report*, Washington, D.C., 1990.
10. U.S. Environmental Protection Agency, *RCRA Orientation Manual — 1990 Edition*, Washington, D.C., 1990.
11. Lee, Robert T., Comprehensive Environmental Response, Compensation, and Liability Act, *Environmental Law Handbook*, Government Institutes Inc., Rockville, MD, 1993, 267–320.
12. U.S. Environmental Protection Agency, *Guidance on the Use of Stipulated Penalties in Hazardous Waste Consent Decrees*, Washington, D.C., 1987.
13. Travis, Curtis C. and Doty, Carolyn B., Remedial Action Decision Process, in Richard B. Gammage and Barry A. Berven, eds., *Hazardous Waste Site Investigations — Toward Better Decisions*, Lewis Publishers, Boca Raton, FL, 1992.
14. Ronan, John T., III, New Horizons for Superfund Enforcement, *Hazmat World*, February, 1990, 42–54.
15. Carr, F. Housley, What a Dump! *Resources*, January, 1991, 12–14.

111

16. Lautenberg, Dingell, Blast Superfund Studies; Reilly Focuses on Accomplishments of Program, *Environment Reporter*, October 11, 1991, 1531.

17. Burke, Thomas A., Overview: Refining Hazardous Waste Site Policies, in Richard B. Gammage and Barry A. Berven, eds., *Hazardous Waste Site Investigations — Toward Better Decisions*, Lewis Publishers, Boca Raton, FL, 1992.

18. Russell, Milton, Colglazier, E. William, and English, Mary R., *Hazardous Waste Remediation — The Task Ahead*, University of Tennessee, Knoxville, 1991.

19. Hope, Michael R., RCRA/CERCLA Interaction in Kenneth M. Kastner, ed., *Current Developments in RCRA*, Government Institutes Inc., Rockville, MD, 1993, 213–223.

20. Hall, Ridgway M. Jr. et al., *RCRA Hazardous Waste Handbook*, 9th Edition, Government Institutes Inc., Rockville, MD, 1991.

21. Briggum, Sue M. et al., *Hazardous Waste Regulation Handbook, A Practical Guide to RCRA and Superfund*, Revised Edition, Executive Enterprises Publications Co. Inc., New York, 1985.

22. Walsh, Thomas J., *CERCLA and RCRA Requirements Affecting Cleanup of a Hazardous Waste Management Unit at a Superfund Site: A Case Study*, Air and Waste Management Association, 88th Annual Meeting and Exhibition, San Antonio, TX, 1995.

23. Robertson, Thomas D., *Integrating Closure, Corrective Action and CERCLA Requirements — A Case History*, Air and Waste Management Association, 88th Annual Meeting and Exhibition, San Antonio, TX, 1995.

24. Case, David R., Resource Conservation and Recovery Act, in *Environmental Law Handbook*, Government Institutes Inc., Rockville, MD, 1993, 60–93.

25. Hazardous substance regulated under RCRA Subtitle I refers to any material defined as "hazardous substance" under Section 101(14) of the 1980 Comprehensive Environmental Response, Compensation, and Liability Act (CERCLA) of 1980.

26. U.S. Environmental Protection Agency, *RCRA Orentation Manual 1990 Edition*, Washington, D.C., 1990.

27. Wagner, Travis P., *Hazardous Waste Identification and Classification Manual*, van Nostrand Reinhold, New York, 1990.

28. Letter from Matthew Straus, EPA's Waste Characterization Branch Chief to K. Seiler, Department of Ecology, Washington, D.C., dated April 8, 1987.

29. McCoy and Associates, Inc., Contaminated Soil — Regulatory Issues and Treatment Technologies, *The Hazardous Waste Consultant,* Vol. 9, No. 5, 1991, 4.1–4.24.

30. Corrective Action — How It Works, *Superfund Report*, July 18, 1990, 3.

31. Davis, Robert C., RCRA Corrective Action Developments, in Kenneth M. Kastner, ed., *Current Developments in RCRA*, Government Institutes Inc., Rockville, MD, 1993, 184–201.

32. McCarrol, John, Introduction, in *Corrective Action Training Course*, U.S. EPA Region 9, San Francisco, 1993.

33. Corbaley, Susan, Corrective Action and the RCRA Facility Assessment Program, in *RCRA Facility Assessment Training for Arizona Department of Environmental Quality*, Science Applications International Corporation, San Francisco, 1993.

34. Wesling, Mary, Conducting the Visual Site Inspection, in *RCRA Facility Assessment Training for Arizona Department of Environmental Quality*, Science Applications International Corporation, San Francisco, 1993.

35. Corbaley, Susan, Preparing the RFA Report Under the RCRA Facility Assessment Program, in *RCRA Facility Assessment Training for Arizona Department of Environmental Quality*, Science Applications International Corporation, San Francisco, 1993.

36. Saracino, Ray, RCRA Facility Investigation, in *Corrective Action Training Course 1993*, U.S. EPA, Region 9, San Francisco, 1993.

37. U.S. Environmental Protection Agency, *RCRA Corrective Action Plan — Interim Final*, Washington, D.C., 1988.

38. Soesilo, Joseph A. and Curry, Thomas E., A Comprehensive Approach to Addressing RCRA Contaminated Sites, (draft) 1994.

39. Upon discovery of contamination, RCRA Subtitle C requires a facility to perform cleanup immediately. In this context, the term immediate is not precisely defined by EPA. This flexibility enables the state agency to approve voluntary cleanup without imposing any permit requirements.

40. Power, Jan G. and Leifer, Steven, Expediting Site Cleanups, *Waste Age*, June, 1993, 177–184.

41. Under Section 4004(u) of the 1984 HSWA, EPA must require corrective action for all releases of hazardous waste to facilities seeking closure or post-closure permits as well as those seeking ordinary operating permits.

42. EPA states that if a facility fails to utilize the opportunity to immediately clean up the contamination, it must obtain a RCRA permit, or conduct a cleanup under a comparable authority, i.e., through consent or compliance orders (55 FR 30807).

43. Case, David R., Resource Conservation and Recovery Act, in *Environmental Law Handbook*, Government Institutes Inc., Rockville, MD, 1993, 60–93.

44. McCoy and Associates, *The RCRA Land Disposal Restrictions: A Guide to Compliance*, Lakewood, CO, 1993.

45. U.S. Environmental Protection Agency, *Waste Analysis at Facilities that Generate, Treat, Store, and Dispose of Hazardous Waste — A Guidance Manual*, Washington, D.C., 1992.

46. U.S. Environmental Protection Agency, *Land Disposal Restrictions — Summary of Requirements*, Washington, D.C., 1991.

47. The wastes are known as the *"California List"* wastes.

48. Environmental Information Ltd., *Underground Storage Tank Cleanup: Status and Outlook*, Minneapolis, 1995.

49. Underground Tanks: Number of Sites Awaiting Remediation Increases by More Than 5,000 in Six Months, Agency Reports, *Environment Reporter*, June 2, 1995, 280.

50. Hall, Ridgway M., Jr. et al., *RCRA Hazardous Wastes Handbook, 9th Edition*, Government Institutes, Rockville, MD, 1991.

51. Nardi, Karen J., Underground Storage Tanks, in J. Gordon Arbuckle et al., eds., *Environmental Law Handbook — Twelfth Edition*, Government Institutes Inc., Rockville, MD, 1993, 94–119.

52. Arizona Department of Environmental Quality, *Arizona Underground Storage Tank State Assurance Fund*, Phoenix, 1993.

53. U.S. Environmental Protection Agency, *Don't Wait Until 1998 — Spill, Overfill, and Corrosion Protection for Underground Storage Tanks*, Washington, D.C., 1994.

54. Underground Tanks: Statement Promoting Risk-Based Cleanup of Sites to be Sent to States, EPA Says, *Environment Reporter*, April 7, 1995, 2416.

55. American Society for Testing and Materials, Emergency Standard Guide for Risk-Based Corrective Action Applied at Petroleum Release Sites, *1994 Annual Book of ASTM Standards*, Philadelphia, 1994, 1673–1714.

56. Jurczyk, Nicole, Marcussen, Claire, and Tucker, William, Risk Assessment: A Tiered Approach, *The National Environmental Journal*, Vol. 5, Issue 5, 1995, 20–22.

PART III
SITE ASSESSMENT
PLANNING

The major components of the site assessment planning process are described in this part. Chapter 6, the first chapter of the five chapters that explain this process, discusses the basic principles of statistics, chemistry, and hydrogeology that are pertinent to site assessment. In particular, the chapter examines important terms and the applicability of various scientific concepts and methods for the assessment of a site.

As an initial stage in the site assessment, environmental sampling and analysis are addressed in the following chapter. Chapter 7 identifies activities that are necessary prior to performing sampling activities. It discusses the issue of goal formulation and the rationale for initiating a sampling program. Included in the discussion are the requirements for achieving acceptable sampling results, the role of laboratory analysis, and the development of a health and safety plan as a supplement to the sampling plan.

Proper site characterization is critical in the assessment process. It is essential to know the type of contaminants that lie in the subsurface, the characteristics of the subsurface's soil and groundwater, and the fate and movement of the contaminants in the subsurface. Various steps in characterizing a site are discussed in Chapter 8. The chapter also identifies the types of models and the utilization of mathematical models to help planners understand the way site contaminants move in the subsurface.

Chapter 9 is specifically devoted to the discussion of risk assessment. The three major topics addressed in this chapter are the human health risk assessment, risk management, and the ecological risk assessment. The chapter discusses the major components of a risk assessment, examines how these components relate to each other, and describes the approaches to quantify them.

The final chapter of Part III (Chapter 10) identifies a variety of remediation standards that may be applicable to a contaminated site. It examines the cleanup standard from the ability of the analytical instrumentation to detect a contaminant, from the risk assessment perspective, technological aspects, and legal requirements, as well as the remediation level that is determined by the concentration of similar contaminants in the surrounding clean areas.

6

INTRODUCTION TO SAP AND RAP

Site remediation planning, whether conducted under the Superfund, hazardous waste, UST, or other program with a remediation requirement, consists of two planning components: site assessment planning (SAP) and remedial action planning (RAP). The goals of site assessment planning are to examine the characteristics of a contaminated site in terms of the extent of contamination and to understand how the contamination threatens human health and/or the environment.

Based on the findings of the site assessment, remedial action planning evaluates and selects remedial alternatives that are technically feasible, cost effective and acceptable to the community living in the vicinity of the site. The RAP represents a continuation and logical sequence of the SAP. Structurally, remedial action planning consists of the development of strategies that establish remedial action objectives, the evaluation of remedial alternatives, and the scheduling and monitoring of remedial action plan implementation.

In the real world, a site may contain a few chemicals in its subsurface soil and groundwater. If the concentration of an individual chemical is below the acceptable level, then the site is considered to be clean. If one (or more) of the chemicals exceeds the acceptable level of concentration, the property is classified as a contaminated site and requires remediation to the appropriate clean level.

Conceptually, site assessment planning is intended to assess the three major components of site contamination:

1. the origin of contamination
2. the contaminated media
3. the receptors of contamination

Assessing the origin of contamination means both identifying the sources of contamination and characterizing the nature of the released contaminants. To perform this particular task, planners need to have a general understanding of fundamental contaminant characteristics. Assessing the media means identifying the extent of the contaminant releases to the impacted environmental media and evaluating the physical features of the media in which the release occurred. The characterization of the contaminant release and the assessment of the extent of the release involve both sampling and site characterization.

Environmental sampling and analysis are aimed at obtaining accurate information from samples collected from the site. Familiarity with representativeness, validity, bias, error, confidence levels, and other statistical concepts are paramount to understanding the sampling procedure. As U.S. Environmental Protection Agency (EPA)[1] asserts, discussing a sampling event is basically talking about statistics.

The site characterization decision-making process relies heavily on geology and hydrogeology. As LaGrega et al.[2] indicate, a few topical discussions on hydrogeology will not make one a geologist or hydrogeologist. However, understanding the concept and application of this branch of science will certainly form a basis for planners to discuss issues and alternative solutions with a variety of technical experts. Given the entirely multi-disciplinary nature of site remediation, such interaction between specialists is critical to the successful outcome of the project.

The receptors of contamination consist of both humans and the surrounding environment. The impact of site contamination on the receptors is analyzed in terms of risk assessment. To facilitate an understanding of risk assessment, a brief description of the health effects caused by human exposure to hazardous contaminants is furnished as part of an overview of contaminant chemistry. Also provided in this chapter is a short discussion of basic statistics and a review of some geological and hydrogeological concepts.

1. CONTAMINANT CHEMISTRY

Chemistry is described as the study of the structure and composition of substances. In chemistry, substances are differentiated as elements, mixtures, or compounds. Substances which cannot be further decomposed by ordinary chemical means are defined as elements. Examples of elements are oxygen, silicon, aluminum, and iron. A material such as air or concrete, which consists of two or more kinds of substances and each retaining its own characteristic properties, is called a mixture. A compound is defined as a substance which can be decomposed into two or more simpler substances; carbon dioxide is an example of a compound.

Site contaminants can generally be described as consisting of both inorganic and organic compounds. The inorganics are mostly metals, such as cadmium, chromium, copper, lead, mercury, nickel, and silver. Metals are generally solid, insoluble in water, and noncombustible in bulk form but will burn in powder form. The physical characteristics of some metals are as follows:

- Cadmium metal (Cd) is a silver-white, blue-tinged, lustrous, odorless solid.
- Chromium metal (Cr) is a blue-white to steel-gray, lustrous, brittle, hard, odorless solid.
- Copper (Cu) is a reddish, lustrous, malleable, odorless solid.
- Lead (Pb) is a heavy, ductile, soft, gray solid.
- Mercury (Hg) is a heavy, silver-white, noncombustible, odorless liquid.
- Nickel (Ni) is a lustrous, silvery, odorless solid.
- Silver (Ag) is a white, lustrous solid.

The organic compounds are more complex, have more variations and are the subject matter of organic chemistry. This branch of chemistry was originally defined as the study of materials derived only from living organisms. Today, organic chemistry is known as the study of carbon compounds, whether or not they are produced by living organisms.

Carbon is extremely abundant and widespread. It is present in the tissues of humans and in all other living things on earth. The reasons for the existence of so many carbon compounds are twofold: the carbon atoms readily form covalent bonds with other carbon atoms, and the same atoms may be arranged in several different ways.[3]

A study of organic chemistry usually starts with a study of the hydrocarbons because hydrocarbons have the basic structures from which other organic compounds are derived. Hydrocarbons are compounds which are composed of hydrogen and carbon. They can be grouped into several different series of compounds based on the type of bonding between carbon atoms. Bonds are typically represented by dashes connecting atoms in the molecule.

There are aliphatic and aromatic hydrocarbons. The aliphatic hydrocarbons are straight-chain or branched-chain hydrocarbons. Methane and propane are examples of the aliphatic hydrocarbons. The aliphatic hydrocarbons can be divided into alkanes, alkenes, alkynes, and alkadienes (see Figure 6.1). The figure shows that alkanes, also known as paraffin, are hydrocarbons in which the carbon atoms are connected by a single covalent bond. Covalent means that electrons are shared between atoms. Alkanes are represented as C_nH_{2n+2}, where n is an integer. For n = 1, the compound is called methane (CH_4); for n = 2, it is called ethane (C_2H_6); and n = 7 is heptane (C_7H_{16}).

The alkenes, formulated as C_nH_{2n}, are hydrocarbons in which two carbon atoms are connected by a double covalent bond. Alkenes are also called olefins. The alkynes are hydrocarbons in which two carbon atoms are connected by a triple covalent bond. The formula for alkynes is C_nH_{2n-2} and an example of alkyne is C_2H_2, which is called ethyne and also known as acetylene. The alkadienes are hydrocarbons which have two double bonds between carbon atoms.

Aliphatic hydrocarbons containing less than 4 carbon atoms are usually gases at room temperature, those containing 5 to 16 are liquid and those having

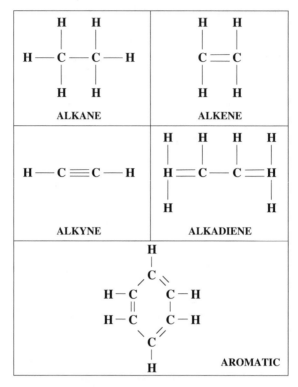

Figure 6.1. Aliphatic and aromatic hydrocarbons.

more than 16 are usually solids. Aliphatic hydrocarbons are used primarily as fuels, refrigerants, propellants, dry-cleaning agents, lubricants and solvents.

Aliphatic halogenated hydrocarbons are aliphatic hydrocarbons in which one or more hydrogen atoms have been replaced by one or more halogens (i.e., fluorine, chlorine, bromine, or iodine). Hydrocarbons containing fluorine tend to be less toxic. Those containing bromine or iodine are more toxic than hydrocarbons containing chlorine (chlorinated hydrocarbons). The halogenated hydrocarbons are used as solvents in degreasing, dewaxing, dry-cleaning agents, and aerosol propellants, refrigerants, fumigants, and insecticides.

Trifluorobromomethane (also known as fluorocarbon 1301, or refrigerant 13B1) is an example of a hydrocarbon containing fluorine. It is a nonflammable, colorless, odorless gas. Trichloroethylene (popularly known as TCE, a chlorinated hydrocarbon) is a colorless liquid with a chloroform-like odor. It is a combustible liquid but burns with difficulty. Dibromoethane, a hydrocarbon containing bromine and known as ethylene dibromide, is a noncombustible, colorless liquid with a sweet odor.

The aromatic hydrocarbons are characterized by alternate single and double covalent bonds in hexagonal carbon rings. Examples of the aromatic hydrocarbons are benzene, toluene, and dioxin. The hydrogen atoms on the aromatic nucleus may be replaced by other univalent elements or groups; however, aromatic hydrocarbons encompass compounds that include only carbon and hydrogen. The aromatic hydrocarbons are used as solvents and as feedstock for many organic compounds. Benzene is the best known aromatic hydrocarbon. It is a flammable, colorless to light-yellow liquid with an aromatic odor. Other examples are naphthalene and toluene. Naphthalene is a colorless to brown solid with an odor of mothballs. It is combustible but is difficult to ignite. Toluene is a flammable, colorless liquid with a sweet, pungent, benzene-like odor.

Naphthalene is a hydrocarbon having two benzene rings. Anthracene, or coal tar, has three rings. Others may have four rings, etc. Aromatic rings containing only carbon and hydrogen are called polynuclear aromatic hydrocarbons (PAH).

Aromatic compounds having a halogen-bearing side chain are called aromatic halogenated hydrocarbons. Generally, the more highly chlorinated the compound, the greater the toxicity. Benzyl chloride and chlorodiphenyls are examples of this group. Benzyl chloride is a combustible, colorless to slightly yellow liquid with a pungent, aromatic odor. Chlorodiphenyl (widely known as PCB, or arochlor 1254) is a nonflammable, colorless to pale-yellow, viscous liquid with a mild hydrocarbon odor.

Other hydrogen substitution compounds include, for example, aldehydes, ketones, alcohol, phenols, esters, and ethers. Aldehydes and ketones are aliphatic or aromatic organic compounds which contain the carbonyl group (C=O). The aldehydes (R–CH=O) are used as chemical feedstock and ketones as industrial solvents. In this formula, the R is used to represent a C atom with three branches. Examples of aldehydes and ketones are acetaldehyde and methyl ethyl ketone (MEK). Acetaldehyde is a flammable, colorless liquid with a pungent, fruity odor. MEK is a combustible, colorless liquid with a unique odor.

Alkanes in which one or more hydroxyl groups (–OH) have been substituted for a like number of hydrogen atoms are called alcohols. Compounds with one hydroxyl group are referred to as alcohols, those with two hydroxyl groups are glycols, and those with three groups are called glycerols. Propyl alcohol, as an example of this group, is a flammable, colorless liquid with a mild alcohol-like odor.

The group of phenols is characterized by the substitution of one or more hydrogens in a benzene ring by hydroxyl (–OH) groups. Phenol (C_6H_5OH) is the simplest of the compounds. It is a combustible, colorless to light-pink, crystalline solid with a sweet, acidic odor. Phenol liquefies when mixed with water.

Ethers are organic molecules that contain a carbon–oxygen–carbon linkage. They are used as industrial solvents and chemical feedstock for organic synthetics. Diethylene ether, or dioxane, is an example of the ethers. It is a flammable, colorless liquid with a mild ether-like odor. Below 53°F, dioxane will be in a solid state.

Esters are organic compounds generated by the reaction between an organic and inorganic acid and alcohol with the elimination of water. Ethyl formate belongs to this group. It is a flammable, colorless liquid with a fruity odor.

In order to understand the danger of site contaminants to human health, it is necessary to recognize the routes of entry for each substance and the target organs that are affected by exposure to the substance. The extent of exposure to a substance and the level of toxicity of the substance are two critical elements in the human health risk assessment. As widely defined, risk assessment is a systematic process for making estimates of risk factors due to the presence of some type of hazard at a contaminated site by evaluating exposure and toxicity.

Table 6.1 displays hazard information for selected chemicals to illustrate the danger of chemicals to human health. The table is constructed from the 1994 Pocket Guide to Chemical Hazards developed by the National Institute for Occupational Safety and Health (NIOSH). The guide is periodically updated.[4]

Organic chemical compounds can be categorized based on their volatility. Volatility is the tendency of liquid or solid organic compounds to vaporize, i.e., to pass into the vapor state. Volatility determines the rate and degree at which organic compounds will evaporate from the solid or liquid state to the vapor state. A volatile organic compound with a higher vapor pressure will vaporize quicker. The categories of organic chemicals are

- volatile organics
- semivolatile organics
- nonvolatile organics

There are no distinct criteria that differentiate between volatile and semivolatile, or between semivolatile and nonvolatile. Volatile organic compounds (VOCs), such as dichloroethane or acetone, evaporate quickly because of their high vapor pressure. Detection of VOCs is conducted by taking advantage of their high vapor pressure through a purge and trap technique. By bubbling helium gas through a water sample, VOCs are removed from the sample, swept away with the evolving gas, and separated from the gas stream for further laboratory analysis.

Nonvolatile organics such as those constituents found in lubricating oils are organic compounds with no volatility even at elevated temperatures. Semivolatile organics are organic compounds with some degree of volatility but not enough to use the purge and trap technique. The chemical analysis of these nonvolatile and semivolatile organics relies on the acidity of the compounds to facilitate their isolation from a water sample into an insoluble solvent. The insoluble solvent phases are separated from the water sample and are ready for laboratory analysis. This technique is known as solvent extraction.

Table 6.1 Health Hazards of Selected Chemicals

Contaminant	Exposure route	Target organs
Acetaldehyde	Inhalation Ingestion Skin contact	Eyes, skin, respiratory system, kidneys, central nervous system, reproductive system
Acetylene	Inhalation Skin contact	Central nervous system, respiratory system
Anthracene	Inhalation Skin contact	Respiratory system, skin, bladder, kidneys (lung, kidney, and skin cancer)
Benzene	Inhalation Skin absorption Ingestion Skin contact	Eyes, skin, respiratory system, blood, central nervous system, bone marrow (leukemia)
Benzyl chloride	Inhalation Ingestion Skin contact	Eyes, skin, respiratory system, central nervous system
Cadmium	Inhalation Ingestion	Respiratory system, kidneys, prostate, blood (prostatic and lung cancer)
Chlorodiphenyls (PCB)	Inhalation Skin absorption Ingestion Skin contact	Skin, eyes, liver, reproductive system
Chromium	Inhalation Ingestion Skin contact	Eyes, skin, respiratory system
Copper	Inhalation Ingestion Skin contact	Eyes, skin, respiratory system, liver, kidneys (increased risk with Wilson's disease)
Dioxane	Inhalation Skin absorption Ingestion Skin contact	Eyes, skin, respiratory system, liver, kidneys
Ethyl formate	Inhalation Ingestion Skin contact	Eyes, respiratory system, central nervous system
Ethylene dibromide	Inhalation Skin absorption Ingestion Skin contact	Eyes, skin, respiratory system, liver, kidneys, reproductive system
Heptane	Inhalation Ingestion Skin contact	Skin, respiratory system, central nervous system
Lead	Inhalation Ingestion Skin contact	Eyes, gastrointestinal tract, central nervous system, kidneys, blood, gingival tissue

Table 6.1 Health Hazards of Selected Chemicals (Continued)

Mercury	Inhalation Skin absorption Ingestion Skin contact	Eyes, skin, respiratory system, central nervous system, kidneys
Methyl ethyl ketone (MEK)	Inhalation Ingestion	Eyes, skin, respiratory system, liver, kidneys
Naphthalene	Inhalation Skin absorption Ingestion Skin contact	Eyes, skin, blood, liver, kidneys, central nervous system
Nickel	Inhalation Ingestion Skin contact	Nasal cavities, lungs, skin (lung and nasal cancer)
Phenol	Inhalation Skin absorption Ingestion Skin contact	Eyes, skin, respiratory system, liver, kidneys
Silver	Inhalation Ingestion Skin contact	Nasal septum, skin, eyes
Toluene	Inhalation Skin absorption Ingestion Skin contact	Eyes, skin, respiratory system, central nervous system, liver, kidneys
Trichloroethylene (TCE)	Inhalation Skin absorption Ingestion Skin contact	Eyes, skin, respiratory system, heart, liver, central nervous system.
Trifluorobromo-methane	Inhalation Skin contact	Central nervous system, heart

Source: U.S. Department of Health and Human Services, *NIOSH Guide to Chemical Hazards,* Washington, D.C., 1994.

The notable differences in the physical and chemical characteristics between volatile organic compounds and semivolatile organics are listed below:

Physical/Chemical Property	**VOC**	**Semivolatile**
Vapor pressure	High	Low
Solubility in water	High	Low
Henry's constant	High	Low
Octanol water partition coefficient (K_{ow})	High	Low
Organic carbon partition coefficient (K_{oc})	Low	High
Bioconcentration factor (BCF)	Low	High

If a liquid is in contact with air, a vapor occurs through evaporation. When the rate of molecules leaving the liquid equals the rate of molecules being redissolved, the equilibrium state is reached. At this equilibrium condition, the pressure exerted by the vapor on the liquid is defined as the vapor pressure. It is measured in atmospheres (atm) and varies with the temperature.

Solubility is the degree to which one substance will dissolve into another. Solubility in water is measured in milligrams per liter (mg/l) and is affected by temperature. The more soluble an organic is, the more volatile the substance.

In the equilibrium state, the partial pressure of a gas above a liquid is proportional to the chemical concentration in the liquid. This proportion is called Henry's constant (H) and can be stated as $H = P_g/C_l$, where P_g = the partial pressure of the gas, and C_l = the chemical concentration in the liquid. The value of H is affected by temperature and the chemical composition of the water.

The octanol water partition coefficient or K_{ow} describes how a chemical divides itself between octanol and water. The coefficient is dimensionless and can be equated as $K_{ow} = C_o/C$, where C_o = the concentration of the chemical in octanol and C = the concentration of the chemical in water.

The organic carbon partition coefficient or K_{oc} is defined as $K_{oc} = X/C \cdot f_{oc}$, where X = the concentration of chemical in the soil, C = the concentration of chemical in water, and f_{oc} = the fraction of organic carbon in the soil, which is dimensionless. The X/C component represents the degree of chemical adsorption by soil or sediment, and is called the soil water partition coefficient (K_p).

The bioconcentration factor (BCF) indicates the portion of a chemical that is likely to accumulate in aquatic organisms. The factor can be stated as $BCF = C_{org}/C$, where C_{org} = the concentration of the chemical in an organism, and C = the concentration of the chemical in water. By knowing the value of BCF for a certain organism (e.g., fish) and the concentration of the chemical in the water, the concentration of the chemical in the fish tissue (C_{org}) can be calculated by using the above equation.

When contaminants move in the subsurface, a complex set of chemical and microbiological processes will affect the fate and movement of the contaminants. These processes may retard the contaminant migration through sorption of organics and ion exchange or chemical precipitation of inorganic substances. Ion exchange is a type of sorption process for inorganic chemicals which involves electrostatic interaction, i.e., the sorption of an ion in solution into an oppositely charged soil particle. Chemical precipitation is the formation of solids out of dissolved constituents due to a change in environmental conditions (e.g., pH, temperature).

The chemical and microbiological processes may also attenuate the concentrations of the contaminants. The attenuation processes include chemical reduction–oxidation (redox), biodegradation, and volatilization. Redox consists of a couple of chemical reactions. Oxidation refers to a reaction in which a substance loses or donates electrons while a reduction reaction involves the acceptance of electrons by a substance. Redox reactions are always coupled because free electrons cannot exist in solution.

Biodegradation is a process by which microorganisms transform the structure of contaminants introduced into the environment through a metabolic process or enzymatic action. Biodegradability is the relative ease with which an organic contaminant will degrade as the result of biological metabolism. Biodegradability normally increases with increasing solubility.

Volatilization is the process of converting volatile chemical constituents from a liquid phase (in groundwater) to a gas phase (or vapor) and, ultimately, transferring the vapor to the atmosphere. The rate of volatilization is determined by solubility, molecular weight, and vapor pressure of the liquid and the nature of the gas–liquid interface.

In many contaminated sites, contamination problems often involve the so-called nonaqueous phase liquids (NAPLs). These are liquids that do not easily dissolve in water and can occur as an independent fluid phase. They can be classified into two groups: the dense NAPLs (DNAPLs) and the light NAPLs (LNAPLs).

The LNAPLs, such as petroleum products, float on water. The presence of LNAPLs can create site investigation problems because of interstitial tension effects; LNAPLs that are present in groundwater at aquifers with relatively low water saturation may not flow into a normally constructed monitoring well.[5]

The dilemma posed by DNAPLs stems from the very characteristics that define them: they are immiscible in and denser than water.[6] These organic liquids include coal tar, creosote, PCB oils, and chlorinated solvents.

The physical nature of DNAPLs can be broadly divided between two classes based on melting point.[7] The first class is the polynuclear aromatic hydrocarbons (PAHs) which have a low melting point and, therefore, are solid at temperatures typical in the subsurface resulting in limited contaminant migration. The second class is the chlorinated solvents which are liquid at normal subsurface temperatures making them extremely mobile in the subsurface environment.

Vance indicates that the properties that make chlorinated solvents so problematic in groundwater are density, solubility, viscosity, surface tension, and dehydration.[7] The problem with the high-density chlorinated solvents is the dimple effect. When released to the subsurface, the solvents will sink through the vadose zone, travel to an aquitard, seek the lowest point on the surface of that aquitard and occupy the depressions and dimples. These are the areas outside of an advective flow regime set up by a pump and treat groundwater remediation system.

The solubility of chlorinated solvents increases the potential for transport of the released solvents in the dissolved phase. Their low viscosity makes chlorinated solvents extremely mobile in the unsaturated zone. The low surface tension characteristics result in significant lateral spreading of a DNAPL release above the capillary fringe in fine-grained soil. Chlorinated solvents dehydrate clays, causing cracking and further migration through what at first appears to be an impermeable layer. These are the characteristics that make remediation of groundwater contaminated with DNAPLs a difficult undertaking.

2. BASIC STATISTICS

Statistics is defined as the branch of mathematics which provides conventions and techniques for describing quantitative data and inferring the state and behavior or environmental variables.[8] The conventions and techniques for describing quantitative data include methods for collecting, compiling, analyzing, interpreting, and presenting data which are collectively referred to as descriptive statistics. The techniques for inferring the state or behavior of variables of the environment are called methods of statistical inference or inductive statistics. In practice, statistical studies typically use both descriptive and inferential techniques.

A field person who collects a specimen of contaminated soil refers to that specimen as a sample; the laboratory chemist, who receives the specimen from the field person for further laboratory analysis, also thinks of the specimen as a sample; the field sampler and the analytical chemist use the term sample in the context of their discipline as an environmental sample. The statistician uses the term sample in another context with a different meaning.[9] The sample in statistics refers to a group of n observations actually available.

The observations represent random variables. When assessing water quality in a river, the concentrations of a pollutant cannot be adequately evaluated by a single measurement. This is because the concentrations of the contaminant (e.g., fecal coliform) in water vary from one location to another. While some of the variations can be attributed to known causes, there still remains a residual component which cannot be fully explained and must be regarded as a matter of chance.[10] It is this random variation that explains why two samples of water of equal volume taken at the same point on the river at the same time give different coliform counts. Statistically, a random variable is defined as the value of the next observation in an experiment.[11]

In descriptive statistics, the distribution of quantitative data is statistically explained. In describing the data, graphical methods and numerical methods are typically used. As an example, consider soil samples taken from the area surrounding a site which shows lead concentrations (in mg/kg) of the following values:

From the above concentration values, it can be seen that one soil sample (i.e., sample #6) concentration falls between 10 and 19 mg/kg, while three samples (sample #5, #10, and #14) have concentrations between 20 and 29 mg/kg. This means that the frequency of lead concentration between 10 and 19 mg/kg is one, and the frequency of the 20–29 mg/kg concentration is three. The following table can be developed for all values. The frequency of the concentrations between 10 and 29 mg/kg can be calculated as 1 + 3, or 4.

Sample #	Concentration	Sample #	Concentration
1	32	11	38
2	47	12	36
3	58	13	49
4	43	14	29
5	24	15	36
6	18	16	42
7	48	17	35
8	33	18	67
9	57	19	45
10	27	20	32

Concentration	Sample #	Frequency	Cumulative frequency
10–19	6	1	1
20–29	5, 10, 14	3	4
30–39	1, 8, 11, 12, 15, 17, 20	7	11
40–49	2, 4, 7, 13, 16, 19	6	17
50–59	3, 9	2	19
60–69	18	1	20

Similarly, the frequency of the concentrations between 10 and 39 is 1 + 3 + 7, or 11. The frequencies of 4 and 11 represent part of the cumulative frequency of the lead concentrations.

The frequency distribution of the soil samples tabulated above is displayed in Figure 6.2. The figure also shows the cumulative frequency of the distribution. This graphical presentation provides a quick general description of the collected soil samples but still needs further numerical descriptive measures for a better understanding of the collected soil samples. The measures include the analysis of both central tendency and variability of the distribution.

One of the most common measures of central tendency is the mean, or the arithmetic average of a set of measurements. A mean is calculated as the sum of the measurements divided by the number of measurements. Another measure of central tendency is the median, which is the value of the measurement that falls in the middle when the measurements are arranged in order of magnitude. As an example, the mean of the set of measurements 1, 3, 7, 6, 4 is (1 + 3 + 7 + 6 + 4)/5 = 21/5 = 4.2. The arrangement of the measurements in order of magnitude is 1, 3, 4, 6, 7 and the middle value of the arrangement is 4.0 which represents the median. A mode, which is defined as the value that occurs most frequently in the distribution, is also another measure of central tendency.

The above example illustrates that the measures of central tendency are intended for locating the center of the distribution of data. The center functions as the representative of the data. For a distribution of data that is normal, the frequency polygon will be in the form of a bell shape. The normal distribution

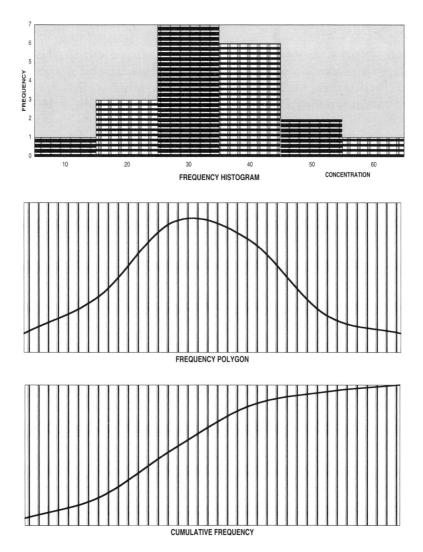

Figure 6.2. Frequency distribution and cumulative frequency.

is also called a Gaussian distribution. In a perfectly symmetrical normal distribution, the values of the mean, the mode, and the median are all the same.

In practice, many geochemical data in nature are not normally distributed. The spread of data often follows the so-called logarithmic normality distribution or simply the lognormal distribution. By utilizing a normalized approach, however, this lognormal distribution can be transformed into a normal distribution.

In addition to a measure that indicates where the center is located on the distribution scale, it is also necessary to have a second measure that provides an indication of how far the data in the distribution scale are scattered. The simplest and easiest way to calculate the measure of spread or variability of the distribution is the range, i.e., the difference between the highest and lowest value in the distribution. Another measure of spread is the standard deviation(s) which is the square root of the variance (s^2). The variance of a sample of n observations is calculated as the mean of the squares of deviations from the mean (see Plate 6.1).

Sample Mea n: $\bar{y} = \dfrac{\sum\limits_{i=1}^{n} y_i}{n}$

Sample Var iance: $s^2 = \dfrac{\sum\limits_{i=1}^{n} \left(y_i - \bar{y}\right)^2}{n}$

Sample Sta ndard Devi ation: $s = \sqrt{s^2}$

Population Mean: $\eta = \dfrac{\sum\limits_{i=1}^{N} y_i}{N}$

Population Variance: $\sigma^2 = \dfrac{\sum\limits_{i=1}^{N} \left(y_i - \eta\right)^2}{N}$

Population Standard Deviation: $\sigma = \sqrt{\sigma^2}$

Plate 6.1. Mean, variance, and standard deviation.

In statistics, it is imperative to distinguish a value that represents a sample and a value that represents a population. A sample is a group of n observations actually collected from the field. A sample is a sub-set of a population. A population is a large set of N observations from which the sample is to come. The mean, variance, and standard deviation of a population are provided in Plate 6.1. The difference between sample mean and population mean is defined as bias.

In a normal distribution, the mean of a population is the center of the distribution and the standard deviation measures the distance from the mean. In describing the normal distribution, the distribution is always characterized by its mean and variance (or standard deviation). Consider the area under the curve in Figure 6.3 as 100%.

In the setting displayed in Figure 6.3, the probability that a positive deviation from the mean will exceed one standard deviation is the area under the curve on the right side of one standard deviation, which is approximately 15.87%. The probability that a positive deviation from the mean will exceed two standard deviations is the area to the right of the two standard deviations

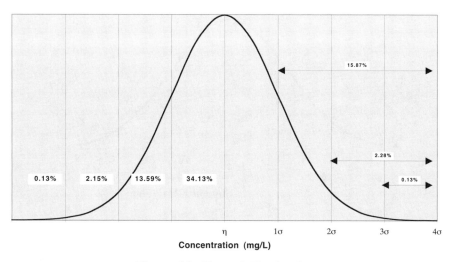

Figure 6.3. Normal distribution.

(2.28%), and the probability that the deviation will exceed three standard deviations is 0.13%.

Because of its symmetrical form, the normal distribution curve shows that the probability that a deviation in either direction will exceed two standard deviations is $2 \times 2.28\% = 4.56\%$. The relationship between standard deviation, the normal curve area, and the tail area are tabulated in Table 6.2. Both areas are indicated by approximate percentages. In addition to the standard deviation, another frequently used measure of spread is the relative standard deviation, or the coefficient of variation (CV), which is defined as the percentage of spread of the data relative to the mean value, which equals the standard deviation divided by the mean and multiplied by 100%.

In order to standardize a normal random variable, both the population mean and standard deviation of the population must be known. In practice, the population standard deviation is unknown, therefore the sample standard deviation is used to calculate the distribution. This distribution is also a bell-shaped, symmetrical curve and is called the Student's t-distribution, or simply the t-distribution. The tail area probability of the Student's t-distribution for some selected values of n is tabulated in Table 6.3. It can be seen from the table that when the value of n is infinite, the Student's t-distribution becomes the standard normal distribution. The t-distribution is useful, for example, to statistically determine whether a site has been remediated to the applicable cleanup level, or, if downgradient monitoring wells with detectable contaminant concentrations are significantly different from upgradient well.

Suppose that random samples of n size are collected from a population and the average is calculated for each sample. The result would be many different average values. These averages, when plotted, will form a probability

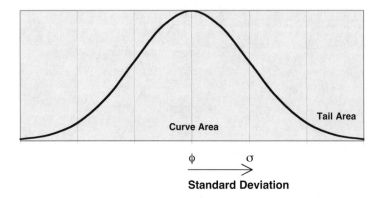

Standard Deviation (σ)	Curve Area (%)	Tail Area (%)
0.26	10.0	40.0
0.53	20.0	30.0
0.84	30.0	20.0
0.88	31.0	19.0
0.92	32.0	18.0
0.96	33.0	17.0
1.00	34.0	16.0
1.04	35.0	15.0
1.08	36.0	14.0
1.13	37.0	13.0
1.18	38.0	12.0
1.23	39.0	11.0
1.28	40.0	10.0
1.34	41.0	9.0
1.41	42.0	8.0
1.48	43.0	7.0
1.55	44.0	6.0
1.65	45.0	5.0
1.75	46.0	4.0
1.88	47.0	3.0
2.06	48.0	2.0
2.33	49.0	1.0
3.08	49.9	0.1

Table 6.2

distribution known as the sampling distribution of the average. The square root of the variance of this distribution is called the standard error of the mean.

A significant test in statistics typically takes the form of a hypothesis test. The hypothesis to be tested is called the null hypothesis and is designated as H_0. A significant statistical level at which the null hypothesis will be rejected is then selected. A significant level (i.e., the tail area of the probability curve) of 0.025 in a two-sided test of the null hypothesis is equal to $2 \times 0.025 = 0.05$ or 5%. This translates to a 95% confidence level.

In testing a hypothesis, there are potentially two types of decision-making errors: type I and type II errors. A type I error is incurred by rejecting the null hypothesis when it is true. The probability of making a type I error is denoted

Table 6.3 Tail Area Probability of the *t*-Distribution

Degree of freedom (n–1)	n	Tail area probability				
		$t_{0.100}$	$t_{0.050}$	$t_{0.025}$	$t_{0.010}$	$t_{0.005}$
2	1	3.078	6.314	12.706	31.821	63.657
3	2	1.886	2.920	4.303	6.965	9.925
4	3	1.638	2.353	3.182	4.541	5.841
5	4	1.533	2.132	2.776	3.747	4.604
6	5	1.476	2.015	2.571	3.365	4.032
7	6	1.440	1.943	2.447	3.143	3.707
8	7	1.415	1.895	2.365	2.998	3.499
9	8	1.397	1.860	2.306	2.896	3.355
10	9	1.383	1.833	2.262	2.821	3.250
20	19	1.328	1.729	2.093	2.539	2.861
30	29	1.311	1.699	2.045	2.462	2.756
Infinite	Inf	1.282	1.645	1.960	2.326	2.576

by an alpha symbol. This type of error is also called a false positive error, e.g., detecting contamination when no contamination is actually present on site.

A type II error is incurred by not rejecting the null hypothesis when it is false and the alternative hypothesis is true. The probability of making a type II error is denoted by a beta symbol. A type II error is also known as a false negative error; an example is failing to detect site contaminant when some contamination exists. Because of this potential for error, the goodness of a statistical test of a hypothesis is measured by the probabilities of making a type I or a type II error.

3. GEOLOGY AND HYDROGEOLOGY

Geology is the science of the earth; it deals with the earth's physical structure, composition, and its changes over time. The three major divisions of the earth are the lithosphere, hydrosphere, and the atmosphere. Lithosphere, the outer crust of the earth, consists of continents and the physical features beneath the ocean's surface. While the hydrosphere is made up of oceans, lakes, streams, and groundwater, the atmosphere consists of gases, dust, and water vapor.

An understanding of geology is imperative in site remediation. Knowledge regarding pertinent geological concepts, earth material and properties is relevant to soil remediation. This knowledge is also critical to the proper characterization of an aquifer in groundwater modeling. In this section, the introductory discussion of geology is focused on a few limited topics. The topics include the discussion of consolidated bedrock, unconsolidated materials, and properties of soil and rock.

Rocks represent the essential component of the earth's crust. They were formed from single minerals and mixtures of several minerals. As the building block of rocks in the lithosphere, minerals are naturally occurring solid elements or components.[12] Although there are many thousands of known minerals, only about 25 minerals are considered abundant in the earth's crust.

The major groups of rock forming minerals include the silicates, oxides, carbonates, sulfates, and ore minerals. The silicates can be classified into aluminum silicates (called feldspar) which are light in color, and ferromagnesian silicates whose colors range dark green to black. An example of a very common oxide is colorless silicon dioxide, better known as quartz. Some gray quartz occurs in nature due to the addition of chemical impurities to the pure colorless quartz. Carbonate, such as calcium carbonate, is an important element because it is the primary constituent of limestone. Gypsum is an example of sulfate, which is the major source of sulfate in groundwater. Ore minerals such as pyrite (iron sulfite) or galena (lead sulfite) are examples of mineral deposits that are commercially mined.

Consolidated rocks were formed from mineral particles that were subjected to a physical force (heat and pressure) or chemical reactions that formed them into a solid mass. This is commonly called bedrock. Bedrock includes sedimentary rocks, unconsolidated materials turned into a solid body, igneous rocks that were formed from a molten state, and metamorphic rocks that were recrystallized by extreme heat and pressure. Examples of some consolidated sedimentary rocks are limestone, dolomite, shale, sandstone, and conglomerate. Examples of igneous rocks are granite and basalt; metamorphic rocks are slate and marble.

Sedimentary rocks are laid down in a series of distinct layers with the oldest layer at the bottom and the youngest layer at the top. The individual layers (strata) can be recognized because of the differences in color, mineral composition, and grain size. In some sedimentary rocks, the rocks may consist of conglomerate covered by sandstone and sandstone covered by shale. In another situation, the sedimentary rocks may consist of shale covered by sandstone and sandstone covered by conglomerate. Conglomerates, however, are less common than sandstone or shale and they may be absent from the rocks.[13]

Most unconsolidated (soil-like) materials consist of material derived from the disintegration of consolidated rocks. Significant unconsolidated materials include soil, clay, silt, sand, and gravel. Also important are fragments of shells of marine organisms.

Different kinds of voids can be found in rocks. The lava tubes in basalt and pores in sand and gravel are examples of voids that were created at the same time as the rock developed. These are called primary openings. If the voids were created after the rock was formed, they are called secondary openings. Voids in limestone that occurred due to the action of water that dissolved the rock, or fractures in granite are examples of secondary openings. Many sedimentary rocks that serve as sources of groundwater can exist in the form of semiconsolidated materials, with both primary openings (pores) and secondary openings (fractures).

Over time, the various types of rocks are subjected to forces that may change their location and their physical characteristics. Although some of these changes may occur deep in the subsurface, many changes also occur on the surface. These surface changes are called weathering. Weathering consists of two processes in which rock is broken up in place: disintegration, a mechanical process that breaks large masses of rock into small fragments without changing the mineral components of the rock; and decomposition, a chemical process that changes the arrangement of atoms and forms new substances. Disintegration and decomposition work together.

Igneous and metamorphic rocks decompose to form clay and soluble forms of silica and carbonates. Quartz is changed very little by the weathering process. Its grains in large quantities are known as sand. Soil is the most familiar product of weathering and consists of a variety of minerals. The mineral composition of a soil with respect to the size and size distribution of the primary particles is referred to as soil texture.[14]

The textural classification of soil is based on the proportion of clay, silt and sand in the soil. Soil consisting of 50% clay, 25% sand, and 25% silt, for example, is called clay. A soil sample of 50% clay, with 40% silt and 10% sand is called silty clay. Soil that has a proportion of sand, silt, and clay of 65:25:10 is called sandy loam. Other textural classes in this classification are sand, silt, loamy sand, loam, silt loam, sandy clay loam, sandy clay, clay loam, silty clay loam, and silt loam. The names of the classes are suggestive of the physical behavior of the respective soil. The clay family, for example, normally has a large surface area per unit mass of particles which makes some classes of clay easy to swell.

The physical behavior of a soil is also influenced by its structure. Decomposed organic matter (humus) and certain inorganics (oxides) bind soil particles together into granules or aggregates. The structure of the aggregate determines the pore size and pore distribution in soil. The size and configuration of soil pores are critical because they affect the water retention ability of a soil as well as the water and air transport properties of the soil. A soil property which is normally used to characterize the structural state of a soil sample is the bulk density, which is defined as the mass of dry soil per unit bulk volume.

The above review of some important geological concepts and terminology is essential for a better understanding of site remediation. Similarly critical is the understanding of basic hydrogeology. Hydrogeology is concerned with groundwater and is defined as the study of the laws governing the movement of subterranean water, the mechanical, chemical, and thermal interaction of this water with the porous earth, and the transport of energy and chemical constituents by the flow.[15] Because groundwater is one component of the hydrologic cycle, the concept of the hydrologic cycle is discussed first before the review of groundwater hydraulics. Finally, the last few paragraphs of this section are dedicated to a description of the contamination processes occurring in subsurface soil and groundwater.

The hydrologic cycle is a continuous process of water circulation between the ocean, the atmosphere, and the land. The cycle can be portrayed as a system of various moisture inputs and outputs on a basin-wide scale where precipitation and water from melting snow are taken as inputs, while evaporation and transpiration along with stream run-off are outputs. Precipitation occurs when water particles fall from the atmosphere and reach the ground surface. Evaporation can take place from both soil and free water surfaces. Evaporation from plants is called transpiration; evaporation and transpiration are referred to as evapotranspiration.

Stream run-off, also termed as overland flow, carries water until it infiltrates the soil, enters a stream or lake, or evaporates. The run-off is augmented by interflow and base flow. Interflow is a process that operates in the subsurface, but above the zone where soils are saturated with water. A base flow is a component of discharge to streams from the saturated zone of the subsurface where groundwater intersects a stream bed. Infiltration of water into the subsurface represents the source of interflow. Infiltration also provides recharge to groundwater. Groundwater is subsurface water below the water table in soils and geologic formations that are completely saturated, i.e., where the pore spaces or soil matrix are filled with water.

Given a volume of soil in the subsurface V_t, this total volume consists of the volume of the solids (V_s) and the volume of the voids (V_v). The voids are capable of containing air or water. The volume of water (V_w) contained in the void indicates the degree of saturation (calculated as V_w/V_v) or moisture content (V_w/V_t) of the soil. Both terms are expressed as a percentage.

The water table is the underground water surface whose pressure equals the atmospheric pressure. Hydrodynamically, any point on the water table has a hydrostatic pressure equal to the atmospheric pressure. The zone below the water table is the saturated zone. The depth to water table is an important parameter measured in the field. The water table represents the upper part of the saturated zone. At the contact between a saturated material and a dry porous material above the water table, the water rises to a certain height in the porous material. In this case, water is adsorbed as a film on the surface of soil and rock particles, and rises in small-diameter pores against gravitational force.

The area where this capillary phenomenon occurs is called the capillary fringe. The capillary phenomenon is caused by surface tension, a force parallel to the water surface in all directions because of an unbalanced molecular attraction of the water at the boundary. The soil above the water table and the capillary zone is collectively called the vadose zone, unsaturated zone, or zone of aeration. In the vadose zone and the capillary fringe, water is under suction due to a negative hydrostatic pressure which may be pulled upward by evapotranspiration or dragged downward by gravity. In the U.S., the unsaturated zone extends from the land surface to depths ranging from very shallow in humid areas to hundreds of feet in the more arid regions.

The water table is not a barrier to flow. Depending on the specific energy relations, flow of water across the water table can occur from the vadose zone to the saturated zone. This is called groundwater recharge. The flow can also occur from the saturated zone to the vadose zone. This is called groundwater discharge.

A porous body is a solid that contains holes. Porosity, which is the proportion of voids in a mass, is defined as $n = V_v/V_t$ and is expressed as a percentage. A related concept is the void ratio which is defined as $e = V_v/V_s$ and expressed as a fraction. Porosity can range from zero to more than 60%. Crystalline rock, for example, has a porosity of 0–5%, limestone 0–20%, coarse sand 31–46%, and clay 34–60%. Permeability is the ease with which a fluid can move through a porous medium and is measured by the rate of flow. The value is approximately proportional to the square of the mean grain diameter, i.e., permeability $= C \cdot d^2$ where C is a dimensionless constant and d is the mean diameter of the grain.

A groundwater reservoir is composed of rock strata or unconsolidated materials. The subsurface unit that is sufficiently permeable to supply water in a usable quantity to wells and springs is called an aquifer. Intervening geological layers between aquifers are called confining beds which have low permeability. Aquifers facilitate groundwater movement from recharge areas to discharge areas. There are two types of aquifers:

1. water table or unconfined aquifers
2. artesian or confined aquifers; a confined aquifer is an aquifer that is situated between two confining beds in which the static water level in a well rises above the top of the water-bearing unit

An imaginary surface that indicates the water level in an artesian aquifer is defined as potentiometric surface. A perched aquifer is an unconfined aquifer separated from an underlying body of groundwater by an unsaturated zone. The top of this saturated subsurface unit is called the perched water table. A perched water table may be either permanent or temporary.

Aquitards are beds or rock strata of lower permeability that contain water but do not readily yield water to pumping wells. A hydrostratigraphic unit is a formation, part of a formation, or a group of formations in which there are similar hydrologic characteristics that allow for a grouping into aquifers and associated confining layers.

When groundwater flows, it may reappear on the surface and form a seep or a spring. Groundwater may flow hundreds of miles before it reappears on the surface. Groundwater discharge to streams constitutes the base flow of many perennial streams. Slow groundwater movement and the large volume of aquifers result in detention times of many years. This makes the annual renewal rate of groundwater very small. Both the velocity and the direction of groundwater movement represent important factors in understanding the nature of groundwater contamination. To comprehend the behavior of groundwater movement, it is necessary to understand Darcy's law. This law states that the flux specific discharge (or flow discharge) q is equal to the volumetric flow Q per unit surface area A, or q = Q/A.

In hydraulics, the flow discharge can be defined as a function of the change in water level elevation (h) over the distance (l) through which the change takes place. This means that as groundwater moves from one point to another, friction must be overcome which results in an energy loss over the distance traveled. This is expressed as dh/dl, or the hydraulic gradient. Darcy's law can thus be stated as q = Q/A = –K · dh/dl where K represents a constant of proportionality. The equation implies that the discharge or the velocity of flow is proportional to the hydraulic gradient. The hydraulic gradient is always negative because groundwater flows in the direction of decreasing elevation. This relationship holds true as long as the flow is laminar. Under turbulent flow, the water particles take more circuitous paths. Darcy's law is applicable for flow through most granular material.

The K coefficient in Darcy's law is also referred to as hydraulic conductivity, i.e., the capacity of a medium to transmit water. K is, therefore, a function of both the characteristics of the medium and type of fluid. For example, the hydraulic conductivity of water is greater than that of diesel. The higher the conductivity, the better the aquifer transmits water. Permeability is a rather qualitative term used to describe the capacity of a medium to transmit any fluid. It is used in the petroleum industry where the fluids of interest are oil, gas, and water. Hydraulic conductivity is used most frequently in groundwater literature when dealing with a single water phase. Hydraulic barriers to flow are those beds having low permeability characteristics, e.g., a fine-grained layer that retards the flow of water. The capacity of an aquifer to transmit fluid of the prevailing viscosity is called transmissivity (T). It is defined as T = K × b where K is the hydraulic conductivity and b is the aquifer thickness.

The work needed to move a unit mass of fluid from any point to the designated point in a flow system is defined as the fluid potential. In groundwater hydrology, potentials are defined in terms of head, which can be expressed simply as an elevation.[16] The total head, or hydraulic head (h) is the sum of the fluid potentials at a point. It is the sum of the elevational head, the pressure head, and the velocity head. The hydraulic head is the driving force influencing the movement of groundwater; groundwater moves in the direction of decreasing total head.

Because groundwater movement through porous media is uniformly slow, the velocity head can be disregarded. As a result, under laminar flow conditions, the hydraulic head is calculated as the sum of the elevation head and the pressure head (see Figure 6.4). By measuring the depth to water in a confined aquifer from the top of the well casing and subtracting the height of the well casing above a given datum (normally sea level), the hydraulic head relative to the given datum can be determined.

Equipotential lines are a set of lines that connect points of equal hydraulic head. This set of lines represents the height of the water table or, in the case of a confined aquifer, the potentiometric surface. A set of idealized paths that

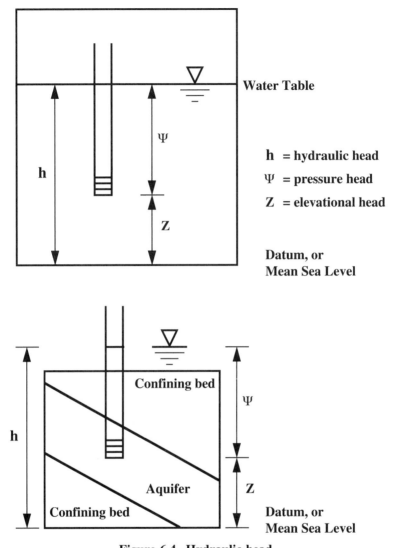

h = hydraulic head

Ψ = pressure head

Z = elevational head

Figure 6.4. Hydraulic head.

are followed by particles of water as they move through an aquifer is referred to as flow lines. Flow lines in isotropic aquifers are perpendicular to equipotential lines. If the hydraulic conductivity at a point in an aquifer is the same in all directions, the aquifer is known as an isotropic aquifer. Conversely, if the hydraulic conductivity is directionally dependent, the aquifer is anisotropic.

There are an infinite number of flow lines and equipotential lines in an aquifer. Equipotential lines and flow lines together make up the so-called flow nets. When groundwater moves through permeable material, there is no change in head with time. This is expressed as $dh/dt = 0$ and represents a steady-state flow condition. Although steady-state flow does not perfectly occur in nature, the concept is useful for aquifer modeling. In practice, aquifers may consist of heterogeneous soil properties which can result in more complicated aquifer modeling. This may include the use of an unsteady-state numerical solution of partial differential equations which require three-dimensional modeling of the hydrogeologic properties of the aquifer.

Modeling the movement of contaminants in groundwater involves the analysis of groundwater flow characteristics and geochemical transport. As contaminants travel through the soil to the groundwater system, the contaminants pass through the different hydrologic zones, experiencing different physical and chemical processes. Significant attenuation processes occur in the vadose zone. Some contaminants are adsorbed onto soil particles and, as the contaminants percolate through the vadose zone in response to gravity, they are decomposed through oxidation and aerobic biological degradation due to microbial activity.

In the capillary zone, the lighter contaminants will float on top of the water table. The dissolved contaminants will travel in the groundwater in the saturated zone in both horizontal and vertical directions. The movement is dictated by the value of the hydraulic gradient. Oxidation of contaminants is limited due to lack of oxygen in the saturated zone.

Dissolved contaminants in a laminar flow follow groundwater movement and form distinct plumes. These contaminant plumes can be several miles downstream of the pollution source and move at a speed of less than one foot per day.[17]

The size and shape of a plume depends on the types of contaminants, groundwater movement, geochemical processes, and site hydrogeology. Pumping from wells may affect groundwater flow and can alter the movement of a contaminant plume. These factors make contaminant movement in groundwater difficult to assess. As a general observation, volatile organic contaminants in groundwater are extremely mobile; the lighter contaminants tend to float on top of the water table and the heavier, denser contaminants sink to the bottom, which tends to create a plume that accumulates at the base of the aquifer.

7   ENVIRONMENTAL SAMPLING

Two of the major steps in site remediation are to identify the types of contaminants that are present in the site's soil and groundwater and to assess at various distances the concentrations of the contaminants that have migrated from the source of contamination. These steps are vital to the process of fully characterizing the site. Site characterization is very important in site remediation because it will determine the type of cleanup activities and, therefore, the costs associated with them. While site characterization is described extensively in Chapter 8, this chapter specifically addresses environmental sampling and analysis which represents an initial stage in site characterization.

Prior to any major sampling activity, a preliminary review of existing site documentation (files, reports, aerial photographs, maps) and a visit to the site (visual observation and interviews) should be conducted to assess the general conditions of the site. The review should obtain meaningful information such as past and present data of site owners or operators, chemicals used or stored on-site, waste storage and disposal areas, layout of structures such as buildings and underground tanks, soil types and thickness, and depth to groundwater. The site visit should be used to obtain information on:

- physical features that may limit sampling activity or indicate evidence of contamination such as soil staining or stressed vegetation
- possible site access and sampling points
- possible off-site contamination
- areas where potential hazards may be encountered due to the past burial of explosives or volatile chemicals on-site.
- locations of wells (potable water, industrial, and monitoring wells) placed on-site and surrounding the site
- the location and concentration of the human population in the area

The purpose of sampling is to obtain a small and informative portion of the site features (i.e., the overall soil, groundwater, sludges, or other environmental media) being investigated. Usually representative samples are sought; however, nonrepresentative (also called targeted) samples are sometimes needed. An example might be sampling a hot spot, a term used to refer to a particular area at a contaminated site that appears to be highly polluted.

In general, however, samples that are not representative of the population of interest are of little use. In fact, samples that are not representative may lead to the incorrect statistical inferences about the site features. Because of this reason, planning for representative sampling should be made an integral part of the design of environmental analyses.[18] In order to obtain samples that are unbiased and collectively representative of the site features, a proper sampling strategy must be adopted.

The sampling strategy, sample collection, and sample analyses involve the field staff, the laboratory analyst, and the data user. Close coordination among these three participants is critical for achieving successful results. An effective and efficient sampling and analysis also requires good planning. Without it, the sampling project may produce data that leads to invalid conclusions regarding the site because the samples were not collected and analyzed in an acceptable manner.

Poor sample collection procedures will likely yield nonrepresentative samples and contribute to the uncertainty of the analytical results. Because sampling errors and analytical errors occur independently of each other, sampling related errors cannot be accounted for by sampling quality control measures such as control samples or laboratory blanks.[19] Keith argues that thorough sampling documentation can identify poor sampling protocol errors; however, the errors can rarely be corrected without resampling.

Because of the importance of appropriate field sampling, laboratory analysis, and data management, the planning aspect of these three components is critical to a successful sampling effort. Although each of the components is often planned independently, field sampling, laboratory analysis, and data management are essentially interrelated and the objectives of each component must be known to all of the participants involved with the assessment of a contaminated site.[20]

The discussion of environmental sampling in this chapter starts with goal formulation and the rationale for initiating a sampling project. Included in this discussion is an overview of data quality objectives (DQOs) and the quality assurance project plan (QAPP) to show how these two planning components are related to the sampling project; also included is the examination of laboratory analysis (or analytical methods) required by existing regulatory programs. The analytical method is an integral part of the sample planning process because any decisions made regarding the analytical method directly impact the sampling protocol, such as the mandatory specifications for the type of sampling container, or the appropriate procedure to preserve samples.

1. REQUIREMENTS FOR ACHIEVING QUALITY SAMPLING

Similar to other categories of professional, technical, or engineering projects, sampling activity is governed by certain sets of performance or quality criteria. The criteria are established as qualitative and quantitative statements which specify the quality of data required to support remedial decisions.[21] The statements, known as DQOs, are determined based on the end uses of the data to be collected and established prior to any data collection activity.

DQOs are applicable to all data collection activities during site remediation, including those performed for preliminary assessment, or data collection activities for remedial design and implementation. Data quality (and quantity) requirements should be specified for each data collection activity. The development of DQOs is initiated during the scoping of a remedial project and is completed with the development of a sampling plan for each phase of the project. The process must provide a framework to ensure that all critical issues related to the collection of data with specified quality are addressed. The 1987 EPA document defines the DQO process as consisting of three interrelated steps:[21]

1. the identification of the rationale for requiring data collection activity
2. the description of data utilization
3. the compilation of all data quality objectives

The first step, justifying why new data are needed to make remedial decisions, includes the thorough review of available information including the examination of previous analytical data, if any, and then confirming the information with on-site observations and determining if additional on-site activities are required. Data examination at this phase refers to assessing the adequacy of existing data in terms of their statistical validity and data sufficiency.

The purpose of data validation is to identify invalid data and to qualify the usability of the remaining data. The validation procedure can be more rigorous when the data are intended for litigation compared to that used for general screening of a site. Determining the adequacy of the data reduces to a certain degree, the uncertainty surrounding a decision.

As an integral part of the first phase of the DQO process, a conceptual model is developed which describes a site and its environment and hypothesizes the contaminants present on-site, their routes of migration, and their potential impacts on sensitive receptors. Based on the model description, some hypotheses such as the source of contamination can be contained, the migration pathway can be eliminated, or the receptors are not impacted by migration of contaminants, require testing. These hypotheses are tested and refined throughout the site assessment process.

The hypotheses are also used in conjunction with the level of adequacy of the existing data to specify sampling decisions in a format of clear and concise decision statements, and are consistent with the ultimate objective of selecting remedial alternatives to address the entire site. An example of CER-CLA's (Comprehensive Environmental Response, Compensation, and Liability Act) general remedial investigation-feasibility study (RI/FS) objectives is tabulated in Table 7.1. The table illustrates that with an objective of determining the presence or absence of contaminants, sampling for remedial investigation activity must be conducted at the source of contamination and in all contamination pathways.

The second step of the DQO process is to identify the utilization of the collected data. Utilization of sampling data reflects the general purposes for which data will be collected. Examples of data uses include:

- Health and safety. Data are used to establish the level of protection for anybody working at the site. The workers should be aware of what chemicals are present and how dangerous they are.
- Site characterization. Data are used to determine the nature and extent of contamination at the site. This requires relatively extensive data collection.
- Risk assessment. Data are used to assess the threat of the site to public health and the environment. This is very critical where exposure to humans is potentially high.
- Evaluation of remedial alternatives. Data are used to evaluate remedial technologies that potentially are applicable to the site.
- Determination of responsible parties. Data are used to assist in identifying the parties who were responsible for contaminating the site. The data are very useful in establishing liability at multi-party sites and sometimes useful in establishing the amount of liabilities.

Based on the intended uses of data, the types of data needed and the quality of the data can be identified. This should not be limited to chemical analytical parameters but also include physical parameters such as the soil's porosity and permeability which can be used to evaluate contaminant migration. During the identification of data type, the types of laboratory analysis performed on each sample must be determined. The quality of data needed can be reflected, among other things, by the level of laboratory analysis. This is illustrated in Table 7.2.

The table shows, for example, that if the data are used for monitoring of contaminants during site characterization, field screening to detect total organic/inorganic vapor using portable instruments is required. The quality of data, i.e., whether they can or cannot provide an indication of contamination, is dependent both on whether the instruments used are calibrated appropriately and the collected data are interpreted correctly.

Indicators of data quality such as precision, accuracy, representativeness, completeness, and comparability, known as PARCC parameters, should be considered when developing DQOs.

Table 7.1 General RI/FS Objectives

Objective	RI Activity	FS Activity
Determine presence or absence of contaminants.	Establish presence or absence of contaminants at source and in all pathways.	Evaluate applicability of no-action alternative for source areas and pathways.
Determine types of contaminants.	Establish "nature" of contaminants at source and in pathways; relate contaminants to potentially responsible parties cost recovery.	Evaluate environmental and public health threats, identify applicable remedial technologies.
Determine quantities or concentrations of contaminants.	Establish concentration gradients.	Evaluate cost to achieve applicable or relevant and appropriate standards.
Determine mechanism of contaminant release to pathways.	Establish mechanics of source and pathway(s) interface.	Evaluate effectiveness of containment technologies.
Determine direction of pathway(s) transport.	Establish pathway(s) and transport route(s), identify potential receptors.	Identify most effective points in pathway to control transport of contaminants.
Determine boundaries of source(s) and pathway(s).	Establish vertical and horizontal boundaries of source(s) and pathway(s) of contamination.	Evaluate costs to achieve relevant and applicable standards, identify applicable remedial technologies.
Determine environmental and public health factors.	Establish routes of exposure, and environmental and public health threat.	Evaluate applicable standards or risk, identify applicable remedial technologies.
Determine source and pathway contaminant characteristics with respect to mitigation (bench studies).	Establish range of contaminants and concentrations.	Evaluate treatment schemes.

Source: U.S. Environmental Protection Agency, *Data Quality Objectives for Remedial Response Activities — Development Process,* Washington, D.C., 1987.

Table 7.2 Analytical Levels Appropriate to Data Uses

Data uses	Analysis level	Type of analysis	Limitations	Data quality
Site characterization, monitoring during implementation.	I	Total organic/inorganic vapor detection using portable instruments.	Instruments respond to naturally occurring compounds.	If instruments calibrated and interpreted correctly, can provide indication of contamination.
Site characterization, evaluation of alternatives, engineering, design, monitoring during implementation.	II	Field test kits. Variety of organics by GC, inorganics by AA, XRF. Tentative ID, analyte-specific.	Tentative ID. Techniques and instruments limited mostly to volatiles, metals.	Dependent on quality assurance/quality control (QA/QC) steps employed. Data typically reported in concentration ranges.
Risk assessment, PRP determination, site characterization, evaluation of alternatives, engineering, design, monitoring, during implementation.	III	Detection limits vary from low ppm to low ppb. Organics/inorganics using EPA procedures other than CLP can be analyte- specific. Resource Conservation and Recovery Act (RCRA) characteristic tests.	Tentative ID in some cases. Can provide data of same quality as levels IV, NS.	Similar detection limits to CLP. Less rigorous QA/QC.
Risk assessment PRP determination evaluation of alternatives engineering design.	IV	HSL organics/inorganics by GC/MS, AA, ICP. Low ppb detection limit.	Tentative identification of non-HSL parameters. Some time may be required for validation of packages.	Goal is data of known quality. Rigorous QA/QC.

Risk assessment PRP determination.	V	Nonconventional parameters. Method-specific detection limits. Modification of existing methods. Appendix 8 parameters.	May require method development/modification. Mechanism to obtain services requires special lead time.	Method-specific.

Source: U.S. Environmental Protection Agency, *Data Quality Objectives for Remedial Response Activities — Development Process*, Washington, D.C., 1987.

- Precision measures the reproducibility of measurements under a given set of conditions and can be stated in statistical terms such as standard deviation, coefficient variation, or range.
- Accuracy is reported as bias. It is caused by error in the sampling process, preservation, sample preparation, handling, and laboratory analysis. Sampling accuracy may be assessed using quality assurance/quality control (QA/QC) procedures, such as evaluating sample blanks. The QA/QC procedure will be discussed later in this chapter.
- Representativeness measures whether the collected samples reflect site conditions and whether the sample data accurately and precisely represent the characteristics of the site.
- Completeness refers to the percentage derived from dividing the number of valid measurements by the total number of measurements. This parameter indicates if a sufficient amount of valid data has been generated.
- Comparability is a parameter that indicates the confidence with which one data set can be compared with another. It is used to ensure that sampling results can be compared, for example, with results from past sampling.

The third and last step in the DQO development process is to compile the array of specific DQOs into a detailed comprehensive data collection program. The output of this step is a well-defined sampling plan. In compiling DQOs, a detailed list of all samples to be collected should be provided. The details include:

1. the type of sampling media or matrices (air, water, soil)
2. sample type (composite, grab)
3. number of samples
4. sample location
5. laboratory analytical methods
6. the type and number of QA/QC samples

The three DQO development steps in the previous paragraphs should be viewed as an interactive and iterative process, in that all the DQO elements are continually reviewed and reexamined. As shown in the process of developing them, DQOs define the level of uncertainty that is allowed for site remediation. This level is determined based on sampling variability and limitations, and uncertainty in the laboratory procedures; therefore, DQOs basically determine the degree of error that can be tolerated in the data. DQOs are used to formulate a sampling design that achieves the desired control on uncertainty, thus allowing the remediation decisions to be made with acceptable confidence.[22]

Because of variations in the intended use of the data, the level of detail and quality requirements will vary from one project to another. The variability of site characteristics makes it impossible to apply generic DQOs to all remedial activities. Procedurally, the DQO development process should be integrated with the overall project planning and the results incorporated into the sampling plan and QAPP.

QAPP is a planning tool designed to assure that a certain level of sampling data quality will be achieved during sampling activity. QAPP elements should be incorporated in the sampling plan. QAPP provides a project- or task-specific blueprint of how QA/QC are to be applied to the environmental sampling to assure the results obtained are of the type and quality appropriate to sound administration of the site remediation. In summary, proper implementation of QAPP is an assurance that quality sampling data will be obtained.

2. ASSURING THE QUALITY OF SAMPLING DATA

The QAPP provisions and the DQO requirements are used when developing a sampling plan which details the selected field sampling and laboratory analysis options and statements of the confidence in decisions made during the remedial process. Confidence statements are justified through the use of statistical analysis of the data. Planning a sampling event is part of the QAPP requirements to obtain sampling data that are valid, legally defensible, and cost effective.

A QAPP incorporates major sampling activities including sample collection and handling, laboratory instrumentation and sample management, and document control. The chain of sample collection, analysis, and documentation is composed of many links and the uncertainty of the overall sampling project is comprised of the errors found in each link. Sources of sampling errors include:

- sample collection
- handling and preparation of samples for transport from the site to the laboratory, and sample transportation
- handling and preparation of samples for analysis at the laboratory
- laboratory analysis

The purpose of the QAPP is to be able to identify, measure, and control those sampling errors to a statistically acceptable level. The control can be done by minimizing or correcting individual errors and their cumulative effect through a QA/QC mechanism. A strict QA/QC requirement to sampling might at first seem to be overkill for a site remediation project in terms of cost and schedules, but not when the essential elements and ultimate application of data are considered.[23] Shashidara argues that errors in data collection or analysis that may appear very minor at the discovery stage may turn out to be major factors in an engineering solution being feasible or infeasible, or may jeopardize the acceptability of the final engineered solution in a court of law.

QA/QC procedures should be in accordance with EPA and state requirements, professional technical standards, and site-specific goals. QC consists of specific procedures designed as system checks on field sampling, laboratory analysis, and documentation of all sample movement and chain-of-custody records, i.e., documentation that is designed to trace the custody of a sample from its point of collection to its final disposition. QC also provides supporting information on the methods employed to ensure quality analytical data.

QA consists of an overview check to verify that the QC procedures were properly implemented. It is an overall checking system for ensuring that data are of adequate quality. In other words, QA refers to those operations and procedures which are undertaken to provide measurement data of a stated quality with a stated probability of being right.[24]

There are a number of QC measures that can be incorporated in the QAPP. They include, for example, the introduction of control samples, laboratory coordination, procedures for decontamination, sample transportation, and chain-of-custody. The chain-of-custody procedure traces the custody of a sample to demonstrate that custody remained intact and that tampering or substitutions were precluded.

The transportation of samples from the site to the laboratory requires special attention. To ensure that the chain-of-custody mechanism is properly implemented, the transporter must be authorized to accept hazardous material, and the samples must arrive at the laboratory — not only within the specified allowable time frame (known as sample holding time, which varies from chemical to chemical; it can be 24 hours or 15 days, etc.) but also with enough time to spare that the laboratory personnel can complete the analysis before the holding time expires.

Because sample results can be affected by the use of contaminated sampling equipment, all new and used equipment must be properly decontaminated prior to use. Contamination of sampling equipment and tools can result from exposure to the sample at hand, during transport and storage or at the laboratory. Decontamination of small sampling tools (e.g., soil scoops) is not required if the tool is properly disposed after use. Proper decontamination procedures are described in the QA/QC document.

Coordination between the field sampling staff and the laboratory is critical. Consultation is necessary prior to sample collection to ensure scheduling for sample delivery and analysis. It is also important to know the use of proper containers and sample preservatives, sample volumes, use of appropriate labels, chain-of-custody forms, etc. Field sampling staff are required to inform laboratory personnel of expected contaminant concentrations to aid in the analysis selection and to inform the analyst of any special safety or decontamination procedures that must be performed. The staff and the analyst must both pay attention to the holding time requirements of the sample contaminant.

Control samples are QC samples that are created to monitor the performance of sample collection, transport, and analytical procedures. Control samples include surrogate samples, matrix spikes, QC check samples, blanks, collocated samples, replicate and split samples, background samples, and critical samples. A brief description of each of these control samples is provided in the following paragraphs.

- Surrogate samples contain compounds that are similar to the analytes of interest in terms of chemical composition, extraction and chromatography, but which are not normally found in environmental samples. Surrogates are used to evaluate the capability of laboratory equipment to quantify certain analytes of interest. An analyte is a specific component or element measured in a chemical analysis.
- Matrix spikes (MS) are soil or water samples that have been spiked (i.e., fortified) with known quantities of specific compounds. MS are used to assess the appropriateness of the laboratory method for that matrix (i.e., soil, water). A matrix spike duplicate is a second aliquot (representative sample from a larger quantity of the sample) of the same matrix as the MS and is spiked in an identical fashion. The reproducibility between MS and the duplicate represents an indicator of the precision of the laboratory method.
- A QC check sample is a prepared sample containing the compounds of interest in known concentrations. The QC check sample is processed through the entire sampling processing procedure including sample preparation and analysis. It is used to verify whether the laboratory method was done properly and whether matrix interferences occurred.
- Blanks are prepared samples free of any of the substances of interest. A blank is devised to detect and measure the extraneous material. Types of blanks include field, trip, matrix, equipment, and material.
- Collocated samples are samples collected independently in such a way that they are equally representative of the parameters of interest at a given point in time and space. Side-by-side soil core samples and two water samples collected at the same time and from the same spot in a lake are examples of collocated samples. When collocated samples are analyzed by two different laboratories, a comparison of the results will provide information on the precision of the laboratories' methods.
- Replicate and split samples. Replicates are samples that have been divided into two or more portions at some step (either in the field or at the laboratory) in the analytical process. Split samples are replicates divided into two portions for analysis by two different laboratories.

- Background samples are samples taken from similar environmental media at the site but outside the area of contamination. Data on natural background concentrations of chemicals in the area should be collected and analyzed to determine background conditions.
- Critical samples are samples that must be collected from sample points considered critical for successful sampling. An example of a critical point is an upgradient well in a groundwater contamination study.

When conducting environmental sampling, a decision must be made on the appropriate type and number of QC samples to take in order to meet the DQOs. Questions that are commonly asked to help make the decision are

- Which QC samples are appropriate for estimating bias?
- Which QC samples are good for estimating precision?
- Which QC samples are useful for determining sources of sampling errors?
- Which QC samples are effective in determining laboratory errors?

Even with a perfectly developed QAPP, errors are often times unavoidable. The reason is that not all factors that affect the reliability and representativeness of data are measurable. Maney and Dallas[25] provide a list of some unmeasurable factors that can severely bias data and that are not necessarily readily identified by good data handling and management procedures. The factors are

- sampling the wrong area
- sampling the wrong matrix
- switching samples prior to labeling
- mislabeling sample containers
- incorrectly preserving the samples
- incorrectly weighing samples
- incorrectly diluting or concentrating samples
- incorrectly documenting any procedure
- using the wrong method for analysis

Once an error of this type is identified, a resampling of the site may become a necessity.

In the QAPP, there is a specific QA/QC provision that requires a sampling plan. In fact, the planning of a sampling event constitutes one of the most vital components of quality assurance. This is because analytical results produced from the laboratory are dependent on the quality and integrity of the sample submitted to the laboratory. When the data are to be used in legal proceedings, having a well-written sampling plan is mandatory.

3. SAMPLING PLAN DEVELOPMENT

Similar to any other planning effort, the first step in planning sampling for a contaminated site is to define the objectives. The purpose is to focus valuable resources for obtaining optimum results. Once the objectives are known, the next steps consist of the identification of the appropriate sampling strategy and the determination of sample types, the number of samples to be collected, the appropriate laboratory test methods, and the sampling devices to be used.

Based on the objectives, sampling activities can be divided into five major categories: exploratory, monitoring, preliminary, supplementary, and sampling for site characterization.

- Exploratory or surveillance sampling's purposes are to confirm whether chemicals are present, and evaluate the degree of concentrations in terms of range and variability if contamination is found.
- Monitoring is a sampling activity with the goal of providing information on specific analytes and their concentrations. Monitoring is conducted within a specific geographic area over a specified time period, e.g., 30 years.
- Preliminary or screening sampling is used to help identify a sampling approach before sufficient data have been collected for statistical analysis. Preliminary sampling is an initial part of a phased data collection approach. It represents 10 to 15% of the overall sampling.
- Supplementary sampling is basically a resampling. The goal is to confirm critical findings from previous sampling, and to clarify uncertainties discovered during the monitoring.
- Sampling for site characterization is a sampling activity with the goal of delineating the extent of contamination, vertically and laterally. Included in the characterization goal is the determination of how far and at what level contaminants have migrated from their source.

The determination of sampling objectives will affect the decision on the number of samples to take, sampling equipment required, appropriate QC criteria, proper analytical methods and sampling protocol. Sampling objectives will also affect the decision on sampling method, i.e., the decision on where to locate sampling points in a site. In this regard, the sampling method must meet the sampling objective while complying with the requirements of DQOs. Other criteria that should be examined when deciding the sampling method are cost effectiveness, patterns and variability of contamination, and practical considerations such as site accessibility.[26]

3.1. Determining Sampling Locations

In general, sampling methods are grouped into two major categories: statistical and nonstatistical sampling. As stated at the beginning of this chapter, the ultimate purpose of sampling is to obtain a small number of samples that are representative of a site. To ensure that samples are truly representative, statistical concepts, approaches, and techniques are widely used to select the most appropriate sampling method. The method used is the one that marks the location of sample points in a site. There are four major statistical sampling methods. The methods are displayed in Figure 7.1. and include the following:

1. Simple random. Sampling locations are chosen randomly. The method is used where there is no observable signs of contamination present at a site.
2. Stratified random. The site is divided into sampling areas (strata) based on site data. In each internally homogeneous area (stratum), a simple random sampling is utilized. The method is appropriate for large sites with variations in topographic features, soil types, or land uses.
3. Cluster. Clusters of individual units are chosen at random. Units in each cluster are all measured. This method may result in a biased sampling.
4. Systematic. The site is divided into uniform sampling areas of similar size and form using a grid system. A sample is collected from each sampling area (grid), normally at the center of each grid. This is the most common strategy effective for minimizing bias. The method, however, is relatively costly which generally precludes its use on all but small-size sampling areas.

The nonstatistical sampling method consists of haphazard sampling and judgment sampling. As the name indicates, these sampling methods do not utilize statistical analysis. A brief description of the methods follows:

1. Haphazard sampling. This sampling represents a principle that any sampling location is good because it assumes that the site is completely homogeneous. This assumption, however, is highly suspect in most environmental sampling.[27]
2. Judgment sampling. This is the sampling of units, which are chosen based on subjective selection. Because the selection of the sampling units is subjective, the quality of sampling is dependent on the expertise of those who conduct the selection.

3.2. Sample Types and Number of Samples

The preceding discussion has shown that there are many QC samples that can be collected and analyzed. However, it is not cost effective to require every

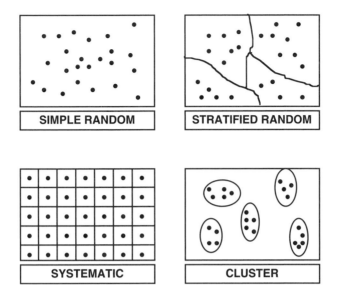

Figure 7.1. Four types of sampling methods.

QC sample at every sampling event. Hence, careful selection of appropriate QC samples will help control project costs and ensure the assessment of the quality of the reported data.[28] It is really important in planning a sampling event to select only those samples that are needed to meet the goals of a project. If the wrong QC samples are chosen, then the goals may be compromised. For those who need assistance, an expert system was developed to provide advice on this subject. Called the QC Advisor, this computer program can be run on IBM-compatible personal computers or Macintosh computers with a DOS simulator.[29]

In addition to QC sample determination, the sampling plan must also indicate whether grab samples or composite samples are to be collected. A grab sample is a single sample collected at a particular spot at a site at a specific time. Grab samples are collected for the purpose of identifying and quantifying contamination at a specific location. When information relating to changes in concentrations of analytes over time is needed, several grab samples may be taken at the same spot over a period of time.

By mixing multiple grab samples, the field sampling staff will generate composite samples. Composite samples can also be collected by using specially designed sampling devices that collect and combine a series of grab samples automatically. Usually, composite samples are established to provide a more representative sample of heterogeneous matrices in which contaminant concentrations may vary over short distances. The rationale of utilizing composite sampling is that the technique may reduce the potential cost of analyzing a large number of samples. This will not be the case if sampling costs are greater than analytical costs. Another compositing drawback is that the practice may dilute low concentrations of contaminants in individual subsamples so that the total concentration of the composite is below the laboratory detection limit, resulting in the existence of the contamination in individual subsamples going unnoticed.

In a number of instances, the sampling event is intended to determine if contaminants found at the site are naturally occurring, or are caused by other sources. With this objective in mind, a sampling plan will differentiate between background samples and control samples. Background samples are samples collected from an area believed to be free from the contaminants of concern. This could be an area adjacent to the site. The area should be similar in topography to the site and have similar soil types with the site.

Control samples are samples collected upgradient from a source of contamination to isolate the effects of the source on the soil. A control sample is used as a control parameter in comparative statistical analysis. The purpose of the analysis is to examine the difference between the observed parameter with the control parameter.

Closely related to the discussion of the types of samples is the issue of sampling size. There is no straightforward, easy answer to the question of sampling size. The fact is that the number of samples that should be taken in a sampling event varies from site to site. Arithmetically, the number of samples is the summation of the test samples, control samples, QC samples, exploratory samples, and supplementary samples. The number of test samples is obviously dependent on the number of analytical methods utilized for the project. For each analytical method, the number of samples needed is determined by the DQOs of the project as well as the size and complexity of the site.

Control samples are estimated from the number of matrices at the sample points. Typically, one or more samples from each matrix is required if a differentiation between contaminated and noncontaminated samples is pursued. The number of QC samples is normally calculated as percentages of the test samples. Exploratory samples represent 5–15% of the test samples. Supplementary samples, which are intended to confirm previous findings, are also estimated to be 5 to 15% of the test samples.

Once the types and number of samples have been determined, the next step in planning a sampling event is to record sampling procedures. This documentation represents the descriptions of the detailed protocols to be followed in collecting, preserving, packaging, labeling, storing, transporting, and documenting samples. The more specific a sampling procedure is, the less chance there will be for errors.

3.3. Recording Sampling Procedures

Sampling procedures must identify sampling locations and all of the equipment needed for sampling. The list of equipment includes all sampling devices, containers, sample preservatives, materials for cleaning the equipment, labels, tape, waterproof pens, packaging materials, sample seals, chain-of-custody forms, protective clothing, and other safety equipment.

Typical sampling procedures contain instructions for each type of sample collection such as number and type of QC samples, exploratory samples, instructions for compositing samples, field preparations such as pH adjustments, maximum holding times of samples, and any special sampling conditions required. Also included are instructions for packaging, transporting, and storing samples; chain-of-custody procedures; and safety precautions.

Selecting the analytical methods is an integral part of the sample planning process. The inclusion of information concerning the analytical methods to be utilized, minimum sample volumes, levels of quantification, analytical bias, and precision limits to the content of sampling procedures helps sampling staff make appropriate decisions when unexpected circumstances require modifications to the sampling procedure.

Sampling protocols ensure that the proper type and quality sample is obtained for subsequent laboratory analysis. If necessary, the sampling protocols also include field testing procedures. The procedures provide assurance that *in situ* information is appropriately gathered. Protocols for environmental sampling, developed by the American Society for Testing and Materials (ASTM) are presented as standard methods and listed under unique ASTM coding numbers.[30] The following list displays selected standard methods sorted by environmental medium:

1. ROCK:
   - D420-93   Guide to Site Characterization for Engineering, Design, and Construction Purposes
   - D2113-83   Practice for Diamond Core Drilling for Site Investigation
   - D5434-93   Guide for Field Logging of Subsurface Explorations of Soil and Rock
2. SOIL:
   - C998-90   Practice for Sampling Surface Soil for Radionuclides
   - D1452-80   Practice for Soil Investigation and Sampling by Auger Borings
   - D1586-84   Test Method for Penetration Test and Split-Barrel Sampling of Soils
3. WATER:
   - D3370-82   Practices for Sampling Water
   - D3856-88   Guide for Good Laboratory Practices in Laboratories Engaged in Sampling and Analysis of Water
   - D4841-88   Practice for Estimation of Holding Time for Water Samples Containing Organic and Inorganic Constituents
4. SEDIMENT:
   - D3213-91   Practices for Handling, Storing, and Preparing Soft Undisturbed Marine Soil
   - D3976-92   Practice for Preparation of Sediment Samples for Chemical Analysis

E1391-94   Guide for Collection, Storage, Characterization, and
           Manipulation of Sediments for Toxicological Testing

5. GROUNDWATER:

D653-90   Terminology Relating to Soil, Rock, and Contained
           Fluid

D4448-85   Guide for Sampling Groundwater Monitoring Wells

D5092-90   Practice for Design and Installation of Groundwater
           Monitoring Wells in Aquifers

6. VADOSE ZONE:

D4696-92   Guide for Pore-Liquid Sampling from the Vadose Zone

D4700-91   Guide for Soil Sampling from the Vadose Zone

D5314-91   Guide for Soil Gas Monitoring in the Vadose Zone

7. WASTE:

D4687-87   Guide for General Planning of Waste Sampling

D4979-89   Test Method for Physical Description Screening Anal-
           ysis in Waste

D5495-94   Practice for Sampling With a Composite Liquid Waste
           Sampler (COLIWASA)

8. ATMOSPHERIC/AIR:

D1357-94   Practice for Planning the Sampling of the Ambient
           Atmosphere

D3614-90   Guide for Laboratories Engaged in Sampling and Anal-
           ysis of Atmospheres and Emissions

D4861-94   Practice for Sampling and Selection of Analytical Tech-
           niques for Pesticides and Polychlorinated Biphenyls in
           Air

An important part of the sampling plan is the selection of the laboratory method to analyze the collected samples. The method chosen will directly affect the sampling protocol. The volume of samples needed to measure an analyte at a given detection level, for example, will depend on the sensitivity of the analytical method. The chosen method also determines the type of sample container, sample preservation and storage requirements.

Procedurally, the selected laboratory method is based on the contaminants of concern at a site. At some sites, the determination of contaminants of concern can be done after analyzing the site history and existing available information. In another situation, a site may be contaminated by a large array of contaminants forcing the planner to initially analyze only for key chemicals and then perform a more thorough analysis later on. The key chemicals are those that are the most toxic, persistent, mobile, or frequently occurring.

Several laboratory or analytical methods are usually available for most environmental analytes of interest. Some of the analytes may have a number of laboratory methods to select from. The appropriate method for sample analysis will depend on several factors:

- the particular regulatory program involved
- the type of sample matrix
- the particular components to be analyzed
- the degree of reliability, precision, and accuracy needed
- the required detection limit
- the potential interferences
- the costs and availability of instrumentation
- the likelihood that the analysis will be used in litigation

To illustrate the issue of laboratory method selection, the Environmental Protection Agency (EPA) Method 502.2, Rev. 2; Method 524.2, Rev. 2; and Method 8240B, Rev. 2 are discussed briefly in the following paragraphs. The first two methods cover 60 volatile organic compounds (VOCs) such as benzene, carbon tetrachloride, chloroform, ethylbenzene, toluene, trichloroethene, and xylene. These two methods are used for drinking water analysis, therefore their utilization is limited to clean water matrices. The third method covers 80 VOCs in groundwater, surface water, soils, or sediments.

The instrumentation needed for Method 502.2 consists of the following components:

- The purge-and-trap system. This requires a water purging device, a sorbent trap, and a thermal desorption apparatus.
- The gas chromatography (GC) system with three capillary columns connected to the purge-and-trap component.
- The photoionization detector (PID) and electrolytic conductivity detector (ELCD) placed in series.

When an inert gas is bubbled through a water sample, purged sample components are trapped in a tube of sorbent materials. The sorbent tube is heated and backflushed with helium to desorb the trapped sample onto a capillary GC system. The column is temperature programmed to separate analytes which are then detected with a PID for aromatic compounds and an ELCD for halogenated compounds.[31]

Method 524.2 is similar to Method 502.2 except that it uses mass spectrometry instead of PID and ELCD. The purge-and-trap technique in both systems requires little in the way of sample preparation. Method 524.2 is more expensive than Method 502.2 but produces fewer false positive errors because of its greater selectivity. This is due to the use of MS which provides excellent specificity for analyte identification and good sensitivity.

Method 8240B uses the purge-and-trap system and a column GC with MS. The 6 ft × 0.1 in PID column glass is packed with 1% SP-1000 on Carbopack-B (60/80 mesh). The method covers 20 more VOCs than the previously mentioned methods, and is not limited to a water matrix. Solid samples must be screened by GC analysis (using EPA Methods 3810 or 3820 as appropriate) prior to the purge-and-trap GS/MS analysis. This prevents expensive instrument cleanup and downtime from contamination of the purge-and-trap system by purgeable organics.

3.4. Selecting Sampling Devices

Currently, hundreds of methods and techniques exist for sampling, monitoring, and field screening at contaminated sites. Existing methods are often refined and new methods are continually being developed. Selecting appropriate sampling devices is vital for successful sampling because it directly affects the sampling costs. The major grouping of devices for soil sampling is the drilling method, hand-driven equipment, and power-driven sampling while groundwater sampling devices can be classified either as portable samplers or *in situ* samplers.[32]

Boulding reports that there are 18 methods available under the category of drilling. The drilling of boreholes is intended to collect soil samples for laboratory analysis, as well as for hydrogeologic characterization using geophysic logging, and installation of monitoring wells or piezometers. The selection of drilling methods is based on cost, suitability for the type of soils at a site (consolidated soil or unconsolidated), and the effects of drilling on sample integrity. These 18 methods can be categorized into three major groups:

1. The hollow-stem auger. This is the most widely used method for unconsolidated soils. The auger is drilled into the soil using a hydraulically powered drill rig. The hollow stem of the auger allows the use of various methods for continuous or intermittent soil sampling (see Plate 7.1).
2. The air rotary with rotary bit/downhole hammer. This method is common for consolidated soils. Compressed air is circulated down the drill rods to cool the tri-cone bit and carries cuttings up the open hole to the surface. A cyclone separator slows the air velocity and allows the soil cuttings to fall into a container. A down-the-hole hammer, which operates with a pounding action as it rotates, can be used to replace the bit.
3. Casing advancement with rotary drill-through or with dual-wall reverse circulation. This is a commonly used method where cross-contamination between aquifers is a concern. This method allows soil cuttings to flow up through the annular space between the drill pipe and the casing. The diameter of the casing is designed to be slightly larger than the bit so that it can be removed once the desired depth is reached.

There are currently 12 methods classified as the hand-held soil sampling devices. They are used for sampling the first two to three meters of soil below the surface. The methods can be grouped into:

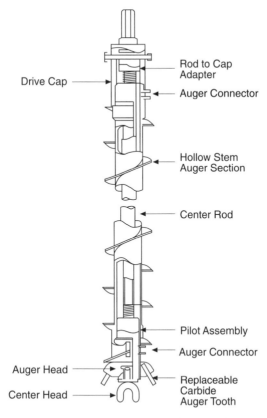

Plate 7.1 **The hollow stem auger consists of auger flights with a hollow interior, a pilot assembly, a center rod and an auger head with replaceable carbide cutting teeth. The interior diameter ranges from 2.5 to 10.25 inches and the standard flight length is 5 feet. Samples brought to the surface can be collected directly from the auger flights. When an undisturbed sample is desired, a tube type sampler can be inserted through the hollow core of the auger flights. (Courtesy of Central Mine Equipment Co.)**

1. Scoops, spoons, and shovels. A spoon has a capacity of 10–100 g; a scoop has a capacity of 300–2000 g. Spoons, scoops, and shovels or shovel-like devices such as trowels can be used separately or in combination to obtain samples. A shovel is normally used to remove the top soil to the desired depth, and spoons and scoops are used for actual sampling. The commonly used materials are stainless steel, plastic, or Teflon-coated.

2. Augers. Hand-held augers consist of an auger bit, a solid or tubular drill rod, and a T handle. As the T handle is rotated, the auger tip bites into the soil. Soil retained on the auger tip is brought to the

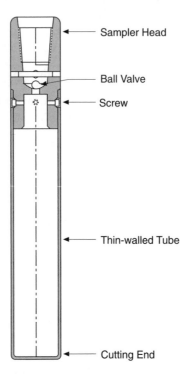

Sampler Head

Ball Valve

Screw

Thin-walled Tube

Cutting End

Plate 7.2.   Thin-wall tubes or Shelby tubes are commonly available in carbon steel. The tube is connected with set screws to the sampler head which in turn is threaded to connect with the extension rods. To collect a sample, the tube is pushed into the soil by hand, with a hydraulic piston or other system. As the sampler is pushed into the soil, the ball valve serves as an air vent. (Courtesy of Mobile Drill Co., Inc.)

surface and used as the sample. When the drill rod is threaded, extensions can be added. Various types of auger bits are available, such as screw-type, bucket-type, or spiral-type.

3. Tubes. The basic equipment is similar to augers. The difference is that a closed or open tube with a cutting tip is attached to the drill rod, and rather than being rotated the tube is pushed into the soil. Types of tube samplers available include thin-wall, Veihmeyer, soil probe, and peat sampler (see Plate 7.2 for thin-wall tube discussion).

The following criteria must be considered when selecting hand-held devices.[33]

1. If an undisturbed core is required, tube samplers must be used, otherwise augers, scoops, spoons, and shovels can be used.

2. Soil conditions at the site (moisture, cohesiveness, rocky). Tube samplers and screw-type augers are not appropriate to use in rocky soils. Other types are suitable. Some devices (e.g., scoops, spoons, shovels, thin-wall tube samplers) are not affected by the moisture content of soils. Some (e.g., peat samplers, wet tip soil probes) are good in moist sample. A dry tip soil probe is only good in dry soils. The cohesiveness of the soil does not affect scoops, spoons, shovels, or soil probes. Most augers only are appropriate for cohesive soils.

3. Size of samples (large or small). Scoops, spoons, shovels, and certain types of augers (sand barrel and mud barrel) and tube samplers (thin-wall and peat samplers) are appropriate for obtaining large samples. The remaining types are good for small samples.

4. Depth of samples (deep or shallow). Shovels and augers are good for obtaining deeper samples. Scoops and spoon are useful for shallow sampling. Some tube samplers (soil probes and peat samplers) are good for deep samples, some (Veihmeyer tube, thin-wall tube) are used for shallow sampling.

5. Number of required operators. Normally one operator is required; however, some tube samplers require two operators, e.g., dry tip soil probes, thin-wall tube samplers, and peat samplers.

Power-driven soil sampling devices are usually operated in conjunction with drill rigs, although thin-wall tube samplers for near-surface sampling can be attached to pickup trucks. The five categories of the devices and their intended uses are

1. Split-spoon samplers are the most common method for the collection of disturbed cores. The sampler is driven by a 140-lb weight dropped through a 30-in. interval. When the split sampler is brought to the surface, it is disassembled and the core removed (see Plate 7.3).

2. Rotating core samplers are used in consolidated geologic material. The rotating bit consists of a tube with a ring fitted to the end of the core barrel. Water is circulated through the bit to cool the cutting surface. The bit cuts through rock, with a solid core remaining in the tube.

3. Thin-wall open tube samplers are most commonly used for collecting undisturbed cores. The sample collection procedure with this method is similar to split-spoon sampling, except that the tube is pushed into the soil using the weight of the drill rig, not driven.

4. Where poor soil cohesiveness is encountered, thin-wall piston samplers should be used. The piston samplers are furnished with internal pistons to generate a vacuum within the sampler as it is withdrawn from the soil.

5. For gravely and very stiff soils, specially designed thin-wall samplers might be required. The denison sampler, or pitcher sampler, for example, has a double-tube core design with an inner tube. The rotating outer tube allows penetration in stiff deposits, while the stationary inner tube collects a minimally disturbed sample.

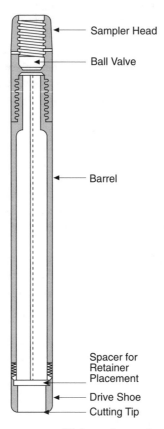

Sampler Head

Ball Valve

Barrel

Spacer for Retainer Placement

Drive Shoe

Cutting Tip

Plate 7.3. **The split-spoon or split-barrel sampler consists of two barrel halves, a drive shoe, and a sampler head containing a ball valve. The sampler is threaded onto the drilling rod and lowered to the bottom of the boring through the flights of the hollow-stem auger. A drop hammer attached to the drill rig is used to drive the sampler into the soil. After the sampler is extracted from the soil, the split-spoon can be opened and the sample removed. A plastic or metal retainer basket can be placed in the drive shoe to prevent the samples from falling out during retrieval. (Courtesy of Diedrich Drilling Equipment Inc.)**

As a subcategory of what are commonly referred to as "direct push" techniques, soil probing refers to tools and sensors that are inserted into the ground without the use of drilling to remove soil and make a path for the tool. Soil-probing equipment is typically used for site investigations to depths of 30 to 60 feet. This range is constantly increasing as better probing equipment is produced. Geoprobe machines, a family of hydraulically powered, soil-probing equipment designed specifically for use in the environmental industry are manufactured by Geoprobe Systems, Salina, KS. The equipment represents one of the popular direct push technologies (see Plate 7.4).

Groundwater sampling devices consist of portable samplers and portable *in situ* samplers. Portable samplers which include submersible pumps, other pumps, and grab samplers are used in permanently installed and screened monitoring wells. *In situ* samplers, on the other hand, do not require permanent wells.

Submersible pumps are placed below the static water level of the well and pump the sample to the surface. Included in this group are bladder pumps which are able to collect samples at a maximum depth of 1,000 ft, gear pumps at 200 ft, and helical rotor pumps at 160 ft. The pumping rate is 1.5 to 3.0 gallons per min. Submersible pumps require a minimum well diameter of 1.5 to 2.0 in.

Other portable pumps that are commonly used include suction pumps. These pumps, operated manually or power driven, apply a vacuum to the well casing/tubing running from the pump to the desired sampling depths. Most pumps are controlled to provide a continuous and variable pumping rate. They fit in wells of any diameter. Sampling using this pump, however, is limited to wells where the water level is less than 25 ft below the surface.

Grab samplers include bailers, mechanical samplers (Kemmerer, Van Dorn and COLIWASA), and pneumatic depth samplers. A bailer is a hollow tube with a single or double check valve at the base. The bailer is attached to a wire and lowered into the water well, with the check valve open allowing water to flow through the bailer. When the bailer reaches the desired depth, it is pulled up. The weight of the water closes the check valve. At the surface, the sample is decanted into a sample container. The bailer can work properly in a well of small diameter (minimum of one half in.) and has no limit in terms of maximum sample depth.

The Kammerer and Van Dorn samplers are tube samplers with end caps that close when triggered by a signal sent down the line. This allows the collection of a water sample at any desired depth. The COLIWASA which is used for sampling fluids in a tank is a tube sampler with neoprene stoppers at each end that are controlled by a rod running through the tube with a locking mechanism (see Plate 7.5).

A pneumatic depth sampler is a sample container which is pressurized before being lowered into the well. Releasing the pressure allows sample to enter the device. Similar to bailers, these samplers can work in small-diameter wells and have no sample depth limit.

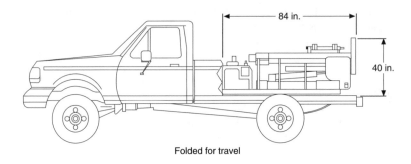

Folded for travel

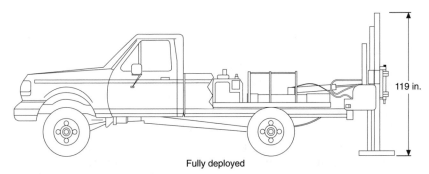

Fully deployed

Plate 7.4.   GEOPROBE® soil probing equipment has been used to depths exceeding 100 feet in many areas of the United States. The equipment is hydraulically powered either from a vehicle or an auxiliary engine. It uses static force and the dynamic percussion force of the GH-40 soil probing hammer to advance small diameter sampling tools without producing soil cutting. The typical penetration rates are 5 to 25 feet per minute. The GEOPROBE can be used for collecting soil cores, groundwater samples, and soil gas samples by using the proper type of samplers. GEOPROBE can be used in 20 to 40 sample locations per day. It can drill through surface pavements 12 inches or more thick (using concrete drill bits) and probe beneath them. (GEOPROBE is a registered trademark of Kejr Engineering, Inc., Salina, KS.)

Portable *in situ* samplers allow fast collection of samples without the installation of permanent wells. A hydropunch, for example, operates in conjunction with conventional cone penetrometer rigs. It is attached to cone penetrator rods and driven into the soil with hydraulic rams. Once the bottom of the probe is five feet below the water table, the outer cylinder is pulled back. Hydrostatic pressure forces groundwater to enter the sample compartment, then the probe is pulled to the surface to collect the sample. Depths of

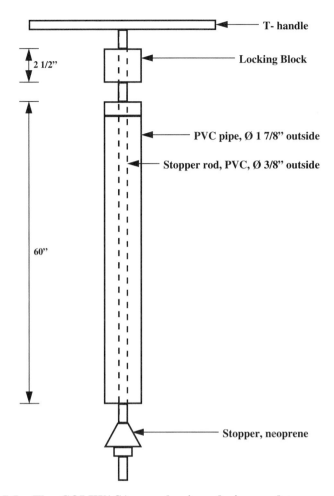

T- handle

Locking Block

2 1/2"

PVC pipe, Ø 1 7/8" outside

Stopper rod, PVC, Ø 3/8" outside

60"

Stopper, neoprene

Plate 7.5. The COLIWASA sampler is a device used to sample free-flowing liquids and slurries contained in shallow tanks or drums. It allows representative sampling of multiphase wastes. Because of its configuration, decontamination is difficult. The coliwasa is not used for sampling high viscosity liquids.

100 to 150 ft may be reached by direct penetration depending on the type of soil materials.

It is important to note that before a groundwater sample from a permanent well is collected using any of the above methods, it is often necessary to conduct well purging, i.e., the pumping of stagnant water from a well before sample collection. Purging is complete when the water quality indicators such as pH, conductance, and temperature have stabilized, indicating that fresh water fills the well. The disadvantage of purging is that it can mobilize colloidal particles upon which contaminants are sorbed.[34] An alternative to purging is to use a dedicated sampling device capable of low pumping rates set at the level of the well screen. The setting will not increase colloid density in the groundwater sample compared to natural colloidal flow through the well screen.

After soil or water samples are collected from a site, they must be directly deposited in containers for shipment to the analytical laboratory. The containers come in a variety of materials and sizes. Containers must be clean and made of material that does not chemically interact with the samples. The type, size, and material of the containers must be specified in the QAPP, including the preparation procedures.

For some types of water samples, certain preservatives are added after the samples have been deposited in the containers. The purpose of adding the preservatives is to retard biological action and chemical hydrolysis, and to reduce volatility and absorption. The following examples illustrate the different types of sample containers for different sample types and concentrations:

1. For water samples containing metals or cyanide at low concentrations, the appropriate sample container is the Type C container, i.e., a 1-l high-density polyethylene bottle with a poly cap.
2. For water samples similar to those in example 1, but with medium concentrations, the proper container is the Type E container, i.e., a 16-oz wide-mouth glass jar with a Teflon-lined black phenolic cap.
3. For water samples containing volatile organics at low or medium concentrations, the appropriate container is the Type B container, i.e., a 40-ml glass vial with a Teflon-lined silicon septum and a black phenolic cap.
4. For the same type and concentration of contaminants as in example 3, but a soil sample, the proper container is the Type D container, i.e., a 120-ml wide-mouth glass vial with a white poly cap.

3.5. Sampling Plan Documentation

Planning and reporting various sampling activities represent an integral part of a sampling event. A comprehensive planning document outlines all aspects of the sampling activities and provides guidance to improve sampling. The major components of a sampling plan include the following elements.[35]

1. scope and objective of the investigation
2. summary of site background information
3. summary assessment of existing data
4. contaminants of interest for sampling
5. types and number of samples to be collected
6. sampling locations and frequency

7. sampling and laboratory testing procedures including decontamination procedures
8. plan of operation and schedule of sampling including the personnel
   to perform each task
9. cost estimates
10. other supporting documents such as QAPP, data management, and
    health and safety plans

When the project is completed, sampling personnel are required to report
the precise conditions under which samples were gathered, including any
deviations from the sampling protocols. Laboratory staff report and describe
analytical data appropriately. Data managers verify and validate the laboratory
data, provide evaluations of data consistency, integrity, and reliability. A good
understanding of the site's contamination problem is possible only after the
sampling data have been completely reviewed and reported. Report formats
will vary, but the sampling report should contain the following components.[36]

1. a summary of the problem being investigated
2. a summary of the DQOs and whether they were met or modified
3. a description of the sampling effort, complete with site maps showing sampling locations
4. a description of the analytical methods and summaries of any analytical problems
5. a summary of the completeness and representativeness of the data
6. interpretation of results and conclusion of findings

Sampling reports must contain sufficient information so that users can
understand the interpretations without having to make their own. Raw data
for each sample, when included in the reports, should be appropriately identified. If average values are reported, then an indicator of the measurement of
precision must be included. Details of the analytical results should be written
with the standard deviation and the mean. Many laboratories tend not to report
all detectable analyte concentration data when they are less than the method
detection limit (MDL) because these data have very poor statistical confidence.
When values are below a detection limit, it is usually suggested that they be
reported as values at the detection limit.

Every laboratory instrument has a minimum detection level. In environmental sampling terminology, the smallest signal above the background level
that an instrument can detect at the 99% confidence level is called the instrument detection limit (IDL). MDL is defined as the lowest concentration of
analyte that is statistically different from the blank. The MDL indicates a 50%
chance of detection. If an analyte is detected, there is a 99% probability that
the true analyte value is greater than zero. Numerically, the MDL is determined
as three times the standard deviation of seven replicate measurements. The

level of six times the standard deviation is defined as the reliable detection limit (RDL), while the ten times standard deviation is called the practical quantification limit (PQL). The PQL represents the lowest concentration of analyte that can be quantified reliably under ordinary laboratory conditions.

4. HEALTH AND SAFETY PLAN

The health and safety plan is required as a supplement to the sampling plan. It is designed for the protection of site workers and the public from acute and chronic exposures to toxic chemicals during site remediation. Exposure to high concentrations of a chemical for a short period of time is acute exposure. If the exposure is to low concentrations over a long period of time, it is called chronic exposure. The exposure routes to the human body are through inhalation, ingestion, skin/eye contact, and skin absorption.

The development of a health and safety plan is also required by Title 29 of the Code of Federal Regulations (29 CFR), promulgated by the Occupational Safety and Health Administration (OSHA). 29 CFR 1910.120 specifically describes standards for employees engaged in hazardous materials operations. The provision, known as OSHA's Hazardous Waste Operations and Emergency Response (HAZWOPER), mandates the development of health and safety programs for site remediation workers; emergency responders; treatment, storage, and disposal (TSD) facility personnel; and underground storage tank (UST) workers.

The health and safety programs require the designation of a safety coordinator responsible for reviewing and approving the site safety plan. A good program requires a written safety standard operating procedure (SOP) for normal work and during an emergency; sufficient employee training; employee medical monitoring; proper safety clothing, monitoring devices, and other personal protective equipment (PPEs). Periodically, the National Institute for Occupational Safety and Health (NIOSH) publishes its Pocket Guide to Chemical Hazards, which provides key information on chemicals normally found in the work environment and which are regulated by OSHA.[37] The guide lists chemicals by name, physical descriptions, chemical and physical properties, potential health hazards, and recommended PPEs.

The health and safety plans are site specific. They are designed to prevent injury and exposure to chemical and work-related hazards. The plan should be prepared by a qualified safety officer and reviewed and revised routinely to keep it current. At a minimum, the site health and safety plan must contain the following components:[38]

1. Evaluate the safety and the health risks associated with the site and with each operation conducted.
2. Identify key personnel and their alternates responsible for site safety and emergency response.

3. Decide the levels of protection appropriate for personnel during various site operations.

4. Divide the area into an exclusion zone (the contaminated area or hot spot), a support zone (the safe area where the command post will be stationed), and a contamination reduction zone (located between the exclusion and support zones). Indicate on the map the location of access control gates into each zone.

5. Determine the number of personnel and equipment necessary in each zone for each operation.

6. Establish emergency procedures for the site, such as emergency notification, communications, and evacuation routes.

7. Determine emergency medical care locations through prior arrangement with local hospitals.

8. Establish action levels for specific atmospheric and radiation conditions based on NIOSH guidance. If the oxygen concentration meter registers the oxygen level at more than 25%, for example, the required actions are to discontinue activity due to fire hazard potential and to consult with a specialist.

9. Implement a program for periodic air monitoring and medical monitoring.

10. Train personnel for nonroutine activities.

11. Consider the effects of weather and site conditions on the health and safety of personnel during normal site operation.

12. Implement strict control procedures for site access.

13. Specify personnel and equipment decontamination procedures and establish a plan to manage the decontamination solutions.

The management of waste solutions as a result of equipment decontamination must be planned carefully. This waste, and several other types of waste generated from soil cuttings, used disposable equipment (gloves, boots), or broken sampling devices (scoops, containers), are called investigative-derived wastes (IDW) and can be defined as hazardous wastes if they meet certain regulatory requirements. As such, they require proper storage, handling, and disposal. Because the management of hazardous waste is relatively expensive, the generation of IDW must be reduced as much as possible. The proper handling of IDW represents one of the issues in sampling management, which is discussed in the following section.

5. SAMPLING MANAGEMENT ISSUES

The management of investigative-derived wastes (IDWs) is mandated by law. Specifically, the National Contingency Plan (NCP) requires that IDW is managed and complies with all applicable or relevant and appropriate requirements (ARARs) to the extent practicable. RCRA regulations, particularly the land disposal restrictions (LDRs), have the potential to be ARARs because they regulate the treatment, storage, and disposal of many hazardous wastes. When managing IDWs, planners are required to choose an option that:[39]

- protects human health and the environment
- complies with ARARs

To identify an IDW, it is necessary to characterize the waste through the use of existing information and best professional judgment. Examples of the existing information are hazardous waste manifests, material safety data sheets (MSDSs), previous test results, and knowledge of the waste generation process. The IDWs include the following wastes:

1. soil cuttings
2. drilling muds
3. purged groundwater
4. decontamination fluid
5. broken sampling devices
6. disposable PPEs

The most important phase of IDW management is planning for its generation and handling before field activity starts. It must be decided early if IDWs will be left on-site or disposed off-site. In this context it is necessary to understand the applicability of the LDRs to IDWs. An important consideration in determining whether LDRs apply is whether land disposal of IDWs has occurred. If IDWs are merely being moved within the same area of concern (AOC), EPA does not consider land disposal to have occurred, so the LDRs are not triggered even if IDWs contain RCRA hazardous wastes. Therefore, if IDWs are being moved only within an AOC, it is unnecessary to determine whether they are subject to LDRs.[40]

If an IDW is a nonhazardous waste, it should be left on-site. The on-site handling options are

1. For soil cuttings and drilling muds, place in the borehole, spread around the well, put into a pit within an AOC, or dispose of at the site's operating treatment/disposal unit.
2. For purged groundwater and decontamination fluids, pour onto ground next to the well to allow for infiltration or dispose of at the site's treatment/disposal unit.
3. For broken sampling devices and disposable PPEs, double bag and deposit in the site's dumpster or a municipal landfill, or dispose of at the site's treatment/disposal unit.

The 1991 EPA document indicates that RCRA hazardous soil also may be left on-site within an AOC unit; however, before the decision is made, it is necessary to consider the proximity of residents and workers in the surrounding area. Hazardous soil that may pose a substantial risk if left on-site, must be disposed off-site. Planning for leaving hazardous soil on-site involves:

- not containerizing and testing wastes designated to be left on-site
- outlining the AOC unit
- choosing pit locations close to the borings within the AOC unit
- covering hazardous soil in the pits with surficial soil

On-site handling of IDWs is relatively inexpensive. The costs for off-site disposal are expensive, consisting of:

1. The cost of containers. Fifty-five-gallon drums are usually used to collect the waste at approximately $50 per drum.
2. The cost of waste testing. Depending on the number of samples and the types of parameters analyzed, the cost is between $40 to $300 per waste sample.
3. The cost of transportation. This varies from $35 to $600 per drum.
4. The cost of disposal. Depending on whether the waste is solid or liquid, hazardous or nonhazardous, the costs range from $12.50 to $615 per drum.

Based on this price scheme, EPA estimates that the cost to manage one drum of IDW off-site may be as little as $350 or as high as $1,650. Clearly, this is not an inexpensive undertaking. For a typical sampling of three soil borings and one groundwater monitoring well, the disposal costs may reach $3,000. Fortunately, there are alternative sampling technologies available that can be used to minimize costs.[41] Saunders illustrates that when push technology is used to collect soil, soil vapor, or groundwater samples, the soil is pushed aside rather than removed. The result is that no unwanted soil is brought up to the ground surface and no purge water is produced. Other waste minimization techniques applicable to IDW include the use of drilling and decontamination methods (such as steam cleaning) that minimize soil cuttings and decontamination fluids, and the use of other materials as substitutes for solvents used for decontamination.

The second sampling management issue deals with the problems in selecting an appropriate testing laboratory. Environmental laboratories have the responsibility of providing sampling results to their clients. These results are often used as the basis of reports required by regulatory agencies. To meet legal requirements, the reports must be legally defensible. If the agency challenges the sampling results, the data may be indefensible and the client, not the laboratory, can be subject to penalties. For this reason, Smith et al.[42] suggest that a prepared client must answer four basic questions before choosing a laboratory:

1. Is the laboratory capable of performing the analysis? Does it have the necessary equipment, facility, and technical expertise?
2. Does the laboratory have and use approved analytical methods?
3. Is the laboratory certified?
4. Is the laboratory reliable?

To answer the first question, the laboratory must have the equipment needed to perform the specified analytical method. Assessing availability of equipment is relatively easy compared to evaluating technical expertise. Looking at resumes can help because a four-year college degree in chemistry or biology is standard for laboratory personnel. Talking to the technicians and managers provides more input. For example, the technicians must know analytical procedures in detail when the analysis fails; they must also know the acceptable range for each QC measure, as well as proper sample containers and preservation for a variety of tests. When talking to laboratory managers, prospective clients should ask about differences among wastewater, drinking water, and solid waste methods. Nonsense answers will indicate the manager's technical incompetence.

Smith et al.[42] underline that there are few signs of technical incompetence that can be easily noticed by a visitor such as the laboratory's level of concern about sample contamination. For example, having instruments for volatile organic analysis in the same room as those for semi-volatile analysis shows that the laboratory ignores potential contamination problems. Another example of sample contamination potential is having samples and sample extracts stored in the same refrigerator as standards and analytical reagents.

To answer the second question, environmental laboratories should have copies of the most current set of regulations. At minimum, a complete set of 40 CFR 1 through 799 must be available. Approved analytical methods found in this set include those for wastewater (40 CFR 136), drinking water (40 CFR 141), solid waste landfills (40 CFR 258), and hazardous waste (40 CFR 261).

To answer the third question, prospective clients should ask if the laboratory is certified. Ask for a copy of the certificate, see the list of parameters for which the certification is awarded, and call the appropriate state agency to check the validity of this certification. Ask for the laboratory's QA manual. This manual is required as part of the certification process. It contains the laboratory organization; DQOs; sampling and analytical protocols; sample custody; equipment calibration; QC methods to assess precision, accuracy, and MDL; validation procedures; and QA reports. The laboratory's standard operating procedures document is also available but it belongs to the laboratory and a copy of it is normally not given to visitors. Check also on the availability of a safety manual or chemical hygiene plan, fire extinguishers, eye-wash stations, and spill and first aid kits. Raw data should be kept in the laboratory for at least three years.

Every environmental laboratory, regardless of its size or reputation, is going to make mistakes. To demonstrate laboratory reliability (the fourth question), clients need to know how the laboratory handles the mistakes. It should explain every aspect of how a sample is handled from the time it is received until the sample is disposed. Clients may test the laboratory by requesting an analysis of a sample with known concentrations of target analytes.

In choosing an environmental laboratory, Smith et al.[42] reminds the prospective clients of the caveat "let the buyer beware". They also suggest that clients visit several laboratories and compare them. Additional questions that can be used to assist with the selection include the laboratory's proximity to the site, whether the laboratory has the capacity to handle all of the samples that are to be delivered within the applicable holding times, and the costs of analysis.[43]

8   SITE CHARACTERIZATION

Lessons learned from the management of contaminated sites suggest that contamination may come from a variety of sources. An old landfill or oil refinery may contaminate a large area for a long period of time. Contamination can also come from a sudden release of chemicals onto the property due to leaking pipelines and valves, or from slowly deteriorating underground storage tanks (UST). To remediate these types of sites, proper characterization of the subsurface conditions and assessment of the contaminants interactions with the surface features are required. It is absolutely essential to know what contaminants lie on and beneath a site, what the sources of the contaminants are, where the contaminants might migrate, what the soil and groundwater characteristics are, and what the likely fate of the contaminants will be.

Stated differently, the characterization evaluates not only the existing distribution of contaminant concentrations, but also the factors that govern their migration and control their fate. The fate of a chemical means the immediate or ultimate change in a chemical as a result of a chemical or biological reaction. The challenge in site remediation lies in the fact that contaminated site conditions are generally too complex and diverse to conform to a standard site characterization plan. In other words, site characterization strategies are very site specific even though they basically involve the characterization of both the sources and media of contamination.

It has become a standard practice to begin site characterization with the generation of planning documents for the establishment of sampling points and the collection and analysis of environmental samples. The parameters that will control contaminant fate and transport are determined during this sampling and analysis process. The location of sampling points is decided on the general knowledge of site conditions derived from topographic maps, aerial photographs, site history related to past waste disposal practices, and other site data. At sites where sufficient information about the subsurface conditions is readily available from previous investigations, it may be possible to reduce the need for the full spectrum of subsurface sampling and testing.

In a subsurface investigation, both direct and indirect methods are normally used. The direct methods such as boring, sampling, and testing more directly measure the parameters of interest. In contrast, an indirect method such as a geophysical survey measures a particular site parameter which can then be related to other parameters of interest. The advantage of indirect measurements is that they can be used for planning the direct methods, interpolation between points where hard data have been developed, and optimization of drilling and sampling costs.[44]

As a common practice, the subsurface conditions at a contaminated site can be roughly evaluated by using geophysical techniques. If it is planned to solve specific site problem areas or simply to provide supportive information rapidly and cheaply, then a geophysical survey can be considered and employed early in the site characterization. Many of the geophysical techniques can operate from the surface, others require boreholes. This latter technique is commonly called geophysical logging. A geophysical survey potentially reduces the number of boreholes necessary for the evaluation of soil and rock strata (stratigraphy) and geotechnical analysis. The technique should be utilized in the initial site characterization when few, if any, boreholes have been drilled.

1. GEOPHYSICAL SURVEY

Geophysical techniques provide information about the subsurface condition through the measurement of certain subsurface physical and electrical properties such as the arrival time of a seismic wave or resistivity measurements. Based on the results of the measurement, parameters of interest, such as depth to bedrock, can then be deducted. Most geophysical techniques provide site characterization information quickly, and at a reasonable cost. This is the main reason why a geophysical survey is attractive for site characterization.

Geophysics is defined as the application of the laws of physics to the study of the earth. It involves the interpretations of fields, such as gravity or magnetism, that can penetrate the subsurface. The direct result of a geophysical survey can take the form of a map of the spatial and temporal distribution of a certain physical property, such as electrical conductivity or resistivity of the site.

When planning a geophysical survey, it is imperative that the objectives of the survey are clearly defined. If it is known that the source of site contamination is an old landfill, and the target is the plume resulting from the landfill contamination, then the objectives of the survey may be to find out the extent of the plume or to investigate the temporal changes of the plume. In this case, the geophysical survey can be used to define groundwater plumes when the plumes contain conductive materials, such as salt.

There are three primary objectives in the use of geophysical techniques in site characterization:[45]

1. The determination of the geometry and properties of the hydrogeo-logic units. The unit refers to the subsurface soil and rock formations or a group of formations that have similar hydrologic characteristics.
2. The detection and mapping of groundwater and soil contaminants.
3. The identification of the location of buried wastes, utility lines, USTs, etc.

Sara and others (e.g., Durant and Myers, 1995[46]) caution that geophysical surveys are intended to supplement, not substitute, for drilling, trenching, and other direct evaluation methods. The surveys are attractive because they can be conducted with a minimum of site disturbance, without causing further environmental damage to the site or posing risks to workers through direct intrusive drilling of contaminated spots.

Related to the formulation of the objective is the scoping of the survey. To meet the stated objectives, the scope of the survey must be carefully delineated, covering the description of the number and types of surveys to be conducted, the size of the area covered, the area of the target, and the cost associated with the survey. Whatever approach is chosen, cost effectiveness must represent the most important decision consideration.

Once the above requirements are completed, planners start to evaluate the choices of techniques available and justify the approaches that are appropriate to address the site problems. Planners must recognize what the geophysics can and cannot do with the equipment. There are a number of geophysical methods that are currently used in site characterization. All are based on elements of physics and geology. They are used to locate subsurface contam-ination, and to delineate subsurface features that influence its movement. The methods respond to the physical properties of the subsurface to infer the geological formations and structure, and to deduce the presence, location, distribution, and size of buried objects.

Geophysical methods fall into six categories. A brief description of each of the methods is provided in the following paragraphs. The categories are[47]

1. electrical, including electromagnetic techniques and direct current resistivity
2. radiometric
3. ground-penetrating radar
4. magnetic
5. gravimetric
6. seismic refraction and reflection

1.1. Electrical Method

Electrical conductivity and resistivity are the two most important geo-physical parameters used in this method. The two related techniques under this category are known as terrain conductivity and subsurface resistivity.

Conductivity is the ratio of the electric current that flows in a medium to the electric force that is applied. With a unit of siemans/meter, electrical conductivity is a measure of the ability of an electrical charge to move through a medium. According to Ohm's law, conductivity (c) is defined as $c = J/E$, where J is the current density, and E is the electric field.

The value of conductivity is not constant. It can change with temperature, pressure, and other physical factors. Buried metallic waste has a high conductivity compared to its surroundings, therefore it can be detected with electromagnetic techniques. Examples of conductivity values for some common materials are

Kerosene	10^{-4}	10^{-3}		Shale	20	70
Basalt	10^{-2}	1.0		Clays	70	80

Resistivity measures the resistance of a medium to current flow and is the reciprocal of conductivity. It has a unit of ohm/meter. For some geophysical surveys, such as electrical soundings using direct current instead of conductivity, the resistivity factor is used to describe the electrical property. There are some tips which can serve as a basis for defining geologic conditions of a site. Examples are that igneous rocks are more resistive than sediments, or that sandy rocks are more resistive than clayey rocks.

Terrain conductivity is an electromagnetic (EM) method that provides a rapid measurement of the electrical conductivity of subsurface soil, rock, and groundwater. Because buried drums and metallic utility lines have a high conductivity, they are easily detected by this technique. The basic principle of the EM method is that contaminants in the subsurface alter the natural conductivity and this alternation can be detected by the appropriate instrument.

Operationally, without ground contact, the EM equipment induces a current in the ground. The equipment must be held above the ground surface. A transmitter mounted in the equipment produces a time-varying electromagnetic field. On penetrating the ground, the EM field induces a voltage that causes an electric current to flow in a conducting subsurface. The subsurface currents create a secondary magnetic field that is picked up by the receiving antenna. This secondary EM field is proportional to the ground current and therefore to the ground conductivity.

Terrain conductivity is influenced by the subsurface's material type, geologic fractures, groundwater chemistry, and the coil spacings and orientation of the instrument. Changes in phase and magnitude of the individual currents are measured by the receiver and by changing the frequency; depth sounding can be produced without changing electrode distance. The EM instruments are usually used for lateral profiling, a technique using several combinations of horizontal and vertical modes and coil spacings to effect a vertical sounding. Profiling measurements are still effective at depths of up to 200 ft.

The EM-31 and EM-34 models, manufactured by Geonics Limited of Missisauga, Ontario, Canada, are popular for terrain conductivity surveys (see Plate 8.1). Geonics recently attached a data logger system to the EM-31 model.

The system allows data to be digitally recorded and stored on-site. Having the advantage of being less expensive and very rapid in delivering results, the limitations of the standard EM equipment is the lack of depth penetration compared to resistivity.

Subsurface resistivity surveys are performed by applying an electric current across a pair of electrodes. When the current is injected into the ground, the patterns of the subsurface current flow reflect the resistivity of the subsurface. These patterns can be mapped on the subsurface by another pair of electrodes which measure the change in electrical potential (voltage) at the surface between the second pair of electrodes. The subsurface resistivity is calculated from the electrode spacing, the applied electric current and the measured voltage. The resistivity value can be used to identify soil type on the subsurface. A resistivity of less than 150 ohm/meter indicates clay, between 100 and 1,500 normally belongs to alluvium and sand, the range of 1,000+ is typical of bedrock.

Resistivity techniques differ from terrain conductivity because resistivity requires the injection of current through electrodes into the ground, while the EM methods induce currents in the ground. One of the advantages of the resistivity method is that the method requires inexpensive equipment. The equipment consists of a power supply, transmitter and receiver, current and potential electrodes, and wire. The new generation equipment records data digitally and can handle more electrodes.

Many resistivity surveys accomplish profiling and sounding simultaneously. Profiling studies the horizontal distribution of resistivity by maintaining a relatively constant depth of analysis through controlling the spacing of the electrodes. Sounding examines the vertical variation of resistivity resulting from the increasing depth of investigation.

Complex resistivity, also known as induced polarization, is another technique that essentially measures the same information as electrical resistivity and electromagnetic techniques. Complex resistivity, however, also provides information on the presence of active chemical processes in the sediment being tested.[48]

1.2.   Radiometric

Surveys are recommended for sites that may be contaminated with radioactive waste. The results of the survey can be used to identify problems early in the site characterization and isolate areas that may require special care in site remediation. Field screening devices are available that will quickly determine the presence of radioactive areas in a site. Called a radiometer, this type of device detects and measures radiation. Once the instruments are appropriately calibrated and all health and safety precautions have been addressed by the team, the survey can start.

Several portable radiometers are commercially available that can detect alpha, beta, and gamma radiation.[49] The alpha counter is a separate unit from the beta/gamma counter. Each unit is battery operated and easily managed by

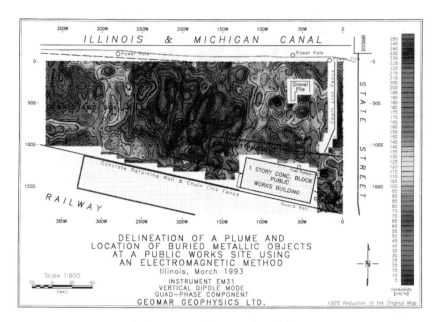

Plate 8.1. **GEONIC EM-31. The Geonics EM31 maps ground conductivity directly in milliSiemans per meter (mS/m). The EM31 also measures the inphase component of the response, i.e., the inphase ratio of the secondary to primary magnetic fields in parts per thousand (ppt), which is useful for detecting shallow ore bodies and for detecting buried metallic drum, pipes, and other ferrous and nonferrous metallic debris. (Courtesy of Geonics Limited.)**

one field staff. The beta/gamma (also known as Geiger) counter has a shield, and when the shield is closed, only gamma rays are measured. The detection is displayed on a meter in millirems per hour, or counts per min. When the shield is opened, both beta and gamma rays are detected. The amount of beta radiation can be calculated by the subtraction of gamma from beta + gamma readings.

A typical radiometric survey consists of two personnel, one with an alpha counter and the other with a Geiger counter. Each sampler would have a log book for recording the readings. The Environmental Protection Agency (EPA) recommends that for screening purposes, any radiation greater than background level must be investigated further to assure a thorough knowledge of the radioactive character of the site.

Another type of radiometer amenable to field use is the portable ion chamber. A hand-held instrument with charged gas in a chamber, this device

Plate 8.1.   (Continued). Two meters on the front panel of the EM31
            simultaneously display conductivity and inphase response.
            These readings can be recorded using the DL720 digital data-
            logger. Data can be collected either manually, with the push
            of a button, or "continuously" at a time interval selected by
            the operator. Stored data are downloaded from the data log-
            ger directly onto the DAF software (IBM compatible). The
            DAT program allows the user to edit, plot and print profiles
            of both the quadrature-phase (conductivity) and inphase
            responses.

is used to detect gamma radiation. For a flat surface, a "pancake" detector is often used for quick screening of beta and gamma radiation. The detector gets its name from its flat, round shape.

1.3. Ground-Penetrating Radar

Ground-penetrating radar (GPR), which was developed in the 1960s as an active remote sensing technique, allows researchers to see through and into otherwise impenetrable solid objects. Unlike X-rays which are hazardous, GPR is a reflection technique that uses nonionizing microwaves and radio waves with very low power requirements. GPR operates at frequencies where the displacement currents dominate and the domination of the displacement currents depends on the dielectric permittivity of the subsurface. GPR is actually an EM method but the underlying principles are quite different from the conventional EM methods, such as terrain conductivity which operates where the conduction currents dominate.

Dielectric permittivity is similar to electrical conductivity. It affects the speed of electromagnetic radiation in earth materials. While conductivity relates to current or charge flow, permittivity describes charge separation or polarization. Dielectric permittivity (E) has a unit of farads/meter and is expressed as a function of vacuum or free space (E_o):

$$E = K \cdot E_o$$

where K is the dielectric constant. K has no unit and its value varies from 1 (for air) to 80 (for water). The value of K for shale is 10. The dielectric constant K is important because it is the major influence on ground-penetrating radar measurements.

The GPR techniques are useful in the mapping of soil stratigraphy; determining depths to the water table and bedrock; and detecting buried drums, underground trenches, or fractured bedrock in the subsurface. Generally, radar penetration is better in dry, sandy, or rocky areas than clay or conductive soils. Penetration of 30 ft is typical for a rocky subsurface; where clays are present in the subsurface, the penetration may be less than 3 ft; under sandy conditions, radar penetration depth may reach 45 ft. The continuous data recorded by GPR provide a large amount of information with substantial detail. The detail is usually difficult to interpret and requires experienced staff to properly analyze the GPR results.

A GPR system consists of the transmitting and receiving units, a control unit, and a display unit. When operated, the transmitting unit generates a short pulse of EM energy that is radiated through an antenna into the ground as radio waves. The energy is reflected to an antenna of the receiving unit, its signal is then amplified, formatted, stored, and displayed. Manufacturers of GPR include Geophysical Survey Systems Inc. (GSSI), North Salem, NH, which manufactures the Subsurface Interface Radar (SIR) equipment. The latest edition of product manufactured by GSSI is called the SIR System-2 (see Plate 8.2).

1.4. Magnetic Susceptibility

Magnetic susceptibility is a measure of the ability of a material to become magnetized. This measure is based on the fact that the geomagnetic field of the earth can induce a magnetization in rocks and earthen materials that have sufficient amounts of ferromagnetic materials such as magnetite. Magnetic susceptibility has no unit and is a function of the magnetization and the intensity of the magnetic fields of the earth. Examples of magnetic susceptibility values for some common materials are
Unconsolidated sediments are generally nonmagnetic. Faults and fracture zones often appear as magnetic lows within the more magnetic background.

Magnetic susceptibility is indicated by a magnetometer. Magnetometers detect magnetic anomalies in the subsurface. Magnetic anomalies are the distortions of the regional magnetic field of the earth caused by materials with magnetic properties in the subsurface. A magnetometer measures the total magnetic field over a site and examines anomalous highs and lows that would indicate ferromagnetic (iron and steel) objects. The vertical gradient of the geomagnetic field is measured by simultaneously recording the magnetic field at two different elevations.

Magnetometer surveys are conducted to locate buried magnetic objects and are very useful for determining the location of metal USTs. The surveys can also provide an estimate of the thickness of nonmagnetic sediments or alluvium overlying magnetic rock. Sources of magnetic interference include power lines, buried metal pipelines, fences, surveyors' big belt buckles, and passing vehicles; therefore, magnetometers generally work better in rural areas.

There are two types of magnetometers: proton procession and fluxgate magnetometers. Proton procession magnetometers measure the amplitude of the magnetic field present at a survey location. Fluxgate magnetometers measure the component of the magnetic field in a particular direction, normally vertical. This type of magnetometer uses the frequency of the magnetic momentum of hydrogen ions after relaxing a strong magnetic field. Modern, portable magnetometers are computerized and digitally record the data.

1.5. Gravimetric

Gravity surveys measure the variations of density in the subsurface. Density is defined as mass per unit volume and is expressed in kilograms per cubic meter. Typical density values for unconsolidated sediment is 2000 kg/m^3 and bedrock is 2500 to 3000 kg/m^3.

The gravity surveys are conducted in areas where cultural noise (i.e., disruptive electromagnetic signals coming from power lines) precludes elec-

Plate 8.2. The SIR System-2 is a portable, digital subsurface interface radar system designed for environmental, geotechnical, geological, and engineering applications. The system is compatible with all GSSI antennas, thus facilitating a broad range of applications. Antennas are the key to quality data; they determine the depth of penetration and the target resolution capabilities. Higher frequency antennas are used for engineering, archaeological, forensic, and environmental studies, and the detection of buried objects. Low frequency antennas are used for fresh water sub-bottom profiling, mining, geotechnical, geological, and hydrogeological studies. The SIR System-2 Digital Control Unit (DC-2) is a self-contained radar system. Data are displayed in real-time on a color display, stored on an internal hard drive, and can be printed on a thermal plotter. Additional features include parallel data transfer to an IBM-compatible PC. (Courtesy of Geophysical Survey Systems, Inc.)

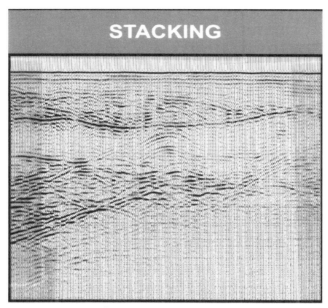

Stratagraphic Profiling

Plate 8.2.   (Continued). The DC-2 system has a variety of data processing, transfer and display options. The data acquisition modes include continuous profiling and point stacking.

Soils	10^{-6}–10^{-3}	Granite	10^{-4}–1.0
Limestone	10^{-5}–10^{-2}	Steel	50–150

tromagnetic, electrical, or seismic surveying. Two types of gravity surveys are available: a standard gravity survey which uses widely spaced stations, up to 1,000-ft intervals, and a micro gravity survey which uses stations spaced at up to 50-ft intervals. Field conditions such as temperature, humidity, and wind will influence accuracy; therefore, meteorological measurement must also be conducted to compensate for these effects.

The gravity survey is designed for preliminary reconnaissance purposes and can be used to delineate faults, buried valleys in the subsurface, or cavities in a karst environment. The results of a gravity survey can be a profile line, or if the data are obtained over a site area a contour map.

1.6.   Seismic Refraction and Reflection

In seismic surveys, a seismic wave is introduced into the subsurface by a small explosive charge or by hitting a steel plate with a sledge hammer into the ground. The ability of a boundary to reflect seismic energy depends on the density and seismic velocity across the boundary. The seismic wave propagates radially in the subsurface from the source. The wave can be detected by a detector termed a geophone. By placing a series of geophones at various distances from the source, the arrival time of the wave at each geophone can be measured and recorded by the seismometer on paper or film strips moving with a constant rate that corresponds to the elapsed time during the motion of the ground.

The time and distance information is then used to calculate the seismic wave velocities which are related to the density and hardness of the soil and rock in the subsurface. Examples of velocity (in ft/sec) for selected materials are

Sand and gravel	1,500–3,000
Clay	3,000–9,000
Water	4,700–5,500
Granite	15,000–19,000

There are two basic seismic exploration methods: reflection and refraction. In the reflection method, the waves are reflected directly back to the surface. Reflection methods are used for deep subsurface investigation. As reported by Benson in 1988, the method is effective from depths of 50 ft to a few thousand feet.[50] In the refraction method, the waves are critically refracted along the boundary and then re-radiated back to the surface. Refraction methods are used in shallow investigations, up to a few hundred feet. When using a sledgehammer, the seismic survey depths are limited to 50 ft.

In practice, it is not uncommon in site characterization to utilize more than one technique. If site characterization requires several geophysical surveys, the following guidelines are suggested by Greenhouse and Gudjurgis:[51]

1. Decide when to survey. In particular, weather can be a deciding factor in the quality of data collected and in the survey cost.
2. Proceed from the general to the particular by looking for broad and general coverage of the area as a guide to later detailed surveys.
3. Start with the simpler and cheaper survey which can be used to delineate the areas where further and more sophisticated work is necessary.

Some of the common technical problems in undertaking a geophysical survey as identified by Greenhouse and Gudjurgis include cultural, electrical, and magnetic noise; vibrations; natural noise; and topography.

- Cultural noise refers to the disruptive electromagnetic signals that come from power lines, buried cables, telephone lines, fences, and buildings. The signals can induce noise in seismic cables, particularly in wet cables, but do not affect magnetic or gravity measurements.

- Vibrations from heavy traffic, nearby drilling, and the roots of trees that are moving in the wind can present problems for seismic and gravity surveys.
- Natural noise refers to EM noise from local and distant electrical storms (sferics) and naturally induced ground currents (tellurics) which can hamper EM and resistivity surveys.
- Topography, particularly rugged topography, can hamper geophysical surveys. In gravity and seismic surveys, site elevation must be appropriately factored in.

Other geophysics limitations include the need to have direct confirmation of results and highly trained personnel for some methods. EPA has developed an expert system program designed to assist and educate nongeophysicists in the use of geophysics at contaminated sites. The 2.0 version of the program is written to run on any IBM MS-DOS compatible computer and is easy to use. It is not meant to replace the expert advice of competent geophysicists.[52]

The computer program asks questions about a site and its contamination problem. Based upon the answers, the program recommends the types of geophysics most likely to be useful at the site to solve problems such as contaminant location and hydrological features of the site. The expert system annotates why various geophysical techniques will likely work or not work at the site.

A geophysical survey requires that planners understand the geological environment in which the contaminants occur and be familiar with the physical properties of the site. To assess contaminant migration at a site, it is essential that planners know the physical characteristics of the subsurface system that govern contaminant movement. The major components of the subsurface system are

- hydrological factors (such as the form and occurrence of precipitation, patterns of surface drainage, vegetation, and land use)
- geological factors (surface topography, stratigraphy, and structural features such as faults, joints, and bedding planes)
- hydrogeological factors including hydraulic gradient, water table depth, and hydraulic properties of the subsurface materials.

The following section discusses the hydrogeological investigation aspect of site characterization. The geochemical investigation is also included in the discussion as well. The discussion addresses the goals of the investigation, the process, methods, tools, and equipment needed for the investigation.

2. HYDROGEOLOGICAL AND GEOCHEMICAL INVESTIGATION

Several steps are required for an adequate assessment of the hydrogeological conditions of a site, including the review of available site documentation, and sampling and environmental analysis. In essence, the purpose of a hydrogeological investigation is[53]

1. to determine the direction and rate of groundwater movement
2. to delineate the principal contaminant pathways, and examine the factors controlling the contaminants migration
3. to quantify the physical parameters that govern groundwater flow and contaminant transport in the subsurface

To achieve this set of goals, a site-specific investigation program must be developed. Because of its specificity, the discussion of the investigation process can only be facilitated by using a general approach. In most situations, the preliminary review of existing data can provide extensive information that enables a more efficient design of the hydrogeological site investigation, ensuring that new field data are continuously incorporated into existing information.

The preliminary review and site reconnaissance is deemed complete when the project is able to provide the description of the general geological and hydrological features of the site, the collection of evidence of source and nature of contamination, the evaluation of the potential site workers' health and safety problems, and the review of the conceptual model of the hydrogeological system developed earlier in the process.

During the site reconnaissance, a shovel and hand auger are often necessary for sampling the surficial materials, and sample bottles for surface water sample collection. The site reconnaissance will also be easier to perform if it is prepared by using the aerial photographs, the Sanborn fire insurance maps, and other technical maps.

- Aerial photos show historical photography from 1920 to the present, and can indicate the past existence and current disappearance of physical objects captured in the photos, such as landfill, surface impoundment, or hazardous waste storage pad.
- The Sanborn maps are available in most public libraries. They provide locations of industries, pipelines, and old dumps in communities with populations of more than 2,000.
- Other useful maps include topographic, geologic, hydrologic, and land use maps.

The site walk-over, or field screening, is often a necessity. The field screening can be conducted by using physical surveys such as geophysical,

X-ray fluorescence, or soil gas. X-ray fluorescence is a field screening technique for identifying metals with field portable equipment. Soil gas primarily identifies volatile organics. It uses a field-portable or truck-mounted instrument. The results can be utilized for delineating shallow groundwater plumes, locating sources of soil contamination, and planning the location of the groundwater monitoring wells. Once the preliminary review and site reconnaissance is completed, the next steps are

1. to conduct the preliminary site investigation
2. to perform the detailed site investigation
3. to report the findings

2.1. Preliminary Site Investigation

This preliminary investigation is conducted for defining the principal hydrostratigraphic unit and general groundwater flow directions on site. Hydrostratigraphic units are a subsurface soil and/or rock formation or a group of formations that have similar hydrologic characteristics. The general subsurface geologic features at the site must be known before an initial test drilling can be conducted. Information such as the types of materials that will likely be intercepted during the test drilling, the depth of the low permeability sediment or bedrock, and the location of buried waste containers is essential for locating potential sources of contamination and avoiding piercing buried containers during drilling, determining the depth of drilling (i.e., reaching the low permeability sediment) and directing the placement of the monitoring wells.

Drilling conducted at three boreholes arranged in a triangular pattern, compiling a detailed record of the subsurface stratigraphy encountered during the drilling in each borehole, and equipping each borehole with a monitoring well will provide enough data to allow the construction of a general map of the groundwater flow field with a contoured map of the hydraulic heads. The initial drilling and analysis should provide the relative position, thickness, physical and geological description of each major stratigraphic unit as well as related soil characteristics such as size, plasticity, porosity, and permeability. If no previous information is available, the borehole is advanced until low permeability sediment such as nonfractured, nonporous bedrock or tight clays is reached. If the low permeability sediments are fractured, deeper drilling will generally be warranted. Selecting the low permeability boundary requires hydrological judgment.

Monitoring wells can be installed after completion of drilling and logging a test hole. At each of the three boreholes, a well should be installed, then monitoring and recording of the recovery and piezometric levels can be started. Obviously, wells in permeable sediments will recover quickly, whereas those in low permeability sediments take longer to reach an equilibrium state. The piezometric level refers to the elevation of the water table or potentiometric surface, measured by a nonpumping well called a piezometer. A piezometer is designed to measure the hydraulic head at a point in the subsurface. The hydraulic head is the height to which water can raise itself above an arbitrary datum level. Plotting the hydraulic heads from the completely recovered monitoring wells on a site map will result in a map containing equipotential lines. An equipotential line is a line that connects points of equal hydraulic heads. For each permeable hydrostratigraphic unit, individual equipotential maps should be drawn.

Drill cuttings that are brought to the surface during the drilling process may be highly contaminated and require special handling and management. During drilling, samples of drill cuttings are collected for on-site chemical analysis. The results of the analysis provide a preliminary indication of the nature of the contamination in the subsurface and the proper disposal method for the drill cuttings. Thorough consideration is required to avoid mobilization of the contaminants by carrying the contamination downward with the drill bit or screening wells to allow migration from contaminated zones to clean zones.

Once the main hydrostratigraphic units have been delineated, the groundwater flow field mapped, and a conceptual model of the hydrogeological system refined, detailed site investigation can be initiated. The benefit of having a detailed site investigation is that vital physical hydraulic properties of a site can be obtained for mathematical modeling of the groundwater flow field as well as chemical data for consideration of remedial options.

2.2. Detailed Site Investigation

This effort requires expansion of the existing monitoring well network, by installing new wells and performing the related field activities and laboratory analysis. At small sites of less than 0.1 square mile where the ground surface is relatively flat, the hydrostratigraphy is normally simple and uniform and the three-well system can provide sufficient information. Physical site features such as springs, rivers, or large excavations, however, can influence patterns of groundwater flow. On larger sites, three boreholes might not be enough because flow directions may change significantly over the site. When selecting additional monitoring points, consider the following features:[54]

1. Recharging conditions and associated mounding of the water table generally occur under areas of higher elevation and groundwater discharge conditions often prevail in the lower topographic areas. These site features impact both the horizontal and vertical movement of groundwater. At least one multi-level monitoring well should be located near each of these features.
2. If a stream flows through the site and the stream receives groundwater discharge (a gaining stream), it is a focus for shallow ground-

water discharge and will certainly affect the groundwater flow direction. If the stream recharges to the groundwater system (a losing stream), mounding on the water table near the stream can be expected. Installing multi-level wells at the edge of the stream will allow the measurement of the vertical hydraulic gradient, and the determination of whether it is a gaining or losing stream. If a pond or a lake is located on the site, installation of monitoring wells along the shore is also necessary.

3. Where upward groundwater flow discharges to the surface, a spring is created. Small creeks may have their headwaters at a spring. Dense vegetation typically grows around a spring. Installing a monitoring well near the spring on the site can provide information on the magnitude of the vertical hydraulic gradients.

4. In response to rain storms or rapid snow melt, the flow in a stream or the level of a lake changes rapidly, resulting in a change of the vertical hydraulic gradients to the point where a gaining stream may become a losing stream creating variations in groundwater flow directions on the site. This means that the nature of surface water bodies must be considered thoroughly during detailed site investigation.

Rudolph recounts that if the data show that the critical hydrostratigraphic units, such as the permeable aquifers, are thin (or quite variable in thickness), a larger number of boreholes may be required. The exact number of additional test holes is difficult to estimate and discussions with the hydrogeologist regarding proper placement of monitoring wells for geochemical mapping is encouraged.

To determine the groundwater flow volumes and velocities, several hydraulic properties of the subsurface materials (such as porosity, permeability, hydraulic conductivity) must be measured. Most measurement techniques provide spot or point measurements. To understand groundwater flow at the field scale, these point measurements must be extrapolated over the entire flow system. Experienced hydrogeologists are needed for planning, performing, and interpreting such measurements.

The procedure of extending the monitoring well network is an iterative process based on the data acquired from each new borehole and a continually refined conceptual model of the site. One of the most difficult decisions of the hydrogeological site investigation is whether sufficient data have been collected.

The hydrogeologist's role in site characterization is to establish a conceptual model of the aquifer; identify the existing distribution of contaminants in the groundwater, soil vapor and subsurface solids; interpret the existing contaminant distribution; estimate future distribution; and propose appropriate remedial action. Identification of the contaminants of concern can be provided in the preliminary review and site visit. Geophysical surveys and the hydrogeological site investigation may have located the contaminants and their sources. To obtain a better understanding of the contaminants on a site, however, a geochemical investigation must be conducted.

2.3. Geochemical Investigation

The purpose of this investigation is to collect additional information on the types, concentration, and spatial distribution of the contaminants on a site. Barker[55] suggests that when additional samples from contaminated areas are needed, they are to be collected for only those target contaminants. If the contaminants are unevenly distributed, reconnaissance sampling of potential contaminated soil suspected from the preliminary review data is appropriate. Systematic sampling is preferred to estimate average and total contamination and to discover patterns of contamination. If the pattern is already anticipated, judgment sampling can be considered. A staged approach similar to the field hydrogeology investigation is also appropriate for defining the distribution of contaminants in the subsurface.

The site characterization can be improved dramatically by a quick determination of the approximate subsurface contaminant distribution. This rapid determination can be accomplished by using a soil gas survey which is intended to identify underlying volatile organic chemicals and X-ray fluorescence for the measurement of inorganics. Field X-ray fluorescence is a site screening technique for soils, using a 20-lb portable instrument that can measure inorganic elements when used appropriately with the proper radioisotope source. The technique is capable of simultaneous analysis of up to six analytes per instrument.[56]

A soil gas survey is a field screening tool which is quick, less expensive than drilling wells, and capable of providing proper plume resolution.[57] EPA[58] indicates that soil gas sampling can be used for site characterization to:

- identify contaminants and their relative concentrations
- identify sources of contamination and indicate the extent of contamination
- guide the placement of subsequent soil borings and wells
- detect leaks through the use of tracer

2.3.1. Soil Gas Survey

The term soil gas refers to the atmosphere present in the soil pore spaces. Volatile compounds introduced into the subsurface can be present in the gas phase or undergo a transition from a liquid to become part of the soil atmosphere. The transformation occurs when the groundwater plume migrates. A contaminant with a high vapor pressure and a low water solubility will volatilize from the groundwater into the unsaturated zone or vadose zone.

In order to be effective, the chemical analyzed by a soil gas survey must be volatile, have a Henry's law constant of at least 0.0005 atm · m³/mol and a vapor pressure of at least 1.0 mmHg at 20°C. Contaminants detectable in soil gases include chlorinated solvents and the lighter fractions of petroleum products as well as some inorganics such as hydrogen sulfide, mercury, and radon. The reliability of a soil gas survey is enhanced when the contaminant does not degrade rapidly into degradation products. One of the major limitations in using a soil gas survey is interference: false negatives are a problem.

Soil gas surveys are conducted to provide chemical information in the vadose zone. The surveys measure a vapor phase plume in the void space of the soil in that zone. This information can then be interpreted in the context of the detection of the contamination source and the delineation of plume.[59] Although the surveys have also been used to assess the nature of the groundwater plume, generally they are restricted to shallow groundwater.

There are two types of soil gas surveys: passive and active methods. A passive soil gas survey refers to leaving in the ground, for a period of time, a sampler containing a sorbent. Charcoal is commonly used as a sorbent. After collection, the sampler is shipped to a laboratory for analysis using gas chromatography (GC) in combination with a detector appropriate for the target contaminant. This method is appropriate when subsequent soil gas data are required to measure the time history of the soil gas contaminant concentrations.[60]

Most soil gas surveys are accomplished by some variant of an active whole-air approach.[61] In active soil gas sampling, soil gas is drawn from the ground through shallow probes and collected in a syringe or tedlar bag. Whole-air samples must be drawn from locations deeper than 3 ft to avoid the variable concentrations found near the surface. After sampling, GC or GC coupled with mass spectrometry (MS) is normally used for the chemical analysis. MS is a three-phase analytical process consisting of ionization, separation, and detection. The detection of an organic contaminant, however, is not always possible by using a low-resolution MS. High-resolution MS offers better detection due to its improved separation capability. It is more dependable, but costly and requires expert interpretation.[62]

Other detection instruments commonly used in site characterization include an organic vapor analyzer and a photoionization detector (PID). The Foxboro Company (Foxboro, MA) manufactures a variety of vapor analyzer equipment. One of the products is OVA 128 GC (see Plate 8.3). HNU Systems Inc. (Newton Highlands, MA) is the manufacturer of portable photoionizers and gas chromatographs. (HNU Model 101 PID, see Plate 8.4; HNU Model 311D GC, see Plate 8.5.)

The analysis of the soil gas survey may be done with field instruments, in a mobile laboratory, or in a remote analytical laboratory. Due to the short holding times, on-site analysis is usually performed. When analysis is performed on-site, results can be expected in a matter of minutes (e.g., 120 min), and therefore the sampling plan can be refined as the survey progresses. Lewis

CENTURY OVA 128GC

Plate 8.3. **The CENTURY Organic Vapor Analyzers (OVA) 128 GC can be used for emergency spill response, initial site discovery and characterization, remedial investigation or ambient air monitoring. In initial surveys, it will deliver immediate warnings of hazardous materials and help determine the level of respiratory protection required for site workers. In site analysis, it can identify the chemicals present and their concentrations in air, soil, or water headscape samples. When equipped with a strip chart recorder, the CENTURY OVA 128 GC will provide continuous readings of organic vapor levels at the site perimeter. Soil gas monitoring is conducted by driving hollow steel probes into the soil and the CENTURY OVA is used to test vapors released. Or soil samples can be collected and heated in sealed vessels for direct headspace analysis. When equipped with a gas chromatograph, it can differentiate those specific vapors present in the headspace samples. (Courtesy of The Foxboro Company.)**

et al.[63] caution that soil gas measurements should be used for screening purposes only, and not for definitive determination of soil-bound volatile organic compounds (VOCs). As Smith et al.[64] have shown, a disparity occurs in concentration of VOCs in soil gas and the VOC concentrations found on the solid phase. In another study, Marks and Singh[65] conclude that the soil gas results correlate well with groundwater sampling results when the soil gas survey is conducted within 5 ft of the water table for clayey soils and 10 ft for sandy soils. In fine grained soils, organic rich subsurfaces, and subsurfaces with a high water content, soil gas surveys will not perform well.

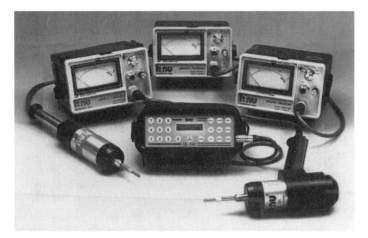

Plate 8.4   HNU's Model 101 portable photoionization detectors (PIDs)
             are used for detecting volatile organic compounds in field sur-
             veys, leak detection, hazardous waste site and industrial
             hygiene applications. The detector works as follows: a gas
             sample is drawn into the ion chamber where it is illuminated
             with an ultra-violet (UV) lamp. Compounds in the gas sample
             with ionization potentials equal to or less than the energy of
             the UV lamp are ionized. An electrical field within the chamber
             forces the ions that are freed to a collection electrode generat-
             ing an electrical signal. The signal is proportional to the con-
             centration of compounds present, to which a value
             representing concentration is assigned and displayed. HNU's
             Model 101 PIDs are available in four types:

             • PI-101 for leak detection and surveys.
             • IS-101 for explosive conditions, i.e., the presence of high
               concentrations of gases and vapors.
             • HW-101 for use in deep wells for soil gas analysis.
             • Dl-101 for microprocessor-controlled, integrated datalog-
               ging. Output data are transferred to a PC for reports.

             Lamp choices are: 9.5, 10.2, and 11.7 eV. The 9.5 eV lamps are
             used to detect aromatics and large molecules; the 10.2 eV
             lamps detect vinyl chloride, MEK, MIBK, TCE and other 2-
             4 C compounds; the 11.7 eV lamps detect halocarbons, meth-
             anol and other single C compounds. (Courtesy of HNU Sys-
             tems, Inc.)

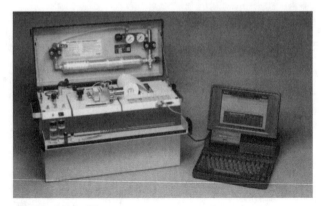

Plate 8.5 **HNU Model 311D GC is a separation technique for identifying and quantifying compounds. A GC consists of five major components:**

1. **The injector — injects a gas sample into the column.**
2. **The carrier gas — moves the injected sample through the column.**
3. **The column — provides separation of the compounds. After separation, the compounds pass into the detection chamber.**
4. **The detection chamber — a detector provides a signal, representing concentration. Types of detectors:**
   - **Photo Ionization Detectors (PIDs) are primarily for volatile organic compounds.**
   - **Electron Capture Detectors (ECDs) detect pesticides, PCBs, chlorinated solvents and other halogenated and nitro-compounds.**
   - **Far Ultra-Violet Detectors (FUVDs) measure inorganic and organic compounds with strong UV absorption signals (O_2, CO_2).**
5. **An output (a chromatogram) — is provided with peaks shown as a function of time. The peak height represents the concentration of a compound. Output results are shown quickly with an on-board printer/plotter or the data are exported directly to a PC with HNU's Peakworks software. (Courtesy of HNU Systems, Inc.)**

2.3.2. Groundwater Sampling

If groundwater monitoring wells are needed for characterizing the geochemistry of a site, determining the appropriate location for the wells is critical. Groundwater monitoring wells can be located in a regular grid pattern, or at specific locations to intersect suspected pathways of contaminated groundwater. The well locations must be selected to accommodate the purpose of the monitoring and the cost must be considered because it is usually

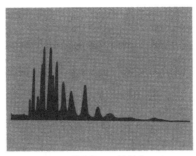

PCBs on an ECD at 180°C

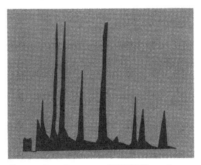

Static headspace of MTBE and BTX on a PID at 60°C

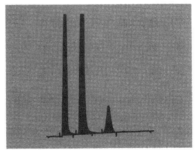

Landfill gas; FUVAD at 130°C, O₂ 18%, CH₄ 1%, CO₂ 2%

Plate 8.5 (Continued).

expensive. Monitoring wells for hydrogeological investigation are normally used for the geochemistry evaluation.

Once the wells are installed, groundwater sampling can be initiated. Barker suggests that the sequence of groundwater sampling consists of the following:

1. Measurement of the water table elevation and the thickness of the nonaqueous phase liquid (NAPL), if any. NAPL is elements or chemical compounds in the liquid phase other than water, such as gasoline. NAPL is immiscible in water. Dense NAPL (DNAPL) may be found at the bottom of a well but it is rare. Light NAPLs (LNAPLs) are more common.
2. Sampling of the LNAPL.
3. Removal of the standing water in the well by purging the well.
4. Sample collection.

It is quite normal to determine the chemistry of the contaminated groundwater early in the site characterization. This early determination, or scanning, is to identify, not quantify, the presence of chemicals.[66] Barker suggests that for inorganic contaminants, the inductively coupled plasma/emission spectroscopy (ICP/ES) technique be used. Volatile organics are scanned by the purge-and-trap GC methods using an MS as the detector. The purge-and-trap is performed by taking advantage of the high vapor pressure of volatile organics. By bubbling a pure gas such as helium through a water sample, volatile organics are removed from the sample, swept away with the evolving gas, and trapped from the gas stream for instrument analysis.

Scanning can include identification of the migration of contaminated groundwater. This requires a nonreactive tracer that moves as fast as the groundwater. A tracer is defined as a substance that is carried by the groundwater and which provides information concerning the direction of the movement of the groundwater and the potential contaminants that might be transported in the groundwater system. Barker provides examples of reliable tracers which include boron, bromide, chloride, phenol, and potassium.

A tracer should have a number of properties in order for its application to be successful.[67] The requirements are

- The chemical and physical behavior of the tracer in groundwater must be understood. The tracer must be able to travel with the same velocity and direction as the groundwater. Most contaminants are retarded and move slower than groundwater, making the selection of a tracer difficult.
- A tracer should be nontoxic, relatively inexpensive, and detectable with simple technology.
- The tracer should be unique to the groundwater environment so that the detection of contaminants in the groundwater can be facilitated.

The artificial introduction of tracers must be carefully considered to prevent possible adverse health effects. It must usually be approved by local and state health authorities. Local citizens must also be informed of the tracer injections.[68]

Chemicals identified in the scanning must be reviewed and selected for subsequent sampling. The need for a hydrogeological investigation and chemical sampling and analysis must be compared. The selection of an analytical method should be based on meeting the QA/QC requirements, among other things. Most methods use GC to separate and identify chemicals. Detectors such as MS are used to quantify the concentrations of each identified chemical.

Barcelona et al.[64] assert that reliable sampling of the subsurface is inherently more difficult than either air or surface water sampling because of the inevitable disturbances which well drilling or pumping can cause and the inaccessibility of the sampling zone to the sampling equipment. To be reliable, sampling protocols must be based on the hydrogeologic setting of the site, and the degree of analytical detail required by the information needs of the project. In this regard, quality control begins with:

- the evaluation of the hydraulic performance of the sampling well
- the proper selection of mechanisms and materials for well purging and sample collection
- the utilization of field blanks, standards, and control samples to account for other variables which can affect data validity

The collected chemical data will be interpreted in lieu of the site characterization objective and in consideration of the nature of the physical hydrogeology of the site. This means that the results of the analysis of soils, rock, and groundwater derived during previous drilling should be used to provide a basis for determining whether additional drilling is required and where to locate additional wells.

The groundwater monitoring strategies have relied on a process of collecting samples from monitoring wells and determining if there is evidence of contamination. Because the purpose of monitoring is to determine the distribution of contamination in the aquifer, the evidence comes after contaminants have reached the resource the planners are trying to protect. This suggests that, whenever possible, prevention of aquifer contamination should be attempted at the vadose zone because vadose zone monitoring can provide early detection of contaminant migration.[70]

In addition to contamination from chemicals, a site may have potential public health risks for pathogens or coliform bacteria particularly if animal or human wastes were disposed on the property. Because it is difficult to attribute coliform counts to the aquifer, Barker argues that collecting such information as part of the site characterization is not warranted.

The results, findings, and conclusions derived from site characterization must be clearly, precisely, and accurately reported. Without proper documentation, the value of the site characterization is reduced significantly. The development of reports is an ongoing effort and should not be left in its entirety to the end of the process.

3. MOVEMENT AND FATE OF CONTAMINANTS

The investigative stage of the site characterization process will provide substantial information that can be used to assess the movement and fate of site contaminants. Knowledge of the types of contaminants, subsurface characteristics, and contaminant locations in the subsurface is needed for assessing the transportation and transformation processes in the subsurface. Generally, site contaminants can be found in five phases (or compartments) in the subsurface:

1. solid grains, rock, or waste
2. ions or molecules which are weakly bound (sorbed) to solids and which can be released into water or vapor
3. gas or vapor in pore spaces around these solids, in the unsaturated zone
4. water in the pore spaces in the saturated zone
5. nonaqueous phase liquid, either as mobile, continuous phase, or immobile, discontinuous phase

The grouping of contaminants based on mobility is necessary because contaminants in the immobile phase, i.e., solid and NAPL, can contaminate subsurface water and vapor. On the other hand, contaminants in the mobile phase, such as subsurface vapor, water, and NAPL, have the potential to migrate. In terms of the migration options, there are three possibilities available for site contaminants:[71]

1. to remain unchanged in the present location
2. to be carried elsewhere by the transport process
3. to be transformed into other chemical species

3.1. Contaminant Transport

The contaminant transport process can occur within a phase. This includes migration under gravity, diffusion, or advection. The transport can also occur between phases. The movement of a contaminant to other phases is facilitated by various processes such as adsorption, volatilization, precipitation, and dissolution. The transport process can occur physically/mechanically (such as diffusion and advection) or chemically. Some of these processes are briefly described in the following paragraphs:

- Diffusion is the movement of a contaminant under the influence of a concentration gradient. Contaminants in air or water have the inclination to physically move from areas of high concentration to areas of lower concentration.

- Advection is a physical process by which a solute (contaminant dissolved in water) is transported in groundwater and moves at the same velocity with the groundwater flow.
- Adsorption is a process in which the sorbates (soluble contaminants) are removed from water by contact with the sorbent (solid surface). This chemical removal of organic substances takes place on a surface. When the sorbate is taken into the sorbent, the process is called absorption. Absorption and adsorption are grouped as sorption. Depending on the type of organics and the sorption mechanism, the rate of sorption can be as fast as 48 hours.
- Volatilization is the transfer of an organic substance from a liquid phase to a gaseous phase. The transfer rate depends on the temperature, the vapor pressure of the chemical, and the difference in the concentrations between the liquid and gas phases. Organics may volatilize from the leachate and move to the ground surface. Emissions may come from the volatilization of dissolved chemicals in contaminated groundwater. Volatilization can be measured by organic vapor analyzers; however, the instruments may not be able to differentiate between varied organic compounds.
- Precipitation occurs when a chemical reaction transfers a solute to a much less soluble form. It is particularly applicable to heavy metals such as nickel, mercury, chromium, and lead. Precipitation is dependent on pH. Most metals precipitate at high pH levels.
- Dissolution is the converse of precipitation. When the concentration of a solute drops below its solubility, dissolution of previous precipitates could occur. This process dissolves a solid or NAPL to form ions or molecules which are uniformly distributed in water or another solvent.

3.2. Transformation Process

Contaminant transformation occurs through biodegradation (aerobic and anaerobic), bioaccumulation, and chemical changes known as weathering. The removal or decrease of groundwater constituents by the sum of mechanical, chemical, and biological processes is known as attenuation. Weathering includes oxidation and reduction, hydration, hydrolysis, and photolysis. In this process, all or a portion of the chemical is transformed to other chemical species. The following paragraphs describe some of these processes:

- Biodegradation is the transformation of organic substances to smaller molecules through the oxidation and reduction mechanism induced by the metabolic activity of native microorganisms. When the process occurs in the presence of oxygen, it is called aerobic biodegradation. Without oxygen present, it is known as anaerobic process.

- Reduction-oxidation (redox) is a chemical reaction that occurs with organic and inorganic chemicals and involves the gain or loss of oxygen. A reaction that adds oxygen to a compound is called oxidation and, likewise, loss of oxygen is termed reduction. The redox process, however, does not necessarily involve oxygen. The oxidation and reduction of organic chemicals is normally a result of biological processes rather than chemical reactions.
- Hydrolysis is a chemical process by which chemical substances react with water molecules. The process represents an exchange between OH⁻ and an anionic group of the chemical compound resulting in decomposition of the compound. It is a significant process for chlorinated organics which normally are not readily transformed by biodegradation.
- Photolysis is a photochemical process that photodegrades chemical substances. In a photochemical reaction, only the light absorbed is effective in producing a photochemical change. The efficiency of a photochemical reaction is measured in terms of quantum yield, i.e., the ratio of the number of molecules of material undergoing that process to the number of quanta of light absorbed.

3.3. Factors Affecting Contaminant Behavior

Contaminant behavior in the subsurface is affected by its own physical and chemical characteristics, as well as the natural characteristics of the subsurface. Some of the properties of a contaminant that are important for predicting its behavior in the environment are solubility, volatility, tendency to adsorb to solids, chemical reactivity, biodegradability, and density. If these properties are not known, they can be evaluated from the chemical structure of the contaminant.

One of the important factors, solubility, is measured as the rate at which one substance will dissolve into another. It is affected by temperature. The solubility of chemicals in water is expressed in mg/l or ppm. The solubility of a gas in a liquid is described by Henry's constant. Henry's constant (H) may be calculated as the ratio of the concentration of a chemical in the gas phase (C_g) to the concentration of the chemical in the liquid phase (C_L):

$$H = C_g/C_L$$

In this equation H is unitless. The constant can also be calculated as $H = P_g/C_L$, where P_g is the partial pressure of the gas expressed in atmospheres (atm). If C_L is expressed as mol/m³, then the units for H will be atm · m³/mol. Because of the different units for Henry's constant, it is critical to check the units carefully before using tabulated values for H.

Certain properties of water and soils also play a significant role in influencing contaminant behavior. The water properties include temperature, pH, alkalinity, hardness, chemical components and concentrations, and the migration pattern. The important properties of soils are adsorption attractions for contaminants, mobility, porosity, and hydraulic conductivity.

- Temperature affects the degree of chemical dissociation of species such as cyanide, sulfide, chlorine, ammonia, and carbonate. Biodegradation of organics is accelerated with increased temperatures which raises the demands on dissolved oxygen.
- pH is a measure of the acidic or basic character of a substance, determined by its concentration of H^+. Based on the equation of pH $= -\log (H^+)$, chemists have developed an equivalent way to express (H^+) as a positive decimal number whose value is between 0 and 14. Lemon juice and soft drinks have a pH of 2 to 4; cow's milk 6 to 7; fresh eggs 8; and household ammonia 10 to 12. When the pH is less than 7, a material is acidic. At a low pH, metals tend to dissolve. When the pH is more than 7, a material is basic. At a high pH, metals tend to precipitate as hydroxides and oxides; however, when the pH gets too high, the precipitates may begin to dissolve again. pH = 7 means that a substance is neutral, e.g., pure water.
- Alkalinity refers to the ability of a substance to neutralize acid. It is measured as the concentration of $CaCO_3$ in mg/l. Alkalinity is closely related to hardness.
- Water hardness was originally used as a measure of the ability of water to precipitate soap. Current practice is to define hardness as the sum of the calcium and magnesium concentrations in mg/l.
- Soil porosity can be defined as a ratio that indicates the volume of the voids divided by the total soil volume. Porosity is inversely related to the retardation process, i.e., the process that obstructs the transport of contaminants by removing or immobilizing the compounds from a free state.
- Hydraulic conductivity of soil (k) indicates the ability of a material to conduct (or discharge) water in response to a given hydraulic gradient. The hydraulic gradient denotes the rate of change in which the head (or energy) is lost as water flows through the material. Stated in algebraic form (known as Darcy's law):

$$k = Q/i \times A$$

  where Q = flow rate (cm^2/sec), i = hydraulic gradient (cm/sec), A = cross-sectional area of flow measured, perpendicular to the flow direction (cm^2). The equation shows that k has units of velocity, i.e., cm/sec. Also known as permeability, k indicates the ease with which water will pass through the soil material. This constant is utilized to estimate the flow rate of fluid through a soil material (Q) and calculate the rate at which contaminants may migrate in the subsurface. The triaxial permeability test is used to determine the value of k.

Assessing the movement and fate of contaminants in the subsurface, as the above paragraphs indicate, is not an easy task. Accurate descriptions of the transport mechanisms, transformation processes, and various interrelated factors that affect these processes are needed. The fundamental processes of groundwater flow and contaminant transport in the subsurface are basically governed by natural laws that can be expressed in mathematical equations. These equations form the principles of all mathematical models. To facilitate the understanding of how contaminants behave in the subsurface, a number of mathematical models have been developed. The following section describes the modeling aspect of site characterization.

3.4. Groundwater Flow and Contaminant Transport Model

Understanding the complex interactions among the physical, chemical, and biochemical processes taking place in the subsurface is essential for the management and proper remediation of contaminated sites. Groundwater flow and contaminant transport models are tools that help planners understand the subsurface. While sampling, testing, and screening provide planners with a limited picture of the subsurface, the vast majority of earthen material that is situated between the measuring points will remain unknown. That is why a model is needed.

Mathematical models are common tools that assist planners in understanding how contaminants move in the subsurface. A site-specific model can also be used to evaluate the effects of remediation. Models can incorporate the physical, chemical, and biochemical processes that are likely to play a role in a given situation.

Modeling is a compromise between the sophistication of theories, the limitations regarding the acquisition of specific data, and economics.[72] The success of a model depends on the accuracy and efficiency in simulating the natural processes that control the behavior of groundwater and the contaminants it transports.

The groundwater transport models rely on the hydrogeological processes that govern the flow and transport of groundwater contaminants. The processes include the transport processes (advection, dispersion, sorption), and the transformation processes. In the subsurface, the organic contaminants can be present in various forms: nonaqueous, dissolved, or gaseous phases. Modeling the transport of organics, therefore, deals with multi-phase flow including its related processes, such as dissolution, biodegradation, volatilization, and vapor transport. Because most of these processes are nonlinear, i.e., each of them depends on the other processes, models to simulate these processes are complex and expensive to develop.

One of the approaches for simplifying the modeling process is to select only a relevant process for the model. If, for example, a model is intended to simulate the effects of a hydrocarbon spill, the model will evaluate the LNAPL migration, independent of the plume of dissolved hydrocarbons although the two are obviously interdependent.

The subsurface can be divided into two zones in relation to the conditions of groundwater flow. They are the saturated zone and the unsaturated zone. Saturated flow conditions in the subsurface exist beneath the water table where the pores are completely filled with water. The subsurface area where the pores are not filled with water is the unsaturated zone. Analysis of the unsaturated zone is necessary when there are important transport processes (volatilization, degradation, vapor phase transport) occurring within it.

The most important aspects of modeling from the planning perspective are the required input parameters and assumptions attached to a model. Planners should always acknowledge the inherent assumptions and limitations of the chosen model. Input parameters are the most critical elements of a model because they represent the foundation upon which the mathematical models rest.

In modeling the saturated zone, Molson et al.[73] state that the geometry of the aquifer base (or its thickness where it is confined) must be identified for developing an areal plane flow model. Other input parameters for this two-dimensional model are the hydraulic conductivity component in the areal plane (i.e., K_x and K_y), recharge distribution, well locations and pumping rates, and boundary conditions including river elevations or locations of flow divides.

Molson et al. also state that in developing a groundwater flux model, the following input parameters are required for this two-dimensional model: the geometry of the aquifer base, the hydraulic conductivity components (K_x, K_z), recharge distribution, and boundary conditions. The groundwater flux refers to the volume of fluid flowing through a specified cross-sectional area per unit of time. The concern for this flow in the vertical dimension is that there may be a potential for vertical contaminant migration to underlying aquifers.

The two groups of input components that are relevant to the transport and fate of contaminant modeling in the vadose zone are

- physical and chemical properties of the contaminants
- soil properties

The properties of contaminants that are commonly required as input parameters are identified by Brusseau and Wilson.[74] They include solubility, diffusion, biodegradability, sorption potential, structure, and vapor pressure. Water solubility represents one of the most important parameters. Contaminants that are soluble in water tend to be readily transported through the vadose zone. The structure of an organic compound often controls the solubility, sorption, and transformations characterizing the compound. The vapor pressure of a contaminant affects its tendency to escape from soils. When a liquid is in contact with air, molecules will leave the liquid as vapor. The vapor pressure of a liquid is defined as the pressure exerted by the vapor on the liquid at an equilibrium state, i.e., at a condition when the rate of molecules leaving the liquid is equal to the rate of molecules being redissolved. Vapor pressure is affected by temperature.

Soil data types required for modeling are inventoried by Breckenridge et al.[75] The properties of soil listed in the inventory are bulk density, pH, texture, depth to groundwater, horizons (soil layering), hydraulic conductivity, water retention, permeability, climate (precipitation), porosity, organic content, cation exchange capacity (CEC), degradation parameters, partition coefficients, oxygen content, temperature, mineralogy, moisture content, and evaporation. Not all of these parameters are needed for a particular model. A model called SESOIL, for example, does not require data on soil texture or soil oxygen content.

Soil property data are normally available from university extension services and the U.S. Soil Conservation Service. Panian[76] prepared a catalog of unsaturated flow properties for many soil types. The data are compiled for depths extending to 60 in. below the soil surface. Data for greater depths require the use of sampling.

The physical and chemical properties of soils generally vary spatially. The variation exhibits a stochastic (random) character; however, the randomness of many soil properties tends to follow classic statistical distributions. Campbell[77] found that properties such as bulk density and soil porosity tend to be normally distributed. Saturated hydraulic conductivity, on the other hand, often exhibits a log-normal statistical distribution.

Advances have been made in understanding the spatial variability of soil properties.[78] Geostatistical sampling is also utilized to statistically describe soil properties relevant to contaminant migration. In this case, representative sampling locations are chosen based on the spatial variability of the soil. Sample points are selected by expert judgment to best cover the site, yet be cost effective. Estimates which are made by a procedure called kriging can be used to create contour maps of the contaminant concentrations.[79]

The application of kriging requires an adequate sample size. As a rule of thumb, at least 50 sampling points are needed. Breckenridge et al.[75] underline that the benefit of using kriging is that it allows sampling points within the domain of interest such as grid points for a computer model. The disadvantage of the method is the difficulty in knowing when various assumptions are acceptable or not acceptable in a particular application.[80]

Geostatistical packages such as GEOPACK or Geo-EAS are available from EPA. General kriging computer codes are also available. A computer code is a complete set of computer language (e.g., Basic, Fortran, Pascal) instructions containing the input data statements, computational algorithm, and output statements.

The list of available transport models has grown steadily.[81] Planners must make an informed decision of which model to use based on the specific nature of the site's problem, the capability of the algorithm and computer code, and the staff's experience. For vadose zone modeling, it is necessary for planners to be familiar with the principle of unsaturated flow which basically consists of gravity-dominated flow and capillary tension-dominated flow. Occurring during drainage of wet soils, gravity flow is directed downward and is relatively simple to predict. Capillary flow which occurs in dry to moist soils, on the other hand, proceeds in the direction of increasing pore water tension. This means that the flow can be in any direction.

The complexity of the subsurface actions in the vadose zone is commonly attributed to the nonlinear interaction between atmospheric, biologic, and geologic factors that creates an extremely complex flow and transport regime in the subsurface.[82] A process is nonlinear when there are at least two unknown entities that depend on each other, such as density-dependent transport and reactive geochemical transport. A nonlinear system requires more computational effort to solve than a linear system.

3.4.1. Types of Vadose Zone Models

Flow and transport models can be categorized based on their capabilities and solution approach.[83] They can also be classified according to their spatial perspective.[84] Kramer and Cullen[84] assert that the simplest and most commonly used models are one-dimensional leaching models which are limited to downward transport through characterized soil columns. Numerous one-dimensional leaching models have been developed by public and private sources. EPA's SESOIL and VLEACH are examples of this category of model that have gained acceptance from state environmental agencies.

SESOIL is a compartmental model designed to describe flow, sediment transport, contaminant transport, and transformation. The model can handle multiple layers, accounts for volatilization, and is widely used in the underground storage tank program. VLEACH is a finite difference model which does not simulate flow or calculate flux, but estimates contaminant leaching through the vadose zone. The model is commonly used in the Superfund program.

Kramer and Cullen maintain that the disadvantages of the leachability models are that they use simplifications to approximate actual processes and can not simulate lateral flow and transport although lateral flow between columns can be approximated by adjusting inputs. The simulation of flow and transport in complex domains for a variety of contaminants and conditions is performed by the two- and three-dimensional models. DYNAMIX is an example of a model in this category.

This model simulates dynamic mixing of multi-species chemical systems to simulate chemical transport.[85] This deterministic model applies to a saturated zone with a fixed hydraulic head condition. The simulation does not consider heterogeneity. In a deterministic model, the coefficients and boundary conditions are always assumed to be known exactly throughout the model domain.

To calculate parameters used in modeling, various computer codes have been established. The codes are furnished to calculate hydraulic properties of soil from limited experimental data for input into flow models. An example of such a code is the RETC code which regresses experimental data to calculate soil moisture characteristics and hydraulic conductivity function.[86] Fitted parameters from this calculation are then used in numerous flow models to calculate continuous hydraulic conductivity functions.

Despite the availability of models that are able to simulate flow in the subsurface, there does not appear to be any multi-dimensional models that can accommodate a multi-constituent contaminant migrating through a heterogeneous subsurface under uncontrolled hydraulic heads.[87] Subsurface flow and transport are in fact difficult to model mathematically due to the nonlinear interdependence between model parameters as described previously in this chapter.

3.4.2. Model Limitations

In order to make a flow and transport model operable, it is necessary to impose certain assumptions on the model. Fogg et al.[87] identify six assumptions that characterize the existing flow and transport models. The assumptions are

1. Most vadose zone models assume steady-state water flow even though this condition is rare near the soil surface in the field.
2. The possibility of variable anisotropy in the hydraulic conductivity has never been considered for predictive vadose zone modeling. Current models assume that porous media is always isotropic, i.e., the hydraulic conductivity is the same in any direction.
3. Most models have been vertical and one-dimensional, ignoring velocity components in the horizontal plane.
4. The water theoretically moves with velocity, v, inside the pore of the porous medium. The v used in the model is an average of local scale velocities. The models also use Fick's law which assumes that deviations in local v from average v are characterized by a normal (Gaussian) distribution. This assumption has been called into question by a number of studies which found that deviations in v from the average would often be non-Gaussian.
5. The use of the distribution coefficient and adsorption isotherms to characterize reactive chemical transport may be appropriate under certain conditions; however, simulation of contaminant movement that involves multi-species ion exchange reactions requires consideration of other chemical reactions.

6. Assumption of instantaneous adsorption is being questioned for a number of field conditions such as the presence of strongly adsorbing species, or water flow under unsteady conditions due to rainfall.

Because the reliability of models for contaminant transport has not been established even for site-specific conditions, Fogg et al.[87] argue that direct monitoring of the constituents in the vadose zone remains a necessity into the foreseeable future. Monitoring provides the opportunity to assess changes in concentrations of specific constituents at particular locations for better management alternatives.

3.4.3. *Flow and Transport Models*

The mathematical models that simulate the groundwater and contaminant transport processes may fall in either analytical, semi-analytical, or numerical models, coupled with a particle-tracking program.[88] The particle-tracking program follows contaminant particles through a pre-computed velocity field. This field is determined by using the numerical analysis.

Basically, the analytical models employ complex differential equations. For any given equation, exact answers or solutions will be provided. Because the properties are generally assumed to be homogeneous and infinite, in addition to other simplifications of actual hydrogeologic conditions, the model can be viewed as less than realistic.

In reality, contaminant transport within natural hydrogeologic systems is a three-dimensional process; this means that contaminants advect, disperse, or diffuse in all three dimensions. Molson et al.[89] suggest that in some site characterization projects, such as defining the lateral extent of the contamination plume originating from a known landfill, groundwater flow can be analyzed in two dimensions rather than three by neglecting the vertical flow and vertical dispersion. This is more cost effective than the full 3-D analysis.

An example of an analytical model is the CAPZONE model. This model computes traveltime-related wellhead-protection areas, i.e., the capture areas of wells, in predominantly two-dimensional flow regimes.[90] Traveltime-related capture zones are computed by placing a series of particles in a circle surrounding a pumping well and tracking their reverse pathline for a specific time. The loci of pathline endpoints encircles the capture zone of the well for the specific time period. EPA[91] defines a wellhead-protection area as the surface and subsurface area surrounding a water well or well field, supplying a public water system, through which contaminants are reasonably likely to move toward and reach such water well or well field.

The numerical models employ relationships between many smaller blocks (or elements) within a larger domain. The relationships represent statements of the original equations specifically developed on a subdomain level. The models solve simpler equations sequentially (one at a time) or simultaneously in matrix form. The numerical solutions are only approximate. The results of this approximation is often verified by analytical solutions.

A numerical, three-dimensional, finite-difference flow model developed by McDonald and Harbaugh[92] and referred to as MODFLOW is used extensively by the United States Geological Survey and others to simulate groundwater flow. The modular design of the MODFLOW model makes it easy to add new packages to broaden the scope and improve the accuracy of the MODFLOW model. Designed to be used with MODFLOW is MODPATH, a three-dimensional particle tracking program.[93]

The semi-analytical method combines an analytical flow model with a numerical transport model. It integrates the calculus- and numerical-based techniques to solve flow and transport equations. RESSQC is an example of the semi-analytical method.[94]

In choosing a suitable flow model, the most critical decision is the selection of a model that simplifies the flow system as much as possible while still preserving the geologic and hydrologic characteristics of the flow system. Springer and Bair[95] found in a comparative analysis that in a complex hydrogeologic setting, only the numerical flow model produces results that are sufficiently accurate and realistic to use. They found that the use of an analytical or semi-analytical flow model in a heterogeneous hydrologic setting turns out to be an oversimplification and can produce errors of sufficient magnitude.

9   RISK ASSESSMENT AND MANAGEMENT

In our daily life, the term risk is frequently related to the word hazard. While hazard is defined as a source of danger, risk embraces the likelihood of the conversion of that source into an actual occurrence of loss, injury, or some form of damage. We acknowledge the fact that humans are in constant interaction with the environment. Technical and social mechanisms enable society to seek in nature that which is useful and to buffer that which is harmful. Kates[96] argues that fulfilling this objective requires a complex set of human adjustments in the "human use system". The study of environmental hazards is a study of this adjustment process. The basic hazard paradigm is to seek answers to how humans adjust to risk and what the understanding of that process implies for public policy and planning.[97] This research paradigm remains the underpinning of most hazard research.[98]

If natural hazards, such as floods, earthquakes, and hurricanes, were once the principal hazards faced by society, today they are accompanied by hazards arising from technology which are manifested in a variety of forms, such as explosions, fires, or other environmental mishaps such as spills and contamination.[99] Zeigler et al.[100] divide technological hazards into public and private hazards. The former presents a threat to the public at large and the latter to that which is assumed by individuals, sometimes unknowingly, as a result of their occupation or their personal susceptibilities to negative side-effects of technologies. Some public hazards originate with the production of primary commodities or manufactured goods, others in the transportation and transmission sector, and still others arise as a result of the consumption of products which the economic systems make available.

Expressing the relationship between risk and hazard, Kaplan and Garrick[101] symbolize the idea in the form of an equation:

$$\text{Risk} = \text{hazard} / \text{safeguard}$$

meaning that by increasing the safeguards we can make risk as small as possible. The hazard-safeguard relationship implies that the risk of a contaminated site will become smaller and less significant with proper remediation of the site.

The variation of public risks and social values as well as the uncertainties in the estimation of risk ensures that there will always be conflict in the management of risk. Policy issues will continue to surface about risk because societal conflict itself is inherent to the problem.[102]

Risk can be viewed in relation to two opposing poles in one continuum: natural vs. man-made, controllable vs. uncontrollable, acceptable vs. unacceptable, voluntary vs. involuntary, strategic vs. tactical, and routine vs. catastrophic. Some people will accept certain risks voluntarily because of economic gains or some other personal reasons. An example is a stunt crew working in the movie-making industry. The crew members assume a relatively high risk voluntarily for high salaries and personal challenges.

With regard to a man-made risk, people normally perceive the man-made risk as catastrophic and uncontrollable. This risk perception, which is not necessarily the real risk, is the determining factor in the acceptance of a risk. Because risk perception does not reflect the real risk, it is a challenge for planners and decision makers to communicate the risk properly to avoid misleading perceptions about a risk. Risk communication and risk perception go hand in hand.

As previously indicated, risk and safety are two distinct sides of the same coin. Lowrance[103] points out that determining safety involves two different kinds of activities:

- measuring risk, i.e., measuring the probability and severity of harm, which is an empirical scientific activity
- judging safety, i.e., judging the acceptability of risks, which is a normative political activity

Hadden[104] argues that the policy decision is the most important part of the process of determining safety even though the political decision depends on the scientific risk assessment. The study of risk today is still shaped by issues of risk estimation and risk acceptability. In the context of site remediation planning, risk estimation is popularly termed risk assessment and risk acceptability is known as risk management.

The study of risk involves not only the uncertainty of occurrence but also the consequent loss or damage. Estimating risk involves determining the probability of consequences (P_c) and the value of each consequence (V_c) to the risk taker.[105] Stated differently,

$$\text{Risk} = f\,(P_c, V_c).$$

In this case, risk estimation is free of value judgment. Value judgments come into play in risk evaluation in that we are able to distinguish between acceptable risks as opposed to unacceptable ones.[106] To determine acceptable risk, the risk analysis helps identify how safe the chosen safeguards should be.[107]

In the field of site remediation, the risk assessment consists of both a human health risk assessment and an ecological risk assessment. A human health risk assessment attempts to determine the impact of a given chemical exposure on the health of the exposed individuals. The exposed individuals can be the permanent employees working at the site, the site remediation workers, or the population residing in close proximity to the site. The National Research Council[108] defines the term human health risk assessment as the characterization of the potential adverse health effects of human exposures to environmental hazards.

1. HUMAN HEALTH RISK ASSESSMENT

The U.S. Environmental Protection Agency (EPA)[109] divides a human health risk assessment (or simply risk assessment) into four distinct components:

- data collection and evaluation
- exposure assessment
- toxicity assessment
- risk characterization

Figure 9.1 displays these components and how they relate to each other. Essentially, a risk assessment is a systematic process for making estimates of risk factors due to the presence of some type of hazard at a site by evaluating exposure and toxicity.

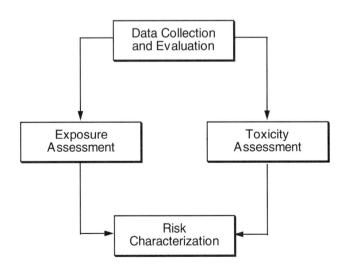

Figure 9.1. Human health risk assessment.

The data collection and evaluation component refers to the gathering and examination of the site data relevant to the human health evaluation, and identification of contaminants present at the site that would be the focus of the risk assessment. An exposure assessment involves an evaluation of the magnitude of human exposure to the contaminants of concern. A toxicity assessment examines the potential for adverse health effects from exposure to the site contaminants. A risk characterization summarizes the results of the exposure and toxicity assessments to characterize the risk of the site to human health.

The purpose of data collection and evaluation is to identify the nature of the hazards at a contaminated site. After sampling (Chapter 7) and site characterization (Chapter 8) activities have been conducted, several analytical results are available for use. At this stage, a list of the chemicals of concern can be developed from the data. A chemical of concern is one that represents the most risk posed by the site. Because chemical risk is determined by the type of chemical (encompassing its hazardous properties) and its amount (or its concentration), these two data components are vital ingredients for hazard identification.

1.1. Hazard Identification

Hazard identification can be described as a process of identifying the chemicals of concern due to their potential adverse health effects on human receptors. A list of chemicals of potential concern, according to a 1989 EPA document, includes chemicals:

1. positively detected in at least one contract laboratory program (CLP) sample
2. detected at levels above levels of the same chemicals detected in blank samples
3. detected at levels above naturally occurring levels of the same chemicals
4. only tentatively identified
5. transformation products of chemicals demonstrated to be present
6. chemicals that are not detected in samples but, based on site history or other knowledge, may be present at the site

For some sites, the list may be lengthy and evaluating all chemicals on the list may consume a considerable amount of time and resources. It is therefore advisable to shorten the list by reducing the number of chemicals through a screening process. EPA recommends the following activities in shortening the list of chemicals:

- group the chemicals into classes
- evaluate the frequency of detection

- assess the usefulness of the chemicals in terms of providing essential nutrients to humans
- rank the chemicals based on their toxicity scores

The grouping of chemicals into classes is necessary when toxicity data for certain chemicals are not available. The grouping can be based on a structure–activity relationship or other similarities. If, for example, some chemicals within the class of polycyclic aromatic hydrocarbons (PAH) are lacking their toxicity information, the toxicity data of benz(a)pyrene (one of the PAH members) can be used to represent the toxicity data of the group. The concentrations of chemicals within a class is calculated as the sum of the individual chemical concentration.

The frequency of chemical detection can be used to eliminate chemicals on the list. Chemicals that are infrequently detected may not be related to the past disposal practice of a site, and may just be artifacts in the data due to sampling or analytical problems. These chemicals can be disregarded if there is no reason to believe that they may be present based on knowledge of site history.

The usefulness of a chemical for human nutrients represents another reason for eliminating chemicals from the list. Examples of chemicals that are essential human nutrients are iron, magnesium, calcium, potassium, and sodium. They will be useful if they are present at concentrations slightly above the naturally occurring levels; therefore, prior to eliminating such chemicals from the list, they must be shown to be present at acceptable levels, i.e., at levels that are not associated with adverse health effects.

The ranking of chemicals based on their toxicity scores is used to evaluate the relative toxicity of one chemical to the others.

The general formula for developing the ranking is

$$TS_{ij} = C_{ij} \times T_{ij}$$

where: TS_{ij} = toxicity score for chemical i in environmental media j, C_{ij} = concentration of chemical i in media j, T_{ij} = toxicity value of chemical i in media j:

- for carcinogenic chemicals the toxicity value is the cancer slope factor (also called cancer potency factor)
- for noncarcinogenic chemicals the toxicity value is the inverse of the chronic reference dose (i.e., 1/RfD)

For example, if the concentration of chloroform in soil is 1.12 mg/kg and the slope factor for chloroform is 0.0061, then the toxicity score of chloroform in soil for carcinogenic toxicity is 1.12 × 0.0061 = 0.0068. If the RfD of chloroform is 0.01, the toxicity score of chloroform in soil for noncarcinogenic toxicity is 1.12/0.01 = 112.

Chlorobenzene found at the same site was detected at a concentration of 1.39 mg/kg. The RfD of chlorobenzene is 0.02, therefore the toxicity score of chlorobenzene in soil for noncarcinonegic toxicity is 1.39/0.02 = 69.5. In terms of ranking, the calculation shows that chloroform (toxic score 112) is more toxic than chlorobenzene (toxic score 69.5) for noncarcinogenic toxicity, even though the concentration of chloroform (1.12 mg/kg) is less than chlorobenzene (1.39 mg/kg). The toxic score can be kept unitless as long as the RfD units of the two chemicals remain the same.

Slope factor information for chlorobenzene is not available; therefore, carcinogenic toxicity between the two chemicals cannot be compared. Chemicals having no toxicity values cannot be ranked using this procedure.

Once the contaminants of concern have been identified, they should be displayed in ways that can easily be used for exposure assessment. Table 9.1 shows an example of the information provided from the hazard identification process. It lists the contaminants of potential concern sorted by environmental medium.

Table 9.1 Summary of Chemicals of Potential Concern at Site X, Location Y

| | Concentration | | | | |
Chemical	Soils (mg/kg)	Groundwater (µg/l)	Surface water (µg/l)	Sediments (µg/l)	Air (µg/m³)
A	5–1,100	—	2–30	—	
B	0.5–64	5–92	—	100–45,000	
C	—	15–890	50–11,000	—	
D	2–12	—	—	—	0.1–940

Source: U.S. Environmental Protection Agency, *Risk Assessment Guidance for Superfund Volume 1 — Human Health Evaluation Manual (Part A), Interim Final,* Washington, D.C., 1989.

The data in the table show that chemical concentrations are likely to vary across a contaminated site. In this case, the basic requirement for risk assessment is to utilize the values that reflect the average concentration to which identified human receptors are expected to be potentially exposed. In other words, given a range of values for concentrations of chemicals at a site, the question arises: What is the representative concentration of that chemical?

Asante-Duah[110] underlines that the statistical procedures used for addressing this issue can significantly affect the results of risk assessment. The Gaussian or normal distribution, for example, is widely used to represent the probability distribution. The average value (or the central tendency of the distribution) is measured by the arithmetic mean. Another procedure, called the log-normal distribution, requires the transformation of the data set $x1$, $x2$, x3, ..., xn into logarithmic distribution ln (x1), ln (x2), ln (x3), ..., ln (xn) that can be expected to be normally distributed. In this distribution the central tendency is measured by the geometric mean. The idea of utilizing the lognormal distribution is because the geometric mean is a much better representation of the central tendency of the sample distribution. Asante-Duah[110] suggests, however, that the appropriateness of any distribution to fit a given data set should be checked using such statistical procedures as the Chi-square test for goodness-of-fit.

1.2. Exposure Assessment

Exposure is defined as the contact of an organism, i.e., humans in the case of health risk assessment, with a chemical/physical agent.[111] The purpose of an exposure assessment is to estimate the magnitude of human exposure to contaminants. The magnitude of exposure is assessed in terms of frequency and duration. Specific questions to be answered by the exposure assessment are who is exposed to substances related to the site, through what pathways are they being exposed, how often and how long are they exposed through these pathways?

This assessment applies to current as well as potential exposure. Exposure pathways are specific routes through which the human population may come in contact with site contaminants under a specified land use. A future exposure assessment involves assumptions as to what the future land use will be. A typical land use for a site would be either residential or industrial. Residential use assumptions generally result in higher risks than industrial use assumptions because exposure in a residential area is usually more frequent and of longer duration; only in some particular cases, such as excavation, may there be exposure in an occupational scenario resulting in higher exposure than in a residential area.

For exposure to occur, a complete pathway is necessary. Failure to identify an important pathway may seriously affect the results of the risk assessment. Contaminant exposure is determined by three factors:

1. the concentrations of contaminants
2. the exposed population or potential receptors
3. the exposure pathways

The identification of the potential contaminants of concern is addressed in the hazard identification stage of the risk assessment. Information needed for the exposed population is related to the size, location (in terms of distance to the site), and nature of the population. The latter refers to activity patterns and different risk groups (e.g., nearby schools, off-site residents, facility employees, site remediation workers, etc.) that are distributed within the exposed population.

- The location of the current population is important to know because those that are closest to the site may have the greatest potential for exposure. Populations at a distance from the site may also be impacted in the future by those contaminants that have migrated from the site.
- The activity pattern will suggest the duration of exposure. Off-site residents, although located farther from the site than the facility employees, may be exposed to a maximum daily exposure period of 24 hours while a reasonable daily exposure period for facility employees may only be eight hours. This exposure period may change in the future; for example, if the site could be used for a residential area instead of an industrial/commercial use.

The exposure pathways describe how the contaminants migrate in the environmental media. The pathways are determined by combining data from the site characterization with knowledge about potential receptors and their behavior. The information needed to achieve an evaluation of a complete exposure pathway includes:

- sources of contaminants (e.g., a leaking tank or soil contaminated from a previous chemical spill could be a source of surface water or groundwater contamination)
- mechanisms of releases (e.g., volatilization, leachate, infiltration)
- release pathways, i.e., environmental transport medium (air, soils, sediments, groundwater, surface water)
- exposure points (e.g., touching contaminated soil, drinking water from a water fountain)
- exposure routes (inhalation, dermal contact, ingestion)

Figure 9.2 displays an illustration of various exposure pathways typically associated with contaminated sites.

Understanding the exposure pathway is essential for the development of an exposure scenario. An exposure scenario is a description of connectivity between a contaminant source to the population receptors through one or more environmental media. If numerous scenarios exist, it will be helpful to use an event tree structure to conceptualize the scenarios. To develop a complete exposure scenario, it is necessary to construct a conceptual model for the site and delineate likely and significant exposure pathways. Table 9.2 displays some typical release sources, release mechanisms, and receiving environmental media at contaminated sites. The table is useful for reference in developing an exposure scenario.

Chemical contaminants entering the environment tend to be partitioned or distributed across various environmental media and biota as a result of several complex processes. A solid understanding of these processes can lead to good predictions of chemical concentrations in the various compartmental

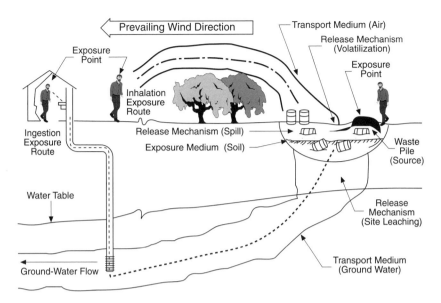

Figure 9.2. Illustrations of exposure pathways. Source: U.S. Environmental Protection Agency, Risk Assessment Guidance for Superfund: Volume I — Human Health Evaluation Manual (Part A) Interim Final, Washington, D.C., 1989.

media. The prediction of chemical transport and fate is necessary in an exposure assessment because the prediction shows the migration of the contaminant to potential receptors. A discussion of contaminant transport and fate is provided in Chapter 8.

Using the predictions of contaminant transport, simplified assessments can be performed for the critical pathways to the potential receptors.[112] Asante-Duah[112] asserts that the common and general types of modeling practices used in the exposure assessment are related to atmospheric, surface water, groundwater and saturated zone, multi-media, and food chain models. The scenarios typically simulated and evaluated using the appropriate models for a contaminated site are

1. infiltration of rainwater
2. erosion/surface run-off release of chemicals
3. emission of vapors and particulates
4. chemical transport and fate through the unsaturated zone
5. chemical transport through the aquifer system
6. mixing of groundwater with surface water

The exposure scenario must be developed based on certain conditions. Examples of the considerations in designing a scenario are listed below:

**Table 9.2 Common Chemical Release Sources
at Sites in the Absence of Remedial Action**

Receiving medium	Release mechanism	Release source
Air	Volatilization	• Surface wastes — lagoons, ponds, pits, spills • Contaminated surface water • Contaminated surface soil • Contaminated wetlands • Leaking drums
	Fugitive dust generation	• Contaminated surface soil • Waste piles
Surface water	Surface run-off	• Contaminated surface soil
	Episodic overland flow	• Lagoon overflow • Spills, leaking containers
	Groundwater seepage	• Contaminated groundwater
Groundwater	Leaching	• Surface or buried wastes • Contaminated soil
Soil	Leaching	• Surface or buried wastes
	Surface run-off	• Contaminated surface soil
	Episodic overland flow	• Lagoon overflow • Spills, leaking containers
	Fugitive dust generation/deposition	• Contaminated surface soil • Waste piles
	Tracking	• Contaminated surface soil
Sediment	Surface run-off, episodic overland flow	• Surface wastes — lagoons, ponds, pits, spills • Contaminated surface soil
	Groundwater seepage	• Contaminated groundwater
	Leaching	• Surface or buried wastes • Contaminated soil
Biota	Uptake (direct contact, ingestion, inhalation)	• Contaminated soil, surface water, sediment, groundwater or air • Other biota

Source: U.S. Environmental Protection Agency, Risk Assessment Guidance for Superfund: Volume I — Human Health Evaluation Manual (Part A) Interim Final, Washington, D.C., 1989.

- How does the contaminant behave in each environmental medium, does it bioaccumulate, is it absorbed by plants?
- How long will the contaminant remain in a medium, does the concentration change over time?
- Does the contaminant react with other chemical compounds in the environment and, if so, may the result be a chemical of concern?
- Is there any possibility of intermedia transfer, e.g., from water to air and, if so, what are the mechanisms for the transfer?

After the complete exposure pathways have been identified, the exposure assessment continues with the determination of exposure point concentrations (EPC) by considering the central tendency maximum exposures that can reasonably be expected to occur at a site. All exposure scenarios for Superfund sites are constructed on an estimate of central tendency and a reasonable maximum exposure (RME), i.e., the highest exposure that is reasonably expected to occur at a site for an individual pathway.[113] If a population is exposed to contaminants at a site through more than one pathway, the combination of exposures must also represent an RME. The RME is supposed to represent the 95th percentile of exposure, or the 95th upper confidence level (UCL).

The degree of exposure to a chemical, known as chemical intake or dose, is a function of a number of variables. Each of these variables can be represented by a range of values. Two exposure measures that are commonly used are the RME and average exposure. The EPCs are estimated from the RME and average exposure scenarios.

- For the RME scenario, the 95th UCL of the arithmetic mean is used to represent the EPC.
- For the average exposure scenario, the arithmetic mean chemical concentration is used to represent the EPC.

The EPCs of particulate chemicals resulting from fugitive dust emissions are estimated using a particulate emission factor (PEF) that relates the chemical concentration in the soil to the concentration of respirable particles in the air, known as PM_{10}, due to fugitive emissions from site soil.[114] An estimate for chemical concentrations in the air can be calculated by dividing the soil chemical concentration by the PEF. EPA suggests the use of the default PEF for soil with unlimited erosion potential as 4.63×10^9 m^3/kg.

For each chemical and each exposure pathway, the chemical exposure in terms of average daily dose or intake must be estimated. Intake is defined as the average amount of chemical systematically absorbed by the human body over a given duration. The definition indicates that the exposure assessment essentially deals with the conversion of the chemical concentration in a medium into a chemical intake. The extent of the chemical intake is determined by various factors:

1. the chemical concentration in the environmental medium over the exposure period in mg/kg, mg/l, or mg/m^3
2. the contact rate, which can be the intake rate (mg/day, l/d) or the inhalation rate (m^3/day)
3. the exposure frequency (days/year)
4. the exposure duration (years)
5. the body weight; the average body weight for estimation of intake is assumed to be 70 kg

6. the averaging time (days); for noncarcinogenic effects, when the intake is averaged over the period of exposure, it is equal to exposure duration; for carcinogens it is assumed to be 70 years

The general equation to estimate the chemical intake for each chemical and exposure pathway is

$$\text{Intake} = \frac{CC \times CR \times EF \times ED}{BW \times AT}$$

where CC = chemical concentration in the environmental medium, CR = contact rate, EF = exposure frequency, ED = exposure duration, BW = body weight, and AT = averaging time. The unit for the chemical intake is mg/kg/day.

The exposure assessment generally assumes that the concentrations of the chemical in the soil remain constant over the entire exposure duration. Chemical concentrations in groundwater are usually assumed to be total, not dissolved concentrations from unfiltered samples. The groundwater concentrations are usually assumed to remain constant throughout the duration of exposure. In addition, site groundwater is assumed to be potable. Individuals are to consume 100% of their drinking water from wells at the home.

The contact rate, exposure frequency, and exposure duration are all determined by the site's land use designation or site-specific information. Examples of exposure assumptions for residential use are:

- People are assumed to have dermal contact with site soils on a daily basis for 30 years.
- Children ages 1 to 6 are assumed to ingest 200 mg of site soils every day, the ingestion for ages 7–70 is half of the ingestion rate for children.
- People are assumed to drink site groundwater on a daily basis for 30 years at a rate of 2 l/d.
- People are assumed to have dermal contact via bathing with site groundwater at a rate of 50 gal/d. They also inhale vapors from showering.

The assumptions show that in a residential land use setting, the exposure assessment applies an exposure frequency of 350 days/year, or daily exposure, and an exposure duration of 30 years. Normally, the exposure duration and frequency for industrial land use are assumed to be 25 years and 250 days, respectively. The contact rates by environmental media for industrial use are generally one third to one half lower than those for residential use.

The assessment of chemical exposure as described in the preceding paragraphs suggests that the accuracy of the exposure calculation depends on the assumptions built into the equation, the exposure pathway determination, and the conclusions regarding the concentrations of chemicals in the soil and groundwater.

- The exposure assumptions are tied to the land use classification of the site.
- The exposure pathway determination is accomplished by using an exposure scenario supported by the contaminant fate and transport models.
- The representative concentrations of chemicals in the soil and groundwater are derived from the sampling data and statistically evaluated to produce values for exposure point concentrations.

At the end of an exposure assessment, a risk assessor must be able to develop an exposure summary table. The table will show the various intake factors (RME and average exposure) for carcinogenic and noncarcinogenic chemicals, for each pathway and receptor. Examples of pathways and receptors are inhalation by site workers, soil ingestion by residents, dermal contact by workers, groundwater ingestion by residents, etc. Table 9.3 illustrates a sample summary of exposure assessment.

Table 9.3 Example of Exposure Assessment at Site Z

| Pathway/ receptor | Health risk | Intake (mg/kg/day) | | Chemical |
		Average	RME	
Soil ingestion	Carcinogens	6.01E-09	1.81E-07	Benzene
by workers	Noncarcinogens	9.55E-08	4.66E-07	Zinc
Soil inhalation	Carcinogens	4.01E-12	2.67E-11	Benzene
by workers	Noncarcinogens	2.99E-11	5.87E-11	Zinc

1.3. Toxicity Assessment

The purpose of a toxicity assessment is to determine what health effects may occur as a result of exposure to the contaminants present at a site. In the toxicity assessment, human health effects are considered in terms of cancerous or acute systemic (noncancerous) effects. Specifically, the purpose of the cancer toxicity assessment is to determine whether a chemical causes cancer in humans under a specified exposure condition. Laboratory animal experiments are widely used to predict the carcinogenicity of a chemical.

Many of the chemicals have been subjected to assessment by EPA to evaluate potential human carcinogenicity. Based on this evaluation, chemicals have been categorized in one of the following cancer groups:

A	Known human carcinogen
B1 or B2	Probable human carcinogen
C	Possible human carcinogen

D   Not classifiable as to human carcinogenicity
E   Evidence of noncarcinogenicity in humans
ND   Evaluation not done

Chemicals in the group of known or probable human carcinogens (groups A, B1, and B2) are carcinogens. Chemicals in the remaining groups (C, D, E, and ND) are considered noncarcinogenic or systemic toxicants. The evidence of carcinogenicity in humans comes from long-term animal studies and epidemiological investigations. The evaluation results are supplemented by short-term toxicological and pharmacokinetic studies. Based on the nature of the evidence collected, EPA determined the cancer grouping as follows:

- Group A is for chemicals with sufficient epidemiologic evidence of carcinogeneity in humans.
- Group B1 is for chemicals with limited epidemiologic evidence in humans and either sufficient or insufficient evidence in animals.
- Group B2 is for chemicals with sufficient evidence in animals but insufficient epidemiologic evidence of carcinogeneity in humans.

The sufficiency of epidemiological data is vital. When data are sufficient, a dose–response relationship for a carcinogen can be established and the slope factor (SF) can be calculated. The SF is expressed as the reciprocal of milligrams per kilogram per day, or 1/(mg/kg/day). In essence, the higher the SF, the more potent the carcinogen (Figure 9.3).

For the systemic toxicants, the toxicity assessment is generally based on the assumption of lifetime exposure. The risk is calculated from the reference dose (RfD) which is obtained from the no-observed-adverse-effect-level (NOAEL) divided by a safety factor. The NOAEL is the dose of a noncarcinogen at which there is no statistically or biologically significant increase in severity or frequency of adverse effect observed between the exposed population and the nonexposed control group. At this dose, effects may be produced but are not considered to be adverse. The RfD is expressed as milligrams per kilogram per day.

Multiplying the RfD by 70 kg (weight of a standard adult) and then dividing it by 2 l (the assumed consumption of water per day) will result in a number which is known as the drinking water equivalent level (DWEL). The DWEL still represents 100% exposure through water and therefore must be allocated to sources which can contribute to the total exposure of a human. The allocation can be done by multiplying the DWEL with an estimate of the relative source contribution (RSC) of a contaminant in the water.

Toxicological data for selected chemicals are provided in Table 9.4. Cited from the 1992 Arizona Department of Environmental Quality document,[115] the table lists the chemicals alphabetically. Included in the table are the cancer group, RfD, slope factor, safety factor, RSC, and the chemical abstract service (CAS) registry number which can be used to verify entries for chemicals having multiple names. CAS numbers are available in standard chemistry

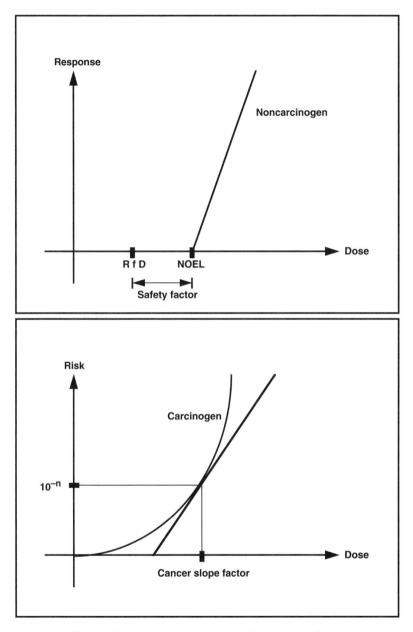

Figure 9.3. The slope factor and reference dose.

references. The unit of measurement for both the RfD and the cancer slope factor is in mg/kg/day. Since June 1995 the list has been updated monthly by the Arizona Department of Health Services.

The primary source for toxicity information for risk assessment is EPA's Integrated Risk Information System (IRIS) database. The phone number for

Table 9.4 Toxicological Data for Listed Chemicals

	Chemical	CAS number	Cancer group	RfD (mg/kg/d)	Slope factor 1/(mg/kg/d)	Safety factor	RSC
1.	Acenaphthylene (PAH)	208-96-8	D	0.06	NA	NA	0.20
2.	Acephate	30560-19-1	C	0.004	NA	10	0.20
3.	Acetone	67-64-1	D	0.1	NA	NA	0.20
4.	Acrolein	107-02-8	C	0.157	NA	10	0.20
5.	Acrrlamide	79-06-1	B2	NA	4.5	NA	NA
6.	Acrylonitrile	107-13-1	B1	NA	0.54	NA	NA
7.	Alachlor	15972-60-8	B2	NA	0.081	NA	NA
8.	Aldicarb	116-06-3	E	0.0013	NA	NA	0.20
9.	Aldicarb sulfone	1646-88-4	D	0.0003	NA	NA	0.20
10.	Aldicarb sulfoxide	1646-87-3	D	0.0013	NA	NA	0.20
11.	Aldrin	309-00-2	B2	NA	17.0	NA	NA
12.	Ametryn	834-12-8	D	0.009	NA	NA	0.20
13.	Ammonium sulfamate	7773-06-0	D	0.2	NA	NA	0.20
14.	Anthracene (PAH)	120-12-7	D	0.3	NA	NA	0.20
15.	Antimony	7440-36-0	D	0.0004	NA	NA	0.20
16.	Arsenic (inorganic)	7440-38-2	A	NA	NA	NA	NA
17.	Asulam	3337-71-1	D	0.05	NA	NA	0.20
18.	Atrazine	1912-24-9	C	0.005	NA	10	0.20
19.	Azinphos-methyl	86-50-0	E	0.0025	NA	NA	0.20
20.	Barium	7440-39-3	D	NA	NA	NA	NA
21.	Benomyl	17804-35-2	D	0.05	NA	NA	0.20
22.	Bentazon	25057-89-0	D	0.0025	NA	NA	0.20
23.	Benz[a]anthracene (PAH)	56-55-3	B2	NA	12.5	NA	NA

#	Compound	CAS					
24.	Benzene	71-43-2	A	NA	0.029	NA	NA
25.	Benzidine	92-87-5	A	NA	230	NA	NA
26.	Benzo[a]pyrene (PAH)	50-32-8	B2	NA	12.5	NA	NA
27.	Benzo[b]fluoranthene (PAH)	205-99-2	B2	NA	12.5	NA	NA
28.	Benzo[k]fluoranthene (PAH)	207-08-9	B2	NA	12.5	NA	NA
29.	Benzyl alcohol	100-51-6	ND	0.3	NA	NA	0.20
30.	Beryllium	7440-41-7	B2	NA	4.3	NA	NA
31.	Bis(2-chloroethyl)ether	111-44-4	B2	NA	1.1	NA	NA
32.	Bis(2-chloroisopropyl)ether	39638-32-9	ND	0.04	NA	NA	0.20
33.	Boron	7440-42-8	D	0.09	NA	NA	0.20
34.	Bromacil	314-40-9	C	0.13	NA	10	0.20
35.	Bromodichloromethane (THM)	75-27-4	B2	NA	0.13	NA	NA
36.	Bromoform (THM)	75-25-2	B2	NA	0.0079	NA	NA
37.	Bromomethane	74-83-9	D	0.0014	NA	NA	0.20
38.	Bromoxynil	1689-84-5	D	0.02	NA	10	0.20
39.	Butyl benzyl phthalate	85-68-7	C	0.16	NA	NA	0.20
40.	Butylate	2008-41-5	D	0.05	NA	10	0.20
41.	Cadmium	7440-43-9	D	0.0005	NA	NA	0.20
42.	Captan	133-06-2	D	0.13	NA	NA	0.20
43.	Carbaryl	63-25-2	D	0.1	NA	NA	0.20
44.	Carbofuran	1563-66-2	E	0.005	NA	NA	0.20
45.	Carbon disulfide	75-15-0	D	0.1	NA	NA	0.20
46.	Carbon tetrachloride	56-23-5	B2	NA	0.13	NA	0.20
47.	Carboxin	5234-68-4	D	0.1	NA	NA	NA
48.	Chloramben	133-90-4	D	0.015	NA	NA	0.20
49.	Chlordane	57-74-9	B2	NA	1.3	NA	0.20
50.	Chlordimeform	6164-98-3	B2	NA	1.17	NA	NA
51.	Chlorobenzene	108-90-7	D	0.02	NA	NA	0.20

Table 9.4 Toxicological Data for Listed Chemicals (Continued)

	Chemical	CAS number	Cancer group	RfD (mg/kg/d)	Slope factor 1/(mg/kg/d)	Safety factor	RSC
52.	Chloroform (THM)	67-66-3	B2	NA	0.0061	NA	NA
53.	Chloromethane	74-87-3	C	0.004	NA	10	0.20
54.	2-Chlorophenol	95-57-8	D	0.005	NA	NA	0.20
55.	Chlorothalonil	1897-45-6	B2	NA	0.0029	NA	NA
56.	o-Chlorotoluene	95-49-8	D	0.02	NA	NA	0.20
57.	Chlorpyrifos	2921-88-2	D	0.003	NA	NA	0.20
58.	Chlorsulfuron	64902-72-3	D	0.05	NA	NA	0.20
59.	Chromium (total)	NA	D	NA	NA	NA	NA
60.	Chrysene (PAH)	218-01-9	B2	NA	12.5	NA	NA
61.	Copper	7440-50-8	D	NA	NA	NA	NA
62.	Cresols (total)	NA	D	0.05	NA	NA	0.20
63.	Cyanazine	21725-46-2	D	0.002	NA	NA	0.20
64.	Cyanide	57-12-5	D	0.022	NA	NA	0.20
65.	Cyromazine	66215-27-8	D	0.0075	NA	NA	0.20
66.	2,4-D	94-75-7	D	0.01	NA	NA	0.20
67.	Dalapon	75-99-0	D	0.03	NA	NA	0.20
68.	DCPA	1861-32-1	D	0.5	NA	NA	0.20
69.	DDD	72-54-8	B2	NA	0.24	NA	NA
70.	DDE	72-55-9	B2	NA	0.34	NA	NA
71.	DDT	50-29-3	B2	NA	0.34	NA	NA
72.	DDT/DDD/DDE (total)	NA	B2	NA	0.34	NA	NA
73.	Diazinon	333-41-5	E	0.0009	NA	NA	0.20
74.	Dibenz[a,h]anthracene (PAH)	53-70-3	B2	NA	12.5	NA	NA

75.	Dibromochloromethane (THM)	124-48-1	C	0.02	NA	10	0.20
76.	1,2-Dibromo-3-chloropropane (DBCP)	96-12-8	B2	NA	1.17	NA	NA
77.	Dibutyl phthalate	84-74-2	D	0.1	NA	NA	0.20
78.	Dicamba	1918-00-9	D	0.03	NA	NA	0.20
79.	Dichlorobenil	1194-65-6	D	0.0005	NA	NA	0.20
80.	1,2-Dichlorobenzene	95-50-1	D	0.089	NA	NA	0.20
81.	1,3-Dichlorobenzene	541-73-1	D	0.089	NA	NA	0.20
82.	1,4-Dichlorobenzene	106-46-7	C	0.1	NA	10	0.20
83.	3,3'-dichlorobenzidine	91-94-1	B2	NA	0.451	NA	NA
84.	Dichlorodifluoromethane	75-71-8	D	0.2	NA	NA	0.20
85.	1,2-Dichloroethane	107-06-2	B2	NA	0.091	NA	NA
86.	1,1-Dichloroethylene	75-35-4	C	0.009	NA	10	0.20
87.	cis-1,2-Dichloroethylene	156-59-2	D	0.01	NA	NA	0.20
88.	trans-1,2-Dichloroethylene	156-60-5	D	0.02	NA	NA	0.20
89.	1,2-Dichloropropane	78-87-5	B2	NA	0.068	NA	NA
90.	2,4-Dichlorophenol	120-83-2	D	0.003	NA	NA	0.20
91.	1,3-Dichloropropene	542-75-6	B2	NA	0.18	NA	NA
92.	Dicloran	99-30-9	E	0.025	NA	NA	0.20
93.	Dicofol	115-32-2	C	NA	0.44	NA	NA
94.	Dieldrin	60-57-1	B2	NA	16.0	NA	NA
95.	Diethyl phthalate	84-66-2	D	0.8	NA	NA	0.20
96.	Di(2-ethylhexyl)adipate	103-23-1	C	0.7	NA	10	0.20
97.	Di(2-ethylhexyl)phthalate	117-81-7	B2	NA	0.014	NA	NA
98.	Difenzoquat	43222-48-6	D	0.08	NA	NA	0.20
99.	Diisopropyl methylphosphonate	1445-75-6	D	0.08	NA	NA	0.20
100.	Dimethoate	60-51-5	D	0.0002	NA	NA	0.20
101.	2,4-Dinitrophenol	51-28-5	ND	0.002	NA	NA	0.20
102.	2,4-Dinitrotoluene	121-14-2	B2	NA	0.68	NA	NA

Table 9.4 Toxicological Data for Listed Chemicals (Continued)

	Chemical	CAS number	Cancer group	RfD (mg/kg/d)	Slope factor 1/(mg/kg/d)	Safety factor	RSC
103.	Dinoseb	88-85-7	D	0.001	NA	NA	0.20
104.	1,4-Dioxane	123-91-1	B2	NA	0.011	NA	NA
105.	Diphenamid	957-51-7	D	0.03	NA	NA	0.20
106.	1,2-Diphenylhydrazine	122-66-7	B2	NA	0.8	NA	NA
107.	Diquat dibromide	85-00-7	D	0.0022	NA	NA	0.20
108.	Disulfoton	298-04-4	E	0.00004	NA	NA	0.20
109.	Diuron	330-54-1	D	0.002	NA	NA	0.20
110.	DPX-M6316	79277-27-3	ND	0.013	NA	NA	0.20
111.	Endosulfan	115-29-7	D	0.00005	NA	NA	0.20
112.	Endothall	145-73-3	D	0.02	NA	NA	0.20
113.	Endrin	72-20-8	E	0.0003	NA	NA	0.20
114.	Epichlorohydrin	106-89-8	B2	NA	0.0099	NA	NA
115.	Ethephon	16672-87-0	D	0.005	NA	NA	0.20
116.	EPIC	759-94-4	D	0.025	NA	NA	0.20
117.	Ethylbenzene	100-41-4	D	0.1	NA	NA	0.20
118.	Ethylene dibromide (EDB)	106-93-4	B2	NA	85.0	NA	NA
119.	Ethylene glycol	107-21-1	D	2.0	NA	NA	0.20
120.	Ethylene thiourea	96-45-7	B2	NA	0.036	NA	NA
121.	N-Ethyltoluene sulfonamide	26914-52-3	ND	0.0025	NA	NA	0.20
122.	Fenamiphos	22224-92-6	D	0.00025	NA	NA	0.20
123.	Fenarimol	60168-88-9	E	0.065	NA	NA	0.20
124.	Fenvalerate	51630-58-1	ND	0.025	NA	NA	0.20
125.	Fluometuron	2164-17-2	D	0.013	NA	NA	0.20

126.	Fluoranthene (PAH)	206-44-0	D	0.04	NA	NA	0.20
127.	Fluorene (PAH)	86-73-7	D	0.04	NA	NA	0.20
128.	Fluoride	7782-41-4	D	0.06	NA	NA	0.20
129.	Fluridone	59756-60-4	D	0.08	NA	NA	0.20
130.	Fluvalinate	69409-94-5	D	0.01	NA	NA	0.20
131.	Fonofos	944-22-9	D	0.002	NA	NA	0.20
132.	Formetanate hydrochloride	23422-53-9	E	0.0015	NA	NA	0.20
133.	Fosetyl-AL	39148-24-8	C	3.0	NA	10	0.20
134.	Glyphosate	1071-83-6	D	0.1	NA	NA	0.20
135.	Heptachlor	76-44-8	B2	NA	4.5	NA	NA
136.	Heptachlor epoxide	1024-57-3	B2	NA	9.1	NA	NA
137.	Hexachloroethane	67-72-1	C	0.001	NA	10	0.20
138.	Hexachlorobenzene	118-74-1	B2	NA	1.6	NA	NA
139.	Hexachlorobutadiene	87-68-3	C	0.002	NA	10	0.20
140.	Hexachlorocyclohexane (alpha-)	319-84-6	B2	NA	6.3	NA	NA
141.	Hexachlorocyclohexane (beta-)	319-85-7	C	NA	1.8	NA	NA
142.	Hexachlorocyclopentadiene	77-47-4	D	0.007	NA	NA	0.20
143.	n-Hexane	110-54-3	D	0.06	NA	NA	0.20
144.	Hexazinone	51235-04-2	D	0.033	NA	NA	0.20
145.	HMX	2691-41-0	D	0.05	NA	NA	0.20
146.	Imazalil	35554-44-0	D	0.013	NA	NA	0.20
147.	Imazaquin	81335-37-7	D	0.25	NA	NA	0.20
148.	Indenopyrene (PAH)	193-39-5	B2	NA	12.5	NA	NA
149.	Isophorone	78-59-1	C	0.2	NA	10	0.20
150.	Lead	7439-92-1	B2	NA	NA	NA	NA
151.	Lindane	58-89-9	C	0.0003	NA	10	0.20
152.	Linuron	330-55-2	C	0.002	NA	10	0.20
153.	Malathion	121-75-5	D	0.02	NA	NA	0.20

Table 9.4 Toxicological Data for Listed Chemicals (Continued)

	Chemical	CAS number	Cancer group	RfD (mg/kg/d)	Slope factor 1/(mg/kg/d)	Safety factor	RSC
154.	Maleic hydrazide	123-33-1	D	0.5	NA	NA	0.20
155.	Mancozeb	8018-01-7	ND	0.03	NA	NA	0.20
156.	Maneb	12427-38-2	D	0.005	NA	NA	0.20
157.	Manganese	7439-96-5	D	0.1	NA	NA	0.20
158.	MCPA	94-74-6	D	0.0005	NA	NA	0.20
159.	Mepiquat chloride	24307-26-4	D	0.03	NA	NA	0.20
160.	Mercury (inorganic)	7439-97-6	D	0.0003	NA	NA	0.20
161.	Metalaxyl	57837-19-1	D	0.06	NA	NA	0.20
162.	Methamidophos	10265-92-6	D	0.00005	NA	NA	0.20
163.	Methiocarb	2032-65-7	E	0.00125	NA	NA	0.20
164.	Methomyl	16752-77-5	D	0.025	NA	NA	0.20
165.	Methoxychlor	72-43-5	D	0.005	NA	NA	0.20
166.	Methyl ethyl ketone	78-93-3	D	0.05	NA	NA	0.20
167.	Methyl parathion	298-00-0	D	0.00025	NA	NA	0.20
168.	Methyl tert butyl ether (MTBE)	1634-04-4	D	0.005	NA	NA	0.20
169.	Methylene chloride	75-09-2	B2	NA	0.0075	NA	NA
170.	Metolachlor	51218-45-2	C	0.15	NA	10	0.20
171.	Metribuzin	21087-64-9	D	0.025	NA	NA	0.20
172.	Metsulfuron-methyl	74223-64-6	D	0.25	NA	NA	0.20
173.	Molybdenum	7439-98-7	D	0.001	NA	NA	0.20
174.	Monocrotophos	6923-22-4	E	0.000045	NA	NA	0.20
175.	Monuron	150-68-5	ND	0.0007	NA	NA	0.20
176.	MSMA (as arsenic)	2163-80-6	A	NA	NA	NA	NA

177.	Myclobutanil	88671-89-0	ND	0.025	NA	0.20
178.	Naled	300-76-5	D	0.002	NA	0.20
179.	Naphthalene (PAH)	91-20-3	D	0.004	NA	0.20
180.	Napropamide	15299-99-7	ND	0.1	NA	0.20
181.	Nickel	7440-02-0	D	0.02	NA	0.20
182.	Nitrate	14797-55-8	D	1.6	NA	0.20
183.	Nitrate/nitrite (total)	NA	D	1.0	NA	0.20
184.	Nitrite	14797-65-0	D	0.1	NA	0.20
185.	Nitrobenzene	98-95-3	D	0.0005	NA	0.20
186.	Nitroguanidine	556-88-7	D	0.1	NA	0.20
187.	N-Nitrosodiphenylamine	86-30-6	B2	NA	0.0049	NA
188.	N-Nitrosodi-n-propylamine	621-64-7	B2	NA	7.0	NA
189.	N-Nitroso-dimethylamine	62-75-9	B2	NA	51.0	NA
190.	N-Nitrosopyrrolidine	930-55-2	B2	NA	2.1	NA
191.	Norflurazon	27314-13-2	D	0.04	NA	0.20
192.	Oryzalin	19044-88-3	C	0.05	10	0.20
193.	Oxamyl	23135-22-0	E	0.025	NA	0.20
194.	Oxydemeton-methyl	301-12-2	D	0.0005	NA	0.20
195.	Paraquat	1910-42-5	C	0.0045	10	0.20
196.	Parathion	56-38-2	C	0.006	10	0.20
197.	Pendimethalin	40487-42-1	D	0.04	NA	0.20
198.	Pentachlorobenzene	608-93-5	D	0.0008	NA	0.20
199.	Pentachlorophenol	87-86-5	B2	NA	0.12	NA
200.	Permethrin	52645-53-1	D	0.05	NA	0.20
201.	Phenol	108-95-2	D	0.6	NA	0.20
202.	Phorate	298-02-2	E	0.0005	NA	0.20
203.	Phosmet	732-11-6	D	0.02	NA	0.20
204.	Phosphamidon	13171-21-6	D	0.00017	NA	0.20

Table 9.4 Toxicological Data for Listed Chemicals (Continued)

	Chemical	CAS number	Cancer group	RfD (mg/kg/d)	Slope factor 1/(mg/kg/d)	Safety factor	RSC
205.	Picloram	1918-02-1	D	0.07	NA	NA	0.20
206.	Polychlorinated biphenyls (PCB)	1336-36-3	B2	NA	7.7	NA	NA
207.	Profenofos	41198-08-7	D	0.00005	NA	NA	0.20
208.	Profluralin	26399-36-0	ND	0.006	NA	NA	0.20
209.	Prometon	1610-18-0	D	0.015	NA	NA	0.20
210.	Prometryn	7287-19-6	D	0.004	NA	NA	0.20
211.	Pronamide	23950-58-5	C	0.075	NA	10	0.20
212.	Propachlor	1918-16-7	D	0.013	NA	NA	0.20
213.	Propargite	2312-35-8	ND	0.02	NA	NA	0.20
214.	Propazine	139-40-2	C	0.02	NA	10	0.20
215.	Propham	122-42-9	D	0.02	NA	NA	0.20
216.	Propiconazole	60207-90-1	D	0.013	NA	10	0.20
217.	Propoxur	114-26-1	C	0.004	NA	10	0.20
218.	Pyrene (PAH)	129-00-0	D	0.03	NA	NA	0.20
219.	RDX	121-82-4	C	0.003	NA	10	0.20
220.	Selenium	7782-49-2	D	NA	NA	NA	NA
221.	Sethoxydim	74051-80-2	D	0.09	NA	NA	0.20
222.	Silver	7440-22-4	D	NA	NA	NA	0.20
223.	Simazine	122-34-9	C	0.002	NA	10	0.20
224.	Strontium	7440-24-6	D	2.5	NA	NA	0.20
225.	Styrene	100-42-5	C	0.2	NA	10	0.20
226.	Sulfate	14808-79-8	D	NA	NA	NA	NA
227.	Sulprofos	35400-43-2	E	0.0025	NA	NA	0.20

228.	2,4,5-T	93-76-5	D	0.01	NA	NA	0.20
229.	2,3,7,8-TCDD	1746-01-6	B2	NA	150000	NA	NA
230.	2,4,5-TP	93-72-1	D	0.008	NA	NA	0.20
231.	Tebuthiuron	34014-18-1	D	0.07	NA	NA	0.20
232.	Terbacil	5902-51-2	E	0.013	NA	NA	0.20
233.	Terbufos	13071-79-9	D	0.0001	NA	NA	0.20
234.	Terbutryn	886-50-0	ND	0.001	NA	NA	0.20
235.	1,2,4,5-Tetrachlorobenzene	95-94-3	D	0.0003	NA	NA	0.20
236.	1,1,1,2-Tetrachloroethane	630-20-6	C	0.03	NA	10	0.20
237.	1,1,2,2-Tetrachloroethane	79-34-5	C	NA	0.2	NA	NA
238.	Tetrachloroethylene (PCE)	127-18-4	B2	NA	0.05	NA	NA
239.	Tetraethyl lead	78-00-2	D	0.0000001	NA	NA	0.20
240.	Thallium	7440-28-0	ND	0.00007	NA	NA	0.20
241.	Thiophanate-methyl	23564-05-8	D	0.08	NA	NA	0.20
242.	Thiram	137-26-8	D	0.005	NA	NA	0.20
243.	Toluene	108-88-3	D	0.2	NA	NA	0.20
244.	Toxaphene	8001-35-2	B2	NA	1.1	NA	NA
245.	Triadimefon	43121-43-3	D	0.03	NA	NA	0.20
246.	Trichlorfon	52-68-6	C	0.0125	NA	10	0.20
247.	1,2,4-Trichlorobenzene	120-82-1	D	0.0013	NA	NA	0.20
248.	1,1,1-Trichloroethane (TCA)	71-55-6	D	0.09	NA	NA	NA
249.	1,1,2-Trichloroethane	79-00-5	C	0.004	NA	10	0.20
250.	Trichloroethylene (TCE)	79-01-6	B2	NA	0.011	NA	NA
251.	Trichlorofluoromethane	75-69-4	D	0.3	NA	NA	0.20
252.	Triclopyr	55335-06-3	E	0.0025	NA	NA	0.20
253.	2,4,5-Trichlorophenol	95-95-4	D	0.1	NA	NA	0.20
254.	2,4,6-Trichlorophenol	88-06-2	B2	NA	0.011	NA	NA
255.	1,2,3-Trichloropropane	96-18-4	D	0.006	NA	NA	0.20

Table 9.4 Toxicological Data for Listed Chemicals (Continued)

	Chemical	CAS number	Cancer group	RfD (mg/kg/d)	Slope factor 1/(mg/kg/d)	Safety factor	RSC
256.	Trichlorotrifluoroethane	76-13-1	D	30.0	NA	NA	0.20
257.	Trifluralin	1582-09-8	C	0.0075	NA	10	0.20
258.	Triforine	26644-46-2	D	0.025	NA	NA	0.20
259.	Trihalomethanes (total THM)	NA	NA	NA	NA	NA	NA
260.	2,4,6-Trinitrotoluene	118-96-7	C	0.0005	NA	10	0.20
261.	Uranium	7440-61-1	A	0.003	ND	NA	0.20
262.	Vanadium	7440-62-2	D	0.007	NA	NA	0.20
263.	Vernolate	1929-77-7	ND	0.001	NA	NA	0.20
264.	Vinclozolin	50471-44-8	D	0.025	NA	NA	0.20
265.	Vinyl chloride	75-01-4	A	NA	1.9	NA	NA
266.	Xylenes (total)	1330-20-7	D	2.0	NA	NA	0.20
267.	Zinc	7440-66-6	ND	0.2	NA	NA	0.20
268.	Zineb	12122-67-7	D	0.05	NA	NA	0.20

Source: Arizona Department of Environmental Quality, Human Health-Based Guidance Level for the Ingestion of Contaminants in Drinking Water and Soil, Phoenix, 1992.
Note: NA = not available; ND = not determined.

IRIS User Support is 513-569-7254. If no verified toxicity value is available through IRIS, then the Health Effects Assessment Summary Tables (HEAST) is the next preferred source. HEAST is a tabular presentation of toxicity information and values for specific chemicals and can be obtained by calling 202-382-3046. If toxicity information of a chemical is not available in both IRIS and HEAST, the assessor may contact the following address:

Superfund Health Risk Assessment Technical Support Center
U.S. Environmental Protection Agency
Mail Stop 114
26 W. Martin Luther King Drive
Cincinnati, OH 45268

The preceding discussion has shown that assumptions play key roles in both the toxicity and exposure assessment portions of a risk assessment. The assumptions are likely to result in overestimates of site risks. EPA's obligation, however, is to make conservative assumptions so that there is a very small chance of the actual health risk being greater than that determined through the risk assessment process.[116] Because of these assumptions, a risk assessment has some uncertainties.

The exposure assumptions and uncertainties associated with each assumption should be described in the risk assessment report which addresses uncertainties in the monitoring data, the exposure models, and the intake parameters. Table 9.5 provides an example of an assumption and the uncertainty table for an exposure assessment at a hypothetical site.

Uncertainties in the toxicity assessment occur because toxicity information for many of the chemicals found at contaminated sites is often limited. A discussion of the strength of the evidence of existing information should be provided in the risk assessment report. The discussion must also address an indication of the extent to which an analysis of the results from different studies gives a consistent and credible picture of toxicity.

1.4. Risk Characterization

The final step in a risk assessment is the risk characterization. It also represents the first input to the risk management process. The characterization integrates the exposure assessment and toxicity assessment to arrive at an estimate of risk to the exposed population. The risks are characterized by the calculation of noncarcinogenic hazard quotients and carcinogenic risks. These risk factors are then compared with applicable standards for risk decisions.

The potential for noncarcinogenic health effects is estimated by summing the hazard quotients for each chemical in each exposure pathway. The sum of the individual hazard quotients from all pathways is called the hazard index. Either the hazard quotient or the hazard index must not exceed a value of 1.00

**Table 9.5. Example of an Uncertainty Table
for Exposure Assessment**

| | Effect on exposure[a] | | |
| | Potential magnitude of exposure for: | | |
Assumption	Over-estimation	Under-estimation	Over or under-estimation
Environmental sampling and analysis			
Sufficient samples may not have been taken to characterize the media being evaluated, especially with respect to currently available soil data.			Moderate
Systematic or random errors in the chemical analyses may yield erroneous data.			Low
Fate and Transport Modeling			
Chemicals in fish will be at equilibrium with chemical concentration in water.	Low		
Use of a Gaussian dispersion model to estimate air concentration off-site.			Low
Use of a box model to estimate air concentration on-site.	Low		
Use of Cowherd's model to estimate vehicle emission factors.		Moderate	
Exposure Parameter Estimation			
The standard assumptions regarding body weight, period exposed, life expectancy, population characteristics, and lifestyle may not be representative of any actual exposure situation.			Moderate
The amount of media intake is assumed to be constant and representative of the exposed population.	Moderate		

Table 9.5. Example of an Uncertainty Table
for Exposure Assessment (Continued)

| | Effect on exposure[a] | | |
| | Potential magnitude of exposure for: | | |
Assumption	Over-estimation	Under-estimation	Over or under-estimation
Assumption of daily lifetime exposure for residents.	Moderate to high		
Use of "hot spot" soil data for upper-bound lifetime exposure.	Moderate to high		

[a] As a general guideline, assumptions marked as "low" may affect estimates of exposure by less than one order-of-magnitude; assumptions marked "moderate" may affect estimates of exposure by between one and two orders-of-magnitude; and assumptions marked "high" may affect estimates of exposure by more than two orders-of-magnitude.

Source: U.S. Environmental Protection Agency, Risk Assessment Guidance for Superfund: Volume I — Human Health Evaluation Manual (Part A) Interim Final, Washington, D.C., 1989.

in order for the contaminants to be considered as not posing significant systemic health effects.

The hazard quotient is calculated from the chemical intake (mg/kg/day) divided by noncarcinogenic RfD (mg/kg/day). Because the units of measurement of both the chemical intake and the RfD are the same, the hazard quotient and hazard index are unitless. The calculation is conducted for each chemical by different pathway and receptor (e.g., soil ingestion by residents, particle inhalation by site workers), and by different intake values (i.e., RME and average values).

Carcinogenic risks are calculated by multiplying the chemical intake (mg/kg/day) with the cancer slope factor (mg/kg/day). Similar to a hazard quotient estimate, carcinogenic risk is also unitless because the chemical intake and the cancer slope factor both have the same unit of measurement. The calculations are performed for each chemical for different pathways and receptors, and for different intake values.

The individual chemical risks derived from these calculations are then added to obtain the estimated cancer risks of the site. In order for the contaminants to be considered of not posing significant cancer risk, the estimated cancer risk of the site must be less than EPA's level of risk.[117] The acceptable range of risk established by EPA is between 1×10^{-4} and 1×10^{-6}. The 1×10^{-6} risk level is often used as the point of departure for establishing remedial goals. Table 9.6 displays a summary of hazard indices and carcinogenic risks for a hypothetical site.

Table 9.6 Hazard Indices and Carcinogenic Risks at Site W

Pathway/receptor	Hazard index		Carcinogenic risk	
	Average	RME	Average	RME
Soil ingestion by workers	0.0145	0.0900		
Soil inhalation by workers			2.01E-10	2.22E-10
Soil dermal contact by workers	0.0002	0.0450		
Total	0.0147	0.1350	2.01E-10	2.22E-10
Soil inhalation by residents			3.60E-10	2.99E-09
Total			3.60E-10	2.99E-09

The table shows that total hazard index is less than 1.00 and the total carcinogenic risk is less than 1×10^{-6}. The findings suggest that the site does not pose a significant risk to the exposed population and, therefore, site remediation is not required. This process of risk assessment as illustrated in the preceding paragraphs is known as a baseline risk assessment.

Typically, the risk assessment process can be applied at a site in three different applications:

- Baseline risk assessment. The purpose of this assessment is to analyze the extent of the risk without site remediation activity.
- Preliminary remedial goal (PRG) assessment. This assessment may be necessary if site remediation is required. The purpose is to analyze an acceptable cleanup standard that will not pose significant risk to the exposed population.
- Risk assessment of remedial options. Similar to the PRG assessment, this effort may be necessary if site remediation is required. The purpose of this assessment is to ensure that the remedial options selected for the site will achieve the intended cleanup level and do not create additional risks to the exposed population.

In a baseline risk assessment the equation for estimating the chemical intake is

$$\text{Intake} = \frac{CC \times CR \times EF \times ED}{BW \times AT}$$

The magnitude of intake in this equation is directly proportional to the chemical concentration (CC). Given the value of the cancer slope factor of the chemical, the carcinogenic risk can be derived by multiplying the chemical intake with the cancer slope factor.

In a PRG risk assessment, the carcinogenic risk is assumed to be 1×10^{-6}. The chemical intake can be calculated by dividing this predetermined carcinogenic risk with the cancer slope factor. For the same chemical, the cancer slope factor will be the same with the slope factor used in the baseline risk assessment. Because the target risk (1×10^{-6}) is lower than the risk calculated from the baseline risk assessment, the chemical intake value will generally be lower than that estimated in the baseline assessment.

By keeping the values of CR, EF, ED, BW, and AT in the equation constant, the decrease in the value of chemical intake will lower the CC value. The values of CR, EF, ED, BW, and AT remain constant when there are no changes in the assumptions of contact rate, exposure frequency, exposure duration, body weight, and averaging time. However, if changes occurred in any of those parameters, appropriate adjustment must be made. Changes in land use, for example, will certainly affect the change in exposure pathway. Table 9.7 illustrates this circumstance.

The target risk of 1×10^{-6} is the National Contingency Plan's point of departure for analysis of remedial alternatives.[118] The target risk 1×10^{-6} is not an actual risk. It is a mathematical model based on scientific assumptions used in risk assessment. In the context of human health risks, 1×10^{-6} is a description for an increased lifetime chance of one chance in a million of developing cancer due to lifetime exposure to a chemical. The number 1×10^{-5} represents one chance in 100,000 and so on.

The use of the cancer-risk point-of-departure for analysis of remedial options reflects EPA's preference for managing risks at the more protective end of the risk range. It should be used as a starting point for discussion of an acceptable target risk at a site, not the ultimate remediation goal. As EPA maintains, the preference does not reflect a presumption that the final remedial action should attain such goals.

Legally, Title 40 of the Code of Federal Regulations (CFR) Section 300.430(e)(2)(i)(A)(2) states that an estimate of excess cancer risk associated with a site that exceeds 1×10^{-4} is considered unacceptable. This forms a sufficient basis for EPA to order cleanup. An estimate of the cancer risk associated with a site that is less than 1×10^{-4} but greater than 1×10^{-6} may be considered unacceptable and may be a sufficient basis for ordering cleanup.

Most states are using the 1×10^{-6} standard; however, some of the states are using various standards based on an excess cancer risk factor of 1×10^{-5}. New York and Texas, for example, use the 1×10^{-5} risk factor for class C carcinogens for soil risk assessment; Montana uses the same risk factor for all carcinogens. In Minnesota, the 1×10^{-5} risk factor is used for the ground-water cleanup goals.

The decision to conduct a risk assessment of remedial options depends on whether the short-term or long-term effectiveness of remedial options is an important consideration in selecting an option, and the perception of risk associated with the selected remedial option. The risk assessment of remedial options follows the same general steps as the baseline assessment. The baseline risk assessment typically is more quantitative and requires a higher level of effort than the risk assessment of remedial options.[119]

Table 9.7 Typical Exposure Pathways by Medium for Residential and Commercial/Industrial Land Uses[a,b]

Medium	Exposure pathways, assuming:	
	Residential land use	**Commercial/industrial land use**
Groundwater	Ingestion from drinking	Ingestion from drinking[d]
	Inhalation of volatiles	Inhalation of volatiles
	Dermal absorption from bathing	Dermal absorption
	Immersion — external[c]	
Surface water	Ingestion from drinking	Ingestion from drinking[d]
	Inhalation of volatiles	Inhalation of volatiles
	Dermal absorption from bathing	Dermal absorption
	Ingestion of contaminated fish	
	Immersion — external[c]	
Soil	Ingestion	Ingestion
	Inhalation of particulates	Inhalation of particulates
	Inhalation of volatiles	Inhalation of volatiles
	Direct external exposure[c]	Direct external exposure[c]
	Exposure to groundwater contaminated by soil leachate	Exposure to groundwater contaminated by soil leachate
	Ingestion via plant uptake	Inhalation of particulates from trucks and heavy equipment
	Dermal absorption from gardening	

[a] Lists of land uses, media, and exposure pathways are not comprehensive.
[b] Exposure pathways included in RAGS/HHEM Part B standard default equations are italicized.
[c] Applies to radionuclides only.
[d] Because the NCP encourages protection of groundwater to maximize its beneficial use, risk-based PRGs generally should be based on residential exposures once groundwater is determined to be suitable for drinking. Similarly, when surface water will be used for drinking, general standards (e.g., ARARs) are to be achieved that define levels protective for the population at large, not simply worker populations. Residential exposure scenarios should guide risk-based PRG development for ingestion and other uses of potable water.
Source: U.S. Environmental Protection Agency, Risk Assessment Guidance for Superfund: Volume I — Human Health Evaluation Manual (Part B, Development of Risk-Based Preliminary Remediation Goals), Washington, D.C., 1991.

In evaluating exposure, the source of the releases for the baseline assessment is the original, untreated site contaminants while the source of releases for the assessment of remedial options is the remedial action itself (e.g., groundwater pump-and-treat may release volatile organic vapors) plus any remaining site contaminants. Figures 9.4 and 9.5 illustrate, respectively, an exposure pathway for a hypothetical remedial action and cumulative exposure from multiple releases.

In evaluating toxicity, the risk assessment of remedial options may include an assessment of chemicals that were not present under baseline conditions. Such chemicals were created as a result of the remediation activities. Similarly, there are additional uncertainties involved in evaluating risks of remedial options that are not considered in the baseline risk assessment.

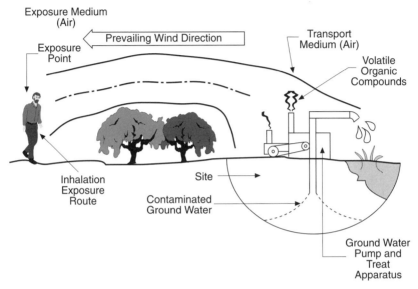

Figure 9.4. An exposure pathway for a remedial action. *Source:* **U.S. Environmental Protection Agency, Risk Assessment Guidance for Superfund: Volume I — Human Health Evaluation Manual (Part C, Risk Evaluation of Remedial Alternatives), Washington, D.C., 1991.**

There are uncertainties in all categories of a risk assessment. Uncertainties exist in the initial selection of the chemicals used to characterize exposures and risk based on the sampling data and available toxicity information. Uncertainties exist in the toxicity values for each chemical used to characterize risk as well as in the exposure assessment for individual chemicals and individual exposures. Uncertainties also exist when exposures to several chemicals across multiple pathways are summed. If necessary, an uncertainty analysis can be conducted using some advanced statistical techniques, such as the Monte Carlo analysis.[120]

For Superfund sites, CERCLA Section 104(i), as amended, requires the Agency for Toxic Substances and Disease Registry (ATSDR) to conduct health assessments for all sites to be listed on the National Priorities List (NPL). ATSDR health assessments consider all health threats, both chemical and physical, to which the exposed population may be subjected. Although the assessments employ quantitative data, ATSDR health assessments are more

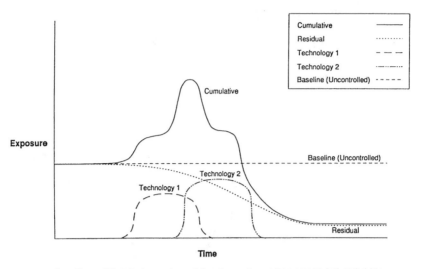

Note: The graph illustrates how nearby populations at some sites could be exposed to both residual risks and risks from remediation technologies. The cumulative exposure illustrated is the sum of residual exposure and exposures associated with releases from Technologies 1 and 2. This exhibit is for illustration purposes only and is not meant to imply that this level of quantitation is neccesary or even desired.

Figure 9.5. Cumulative exposure from multiple releases. *Note:* **The graph illustrates how nearby populations at some sites could be exposed to both residual risks and risks from remediation technologies. The cumulative exposure illustrated is the sum of residual exposure and exposures associated with releases from Technologies 1 and 2. This exhibit is for illustration purposes only and is not meant to imply that this level of quantitation is necessary or even desired.** *Source:* **U.S. Environmental Protection Agency, Risk Assessment Guidance for Superfund: Volume I — Human Health Evaluation Manual (Part C, Risk Evaluation of Remedial Alternatives), Washington, D.C., 1991.**

qualitative in nature. The results can be used to supplement the human health risk assessment in order to give a complete picture of health threats imposed by a contaminated site.

A risk characterization cannot be considered complete unless the numerical expressions of risk are accompanied by a qualitative description of the risk, i.e., the explanatory text interpreting and qualifying the quantitative results. The final discussion of the risk characterization results represents a significant part of the risk characterization process. The discussion provides a mechanism in which the numerical estimates of risk and hazard are placed in the context of what is known and what is not known (or uncertainties) about the site and in the context of decisions to be made about the selection of remedial options.

2. RISK MANAGEMENT

Risk management is a process of weighing policy alternatives and select-ing the most appropriate risk reduction action by integrating the results of risk assessment with engineering data and with social, economic, and political considerations.[121] The process may include a prioritization of risks and an evaluation of the costs and benefits of the proposed risk reduction programs. Deciding to what extent a contaminated site must be remediated, establishing acceptable levels for air contaminant emissions, and setting allowable levels of chemicals in drinking water or food are examples of risk management decisions.

The role of risk assessment in the risk management process is to provide information to help decision makers determine how to protect public health and the environment. While a risk assessment is performed objectively, a risk management decision is conducted in a subjective manner because it involves preferences and attitudes of the decision makers.[122,123] For analytical purposes risk management, risk communication, and risk assessment are often thought of as separate processes although in practice they frequently blend together.[124]

Because health and safety are sensitive political issues, actions and pri-orities are often decided in response to alarming events and perceived risk, rather than actual risks.[125] Many investigators believe that the fundamental causes of miscommunication about environmental health risks derive from differing perceptions of the significance of various risks between risk analysts and the public.[126–128]

Due to the socio-political orientation of risk management, risk commu-nication is becoming a vital and integral part of risk management. Effective communication dictates public perception and, therefore, public acceptance of risk reduction alternatives. Unfortunately, there is no single procedure for good risk communication.

The format used and the messages delivered will vary from site to site. The variations will depend on the characteristics of the public, their level of concern, the complexity of the problems at the site, the type of contaminants of concern, and the nature of options proposed for remediating the site. The format of risk communication can be fact sheets, public meetings, and risk assessment summary documents.

In order to achieve appropriate remedial decisions, a risk manager must have a comprehensive understanding of the risk assessment for the remedial options. EPA[129] developed a list of key questions that should be asked about the risks of remedial alternatives at a site. The questions include the following:

1. Which technologies can readily achieve all preliminary remediation goals in a given environmental medium? What uncertainties are involved in this determination?
2. Which remedial options will clearly not address the significant human exposure pathways identified in the baseline risk assess-ment?

3. Are the estimated residual risks from one technology different from another? How significant are the differences?
4. What other benefits are achieved by selecting one technology over another?
5. Will implementation of specific remedial options create new chemicals of concern or new threats to the surrounding community?
6. Is there a need for precautionary measures to mitigate risks during implementation? Are such measures available? How effective are they?
7. Does the remedial option selected result in on-site residual contamination which necessitates monitoring in the future?

Answers to these questions will certainly enhance the understanding of the site's risks. Such an understanding is necessary because it is one of the main sources of risk perception. The perceived risk includes the professional judgment of the site decision makers and the concerns of the surrounding communities. The following factors are considered by EPA to lead to a higher perceived risk of a site:

1. When the site is in close proximity to a population, particularly when persons in proximity to the site believe that they have little influence with regard to remedial decisions that are made.
2. When the site is contaminated with highly toxic chemicals.
3. When the remedial technologies chosen have a high release potential (e.g., emitting air pollutants), whether the release is planned or accidental.
4. When the remedial technologies chosen are innovative technologies with high uncertainties concerning potential releases, e.g., unknown amount or identity of contaminants released.
5. When multiple contaminants and/or exposure pathways affect the same individuals.
6. When multiple releases occur simultaneously.
7. When multiple releases occur from several units.
8. When releases occur over long periods of time.

If one or more of these conditions apply to a site, the chance of an unfavorable risk perception will be high. EPA suggests that a more quantitative evaluation including emission modeling and detailed treatability studies be conducted to help in the decision-making process. If, for example, the selected remedial option for a site involves extensive excavation very close to a residential population, then a more quantitative evaluation of the short-term risks is necessary to assess the impact of this technology on the exposed population.

As a tool to determine whether and how a site should be remediated, risk assessments have become a source of scrutiny and controversy. Risk assessment has been criticized for its inconsistent estimation practices. In a review of 20 of the approximately 70 risk assessments performed in 1992, the General Accounting Office (GAO)[130] discovered that the risk assessments they reviewed generally followed a consistent approach. The assessments, however, used different assumptions in estimating exposure. The different assumptions occurred in the prediction of how a site might be used in the future, of how much contamination would remain on-site, and of how people absorb contaminants through the skin.

Historically, U.S. risk assessment practice can be characterized by the dominance of carcinogenic chemical concerns while risks from other diseases have been less researched. This is because cancer is such a dreaded disease and it always attracts considerable political interest. From a scientific perspective, emphasis on cancer means that other health effects are downplayed or ignored. According to Locke,[131] this preeminence of cancer risk assessment is an accident of history due to quirks in the development of risk analysis.

The dominance of cancer risk research also results in the fact that ecological degradations are less noticed. Key problems in an ecological assessment have been widely recognized as the lack of availability of relevant data, inadequate understanding of complex ecological processes, and procedural difficulties.[132,133] In practice, an ecological assessment has been used primarily to predict the consequences of development activities on organisms other than humans. Many of the techniques used in ecological assessment were originally developed for purposes of nature reserve selection or wildlife management. Before the techniques can be adapted for purposes of ecological assessment, they require some modifications.[134]

CERCLA and RCRA require EPA to develop a program that protects human health as well as the environment from the releases or potential releases of contaminants. The qualitative and/or quantitative appraisal of the actual or potential effects of a contaminated site on plants and animals other than humans and domesticated species is called ecological assessment.[135]

3. ECOLOGICAL RISK

The ecological assessment is performed by an ecologist. The ecologist must ensure that the ecological aspects of a site are properly considered in the site assessment. Although the human health aspect is excluded from EPA's ecological assessment definition, as information from the ecological assessment becomes available, the newly obtained information may point to new or unexpected exposure pathways for the human population that can be used to refine the human health risk assessment. As a matter of fact, the human health and ecological assessment processes are parallel activities in the risk assessment of a contaminated site. As Figure 9.6 shows, information related to the nature, fate, and transport of a site's contaminants will be used for both evaluations.

Ecological studies normally examine three levels of organism organization: population, community, and ecosystem.

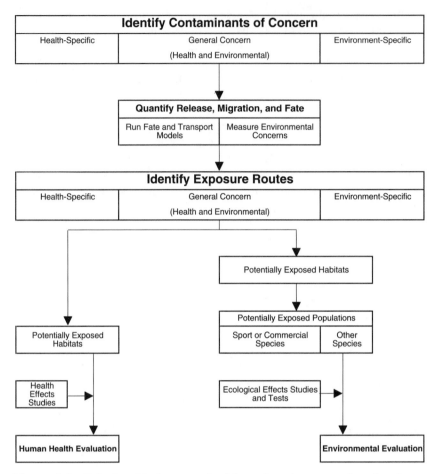

Figure 9.6. **Relationship between health and environmental evaluations.**
Source: **U.S. Environmental Protection Agency, Risk Assessment Guidance for Superfund: Volume II — Environmental Evaluation Manual, Washington, D.C., 1989.**

- Population means a group of organisms of the same species, generally occupying a contiguous area and capable of interbreeding.
- Community is a complex association in which populations of different species live together. The structure of a community is determined by what species are present, in what numbers, and which species eat which other species.
- Ecosystem is a system whereby nonliving components of the system interact with communities. Ecosystems are characterized by species composition, nutrient and energy flow, and rates of production, consumption, and decomposition. The ecosystem is the unit of ecology.

There are five ecosystems in our ecology. The types of ecosystems which are determined by a number of factors such as climate, topography, geology, chemistry and biota include:

1. Terrestrial. This land ecosystem is categorized based on the vegetation types that dominate the plant community.
2. Wetlands. These are areas in which topography and hydrology create a zone of transition between the terrestrial and aquatic environments.
3. Marine. This ocean ecosystem is vital because of its large size and critical ecological function to maintain much of the global environment's capacity to sustain life.
4. Estuaries. These are open bodies of water closely related with the ocean in coastal zones such as river mouths, bays, tidal marshes, and waters behind barrier beaches.
5. Fresh water. This ecosystem is smaller than the marine and terrestrial habitats; however, it is important because the ecosystem represents a major component in the hydrological system, and is a source of water for human use and habitat for wildlife species of value to people.

The release of contaminants into an ecosystem can certainly harm organisms. The exposed organism can differ in its response to contaminants depending on the types of contaminants, the exposure dose, the frequency and duration of the exposure, the physical/chemical characteristics of the environment, and the organism's natural tolerance to the contaminants.

The impact of contaminants to an ecosystem can be in the form of increasing mortality of some organisms and/or decreasing birth rates. This may lead to changes in the community structure due to the extinction of some species and the survival of a few species that are tolerant of the contaminant. A classic example is when an insecticide kills off the target pests along with predatory insects. The pest population can rebound to much higher numbers because only a few predators remain to exterminate them.

In the U.S., preliminary work on guidelines for ecological effects began in the mid-1980s. The continuous efforts of EPA resulted in the issuance of *Framework for Ecological Risk Assessment*[136] in 1992. The use of the document is not an EPA requirement. The publication of the framework is expected to generate discussions for developing concepts, principles, and methods of effective ecological risk assessment. Eventually, the discussion will lead to EPA publishing ecological risk assessment guidelines.

EPA's 1992 ecological risk framework has been used, evaluated, and tested both inside and outside of the agency.[137] One approach to evaluate the framework was to provide case studies at peer review workshops for discussion as to whether the case studies consistently follow EPA's framework.[138–140]

The framework offers a simple, flexible structure for performing and evaluating ecological risk assessment. The term stressor is used in the framework to describe any chemical, physical, or biological entity that can induce adverse effects on organism populations, communities, or ecosystems. Adverse ecological effects include a wide range of disturbances ranging from mortality in an individual organism to a loss in ecosystem function.

Structurally the framework consists of three major phases: problem formulation, analysis, and risk characterization (Figure 9.7). Problem formulation deals with the planning and scoping activities that result in the establishment of the goals, scope and focus of the risk assessment. The end product of the process is a conceptual model that identifies the environmental values to be protected (termed as the assessment endpoints), the data needed, and the analyses to be used.

The focus of the conceptual model is the development of hypotheses regarding how the stressor might affect the ecological components of the natural environment,[141] as well as the description of the ecosystem potentially at risk and the relationship between measurement and assessment endpoints. An endpoint is a characteristic of an ecological component, such as increased mortality in fish, that may be affected by exposure to a stressor. Suter[142] describes assessment endpoints as the explicit expressions of the actual environmental value to be protected and measurement endpoints as measurable responses to a stressor that are related to the assessment endpoints.

The analysis phase of the framework develops profiles of environmental exposures and the effects of the stressor. The profile characterizes the ecosystems in which the stressor may occur and the biota that may be exposed. The exposure is described in terms of the magnitude, and the spatial and temporal variations, while the effects of the stressor are summarized and related to the assessment endpoints.

The risk characterization phase of the framework integrates the exposure and effects profiles. During this phase, the likelihood of adverse effects as a result of exposure to a stressor is examined. This phase consists of risk estimation and a description of risk. Risk estimation compares the exposure and stressor–response profiles and summarizes the associated uncertainties. The risk description discusses confidence in the risk estimation and interprets the ecological significance by describing the magnitude of the identified risks to the assessment endpoint.

Figure 9.7 shows that the framework also recognizes important activities that are related to the risk assessment process. The activities include the continuous discussions between the risk assessor and risk manager, the acquisition of data, and the verification and monitoring studies which can be used to improve future risk assessments.

The distinctive features of the framework are its broad and flexible approaches. An ecological risk assessment can be conducted beyond the individuals of a single species and may assess an organism's population, community, or ecosystem. Secondly, in the assessment, there is no set of ecological values to be protected; rather, the values are chosen from a number of possi-

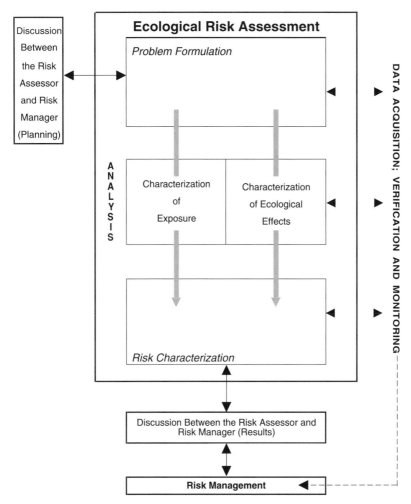

Figure 9.7. Framework for ecological risk assessment. *Source:* **U.S. Environmental Protection Agency, Framework for Ecological Risk Assessment, Washington, D.C., 1992.**

bilities based on scientific and policy considerations. It is also possible to consider nonchemical stressors in the assessment.

A critical element in the ecological risk assessment process calls for distinguishing environmental responses that are effective from those that are not. This implies that the determination of the ecological significance of the risk is vital for a successful assessment of ecological risk.[143] The problem is that what is significant and what is acceptable can only be determined through decision making that takes place in the context of human values.

In this regard everyone agrees that a reduction of risk will correspond to an increment of human welfare; however, no agreed-upon methods currently exist to connect changes in ecosystems with measures of individual welfare.[144] As Sagoff[145] states, while it is clear that ecological risk-based decision making requires input regarding public values, it is less clear how those values are to be determined, measured, and considered in decision making.

One of the critical limitations in an ecological risk assessment is the concern about the uncertainty involved in the assessment. The general lack of knowledge and lack of certainty are referred to as uncertainty.[146] Lack of knowledge about how the particular system functions, how the stressor relates to the system, and how the system affects the stressor will generate uncertainty. Lack of certainty in ecological risk is frequently associated with variability. Variability is the difference between what is expected and what actually happens.

In a number of circumstances it may not be possible to reduce uncertainty enough to allow reasonable risk estimates to be made. In many instances, however, good scientific practices can minimize uncertainty. The practices include the proper use of statistical principles and adherence to quality assurance methods.

10   REMEDIATION STANDARDS

Remediation standards are cleanup levels that are established to indicate that the concentration of a contaminant at a cleanup level is considered not to pose a threat to human health and the environment. The standards represent a measure of the cleanliness of a site. Because the concept of "clean" is ambiguous, relative, and subjective, determining how clean is clean has always been problematic. Since the passage of the Resource Conservation and Recovery Act (RCRA) in 1976 and the Comprehensive Environmental Response, Compensation, and Liability Act (CERCLA) in 1980, remediation criteria for contaminated soils and groundwater have taken many different forms. From this list of criteria, a decision to apply a proper cleanup standard to a site must always be based on sufficient protection of human health and the environment, as well as compliance with the existing laws.

A concentration of a contaminant of concern that is calculated safe from the risk assessment perspective may not necessarily be acceptable to state hazardous waste laws which, for example, rely solely on the toxicity characteristics leaching procedure (TCLP) rules for that particular contaminant. As a practical matter, the decision to apply a standard must also be realistically achievable in terms of cost and the implementation schedule.

After a thorough analysis, Buonicore[147] identifies eight cleanup criteria currently available for use. The criteria include the following:

1. levels established by precedent
2. no-detect levels
3. limits of detection
4. background levels
5. levels determined by standards or guidelines
6. human health risk assessment levels
7. best demonstrated available technologies (BDAT) levels
8. combinations of the above

The applicability of the standards to a particular site depends on the regulatory status of the site, the cleanup policy of the state, or the specific conditions of the site. The level of cleanup for a contaminated site under RCRA corrective action regulations, for example, is determined by RCRA standards. State cleanup policy may require that certain contaminated sites be remediated to limits of detection, or background level. Using the risk assessment approach, a site may also be remediated to a level that is protective of human health based on site-specific conditions.

Because of cost considerations, it is critical for the owner of a contaminated site to negotiate as early as possible with the environmental agency regarding which cleanup level is appropriate for the site. The rule is that the higher the level of cleanup, the lower the cost of remediation. A thorough and systematic analysis that considers all the possible approaches and uncertainties in remediating the site is part of the planning process. This analysis is particularly important because the level of cleanup affects the type of technology used for remediating the site.

Remediation standards can be established by precedents. Precedent can be a powerful tool in negotiating site-specific cleanup levels. Examples of precedent situations are

- similar remediation in another location in the state
- similar remediation in another state
- remediation criteria recorded in the Superfund's record of decision (RODs)

Not only are precedents helpful for site owners to negotiate with the regulatory agency, they are also useful for the agency to justify its remedial decision associated with a site. Buonicore[147] cautions, however, that the situation surrounding the precedent must be analyzed properly to ensure that the site situations are truly similar.

As one of the options, remediation to no-detect means that the site must be remediated until all contaminants are removed from the soil and groundwater. Obviously, this option is very costly and rarely required by the environmental agency. Another option is to remediate a site to a level that is set to the capability of the analytical equipment to detect the contaminants of concern. This option is called the limits of detection and is discussed in the following section, after which the types of remediation standards are discussed, including site remediation to background levels, the regulatory cleanup levels, the human health risk standards, and the technology-based cleanup.

1. LIMITS OF DETECTION

This standard is established through the detection of contaminants of concern by laboratory equipment. The capability of analytical equipment to detect contaminants of concern, however, varies from one laboratory to another. Typical detection limits for contaminants in soils and groundwater are displayed in Table 10.1. The unit of limits of detection for soil in this table is parts per million by weight (ppmw) and milligrams per liter (mg/l) for groundwater.

The detection limits shown in the table can change in the future. Continuous improvement in the field of analytical instrumentation may significantly improve the detection capability and, therefore, lower the limits of detection. Because of the limited capability of analytical equipment at the present time, the detection limit of a contaminant can be higher than the standard imposed by a human health risk assessment. This represents a situation in which a policy decision must be made.

If the state mandates that a cleanup be conducted to the human health risk assessment level, then any contamination above that level is not acceptable, including those sites having contaminants with higher limits of detection. A notice to the deed of the property may be required, and recorded in the county where the site is located. The notice informs the potential buyer that contamination has been remediated to the limits of detection.

2. BACKGROUND REMEDIATION

The remediation of a contaminated site to the condition that existed before the contamination occurred is defined as the cleanup to background levels. The difficulty in this definition is that what constitutes the precontamination condition at a particular site is subject to debate and leads to a policy issue. A soil background level may be determined as those concentrations found in the surrounding, uncontaminated area. A groundwater background may be defined as those concentrations found in the upgradient of a contaminated site.

Even if the surrounding area of a contaminated site is legally defined as a clean area, in reality the area may have long been contaminated with organic chemicals due to the widespread agricultural application of various fertilizers and pesticides in the past. The anthropogenic background, i.e., the background level created by lawfully conducted human activities, can also be found in many sites along a roadway which is contaminated with emissions from motor vehicles. Establishing a cleanup level using these high anthropogenic background concentrations is certainly not a sound policy decision.

Natural background, i.e., the background level created by the naturally occurring inorganics, can be found in many locations. The concentrations of these inorganics can be higher than the human health risk assessment level. Establishing a background remediation standard for this type of contaminant is appropriate, otherwise we may have to unnecessarily remediate a copper-rich mountain due to the higher concentration of copper in the site which actually must be remediated only for its petroleum hydrocarbon contamination.

In addition to the above definitional issue, given an array of contaminant concentrations obtained from sampling results from various background locations, the question is what is the representative value of the background

Table 10.1 Representative Limits of Detection

Constituent	Soil (ppmw)	Groundwater (mg/l)
Acenaphthylene	0.2	0.0003
Acephate	0.6	—
Acetone	0.1	0.01
Acrolein	—	0.005
Acrylonitrile	—	0.005
α-AHC	0.05	0.00005
Alachlor	1.0	0.01
Aldicarb	1.0	0.001
Aldicarb sulfone	1.0	—
Aldicarb sulfoxide	1.0	0.001
Aldrin	0.005	0.00005
Ally	2.0	—
Ametryn	0.1	0.001
Anthracene	0.17	0.005
Antimony	0.15	0.00005
Arsenic	0.10	0.004
Atrazine	1.0	0.01
Azinphos-methyl	0.1	0.0005
Barium	0.5	0.01
Bayleton	0.5	0.005
Bentazon	0.66	0.002
Benzene	0.01	0.0002
Benzidine	0.1	0.02
Benzo(a)anthracene	0.17	0.00001
Benzo(a)pyrene	0.17	0.00001
Benzo(b)fluoranthene	0.17	0.00001
Benzo(k)fluoranthene	0.17	0.00001
Benzyl alcohol	0.07	0.002
Beryllium	0.3	0.0005
bis(2-ethylhexyl)adipate	1.0	0.01
bis(2-chloroethyl)ether	0.17	0.005
bis(2-chloroisopropyl)ether	0.17	0.005
Boron (water soluble)	0.50	0.01
Bromacil	0.33	0.02
Bromoform	0.01	0.0002
Bromomethane	0.01	0.0002
Bromaxynil	0.66	—
Butyl benzyl phthalate	0.17	0.005
Butylate	0.5	0.005
Cadmium	0.3	0.0002
Captan	0.05	0.03
Carbaryl	0.33	0.001
Carbofuran	0.66	0.001
Carbon disulfide	0.05	0.01

Table 10.1 Representative Limits of Detection (Continued)

Constituent	Soil (ppmw)	Groundwater (mg/l)
Carbon tetrachloride	0.01	0.0002
Carboxin	1.0	0.01
Chloramben	0.66	0.001
Chlordane	0.05	0.0005
Chlorobenzene	0.01	0.0002
Chloroform	0.01	0.0002
Chloromethane	0.01	0.0002
2-Chlorophenol	0.3	0.005
Chlorothalonil	0.05	0.001
Chlorpyrifos	0.05	0.01
Chromium	0.5	0.01
Chrysene	0.17	0.00002
Copper	0.5	0.01
Cyanazine	1.0	0.01
Cyanide	0.5	0.01
Cyromazine	6.0	—
2,4-D	0.4	0.0004
Dacthal	1.0	0.010
Dalapon, sodium salt	3.3	0.004
Di(2-ethylhexyl)phthalate	0.17	0.01
Diazinon	0.1	0.0005
Dibenzo(a,h)anthracene	0.2	0.0001
1,2-Dibromo-3-chloropropane	0.003	0.000005
Dibromochloromethane	0.01	0.0002
1,2-Dibromomethane	0.002	0.000005
Dibutylphthalate	0.3	0.005
Dicamba	0.04	0.0005
1,2-Dichlorobenzene	0.025	0.0005
1,3-Dichlorobenzene	0.02	0.0005
1,4-Dichlorobenzene	0.025	0.0005
3,3-Dichlorobenzidine	0.34	0.02
Dichlorobromomethane	0.01	0.0002
Dichlorodifluoromethane	0.01	0.0002
1,2-Dichloroethane	0.01	0.0002
cis-1,2-Dichloroethylene	0.01	0.0002
trans-1,2-Dichloroethylene	0.01	0.0002
Dichloromethane	0.05	0.0005
2,4-Dichlorophenol	0.03	0.005
1,2-Dichloropropane	0.01	0.0002
1,3-Dichloropropylene	0.01	0.0002
Dicloran	2.0	0.01
Dicofol	2.0	0.02
Dieldrin	0.01	0.0001
Diethylphthalate	0.3	0.01

Table 10.1 Representative Limits of Detection (Continued)

Constituent	Soil (ppmw)	Groundwater (mg/l)
Dimethoate	0.13	0.01
2,4-Dinitrophenol	1.67	0.01
2,4-Dinitrotoluene	0.17	0.005
Dinoseb	0.05	0.0002
Diphenamid	1.0	0.01
1,2-Diphenylhydrazine	0.17	0.01
Disulfoton	0.1	0.0005
Diuron	0.17	0.02
Endosulfan	0.01	0.0001
Endrin	0.01	0.0001
s-Ethyldipropylthiocarbamate	0.017	0.01
Ethylbenzene	0.025	0.0005
Ethylene glycol	0.1	0.01
Fenamiphos	0.66	0.01
Fenarimol	1.0	0.01
Fluormeturon	3.0	0.03
Fluoranthene	0.17	0.00003
Fluorene	0.17	0.00004
Fluorine (soluble fluoride)	—	0.05
Fluridone	1.0	0.01
Fluvalinate	1.0	0.01
Heptachlor	0.005	0.00005
Heptachlor epoxide	0.005	0.00005
Hexachloro-1,3-butadiene	0.17	0.01
Hexachlorobenzene	0.17	0.005
β-Hexachlorocyclohexane	0.05	0.00005
Hexachlorocyclopentadiene	0.17	0.01
Hexachloroethane	0.17	0.01
Hexazinone	5.0	0.005
Imazalil	1.0	0.015
Indeno(2,3-cd)pyrene	0.17	0.00003
Isophorone	0.17	0.01
Lead	0.50	0.002
Lindane	0.005	0.00005
Linuron	0.17	0.05
Malathion	0.1	0.0005
Manganese	0.5	0.01
Mercury	0.001	0.0002
Metalaxyl	0.5	0.005
Methiocarb	1.0	0.001
Methomyl	0.83	0.001
Methoxychlor	0.05	0.0005
Methyl ethyl ketone	0.1	0.01
Methyl parathion	0.03	0.0005

Table 10.1 Representative Limits of Detection (Continued)

Constituent	Soil (ppmw)	Groundwater (mg/l)
Methyl *tert*-butyl ether	0.25	0.0025
2-Methyl-4-chlorophenoxyacetic acid	2.5	0.1
Metolachlor	1.0	0.002
Metribuzin	1.0	0.01
Molybdenum	10	0.01
Monocrotophos	0.17	0.04
n-Hexane	0.05	0.005
N-Nitrosodi-*n*-propylamine	0.17	0.005
N-Nitrosodimethylamine	0.17	0.005
N-Nitrosodiphenylamine	0.17	0.005
N-Nitrosopyrrolidine	0.17	0.01
Naled	0.1	0.0005
Naphthalene	0.2	0.0003
Napropamide	1.0	0.01
Nickel	0.5	0.02
Nitrate (as N)	—	0.06
Nitrite (as N)	—	0.06
Nitrate/nitrite (total)	—	0.06
Nitrobenzene	0.17	0.005
Norflurazon	1.0	0.01
o-Chlorotoluene	0.05	0.0005
Oryzalin	0.8	—
Oxamyl (vydate)	0.83	0.001
Parathion	0.03	0.0005
Pendimethalin	1.0	0.01
Pentachlorophenol	2.0	0.01
Permethrin	0.5	0.00004
Phenol	0.3	0.005
Phorate	0.1	0.0005
Phosmet	1.0	0.01
Picloram	—	0.0004
Polychlorinated biphenyls		
p,p'-dichlorodiphenyl dichloroethane	0.2	0.0005
p,p'-dichlorodiphenyl dichloroethylene	0.01	0.0001
p,p'-dichlorodiphenyltrichloroethane	0.01	0.0001
Profluralin	0.5	—
Prometon	0.1	0.003
Prometryn	1.0	0.002
Pronamide	0.5	0.005
Propachlor	—	0.005
Propazine	0.1	0.0013
Propham	0.33	—
Propoxur	1.7	0.01
Pydrin	1.0	0.01

Table 10.1 Representative Limits of Detection (Continued)

Constituent	Soil (ppmw)	Groundwater (mg/l)
Pyrene	0.17	0.00004
Selenium	0.1	0.002
Sethoxydim	1.0	0.01
Silver	0.5	0.001
Simazine	2.0	0.01
Styrene	0.05	0.0005
Sulprofos	0.03	0.0005
Systhane	1.0	0.01
Tebuthiuron	3.0	0.015
Terbacil	2.0	0.01
Terbufos	1.3	0.005
Terbutryn	0.1	0.003
1,2,4,5-Tetrachlorobenzene	—	0.01
1,1,1,2-Tetrachloroethane	0.05	0.0005
1,1,2,2-Tetrachloroethane	0.05	0.0002
Tetrachloroethylene	0.01	0.0002
Tetraethyl lead	1.0	0.05
Thallium	0.5	0.001
Toluene	0.01	0.0002
Toxaphene	0.05	0.0005
1,2,4-Trichlorobenzene	0.17	0.001
1,1,1-Trichloroethane	0.01	0.0002
1,1,2-Trichloroethane	0.01	0.0002
Trichloroethylene	0.01	0.0002
Trichlorofluoromethane	0.025	0.0005
2,4,5-Trichlorophenol	0.2	0.005
2,4,6-Trichlorophenol	0.17	0.005
2,2,4,5-Trichlorophenoxy propionic acid	0.02	0.0002
2,4,5-Trichlorophenoxyacetic acid	0.02	0.0002
1,2,3-Trichloropropane	0.05	0.0005
Trichlorotrifluoroethane	0.25	0.0005
Trifluralin	0.005	0.00005
Trihalomethanes (total)	—	0.004
Vanadium (fume or dust)	0.50	0.01
Vernam	2.0	0.002
Vinclozolin	0.5	0.005
Vinyl chloride	0.05	0.0005
Vinylidene chloride	0.01	0.0002
Xylenes (mixed isomers)	0.01	0.0002
Zinc (fume or dust)	0.025	0.0005

Source: Buonicore, Anthony J., ed., *Cleanup Criteria for Contaminated Soil and Groundwater,* ASTM, Philadelphia, 1995.

concentrations that will be used to remediate the site. Should the highest concentration be used to represent the background? Should it be the lowest concentration? Or perhaps the average value? In this situation, statistics are normally used to characterize a background level by estimating the mean concentration and measuring the standard deviation.

The background concentrations of elements in soils have been regularly evaluated by the U.S. Geological Survey (USGS). Table 10.2 provides background concentrations in terms of average and range. These numerical values, stated in ppmw, are obtained from the results of soil samples taken throughout the U.S. at a depth of approximately 8 in.[148,149] The table shows how large the range of concentrations is for some elements such as aluminum, calcium, iron, and sodium.

Table 10.2 Background Concentrations of Elements in Soils

Constituent	CAS number	Average (ppmw)	Range (ppmw)
Aluminum (fume or dust)	7429-90-5	66,000	700–100,000
Antimony	7440-36-0	0.67	<1–8.8
Arsenic	7440-38-2	7.2	<0.1–97
Barium	7440-39-3	580	10–5,000
Beryllium	7440-41-7	0.92	<1–15
Boron (watersoluble)	7440-42-8	34	<20–300
Cadmium	7440-43-9	0.06	0.01–0.7
Calcium	7440-70-2	24,000	<150–320,000
Cerium	7440-45-1	86	<150–300
Chromium	7440-47-3	54	1.0–2,000
Cobalt	7440-48-4	10	<3–70
Copper	7440-50-8	25	<1–700
Gallium	7440-55-3	19	<5–70
Iron	7439-89-6	25,000	100–>100,000
Lanthanum	7439-91-0	41	<30–200
Lead	7439-92-1	19	<10–700
Manganese	7439-96-5	560	<1–7,000
Mercury	7439-97-6	0.089	<0.01–4.6
Molybdenum	7439-98-7		<3–7.0
Nickel	7440-02-0	19	<5–700
Phosphorus (yellow or white)	7723-14-0	420	20–6,000
Potassium	7440-09-7	23,000	50–70,000
Selenium	7782-49-2	0.39	<0.1–4.3
Sodium	7440-23-5	12,000	<500–100,000
Strontium	7440-24-6	240	<5–3,000
Vanadium (fume or dust)	7440-62-2	76	<7–500
Zinc (fume or dust)	7440-66-6	60	<5–2,900

Source: Buonicore, Anthony J., ed., *Cleanup Criteria for Contaminated Soil and Groundwater,* ASTM, Philadelphia, 1995.

In its 1988 document, the Environmental Protection Agency (EPA)[150] recognized that groundwater quality measurements do not vary only as a result of random occurrences. Whereas the predictable components of groundwater quality are associated with groundwater recharges and quality differences related to the location of monitoring wells, the groundwater quality is basically a result of a dynamic natural system that behaves in a semi-predictable way. This unique behavior particularly represents one of the difficulties in establishing the background level for groundwater quality.

3. REGULATORY CLEANUP LEVEL

Cleanup levels established by regulations and guidelines define acceptable levels of contamination in soil and groundwater. The current regulatory cleanup levels include those established by the Safe Drinking Water Act (SDWA), RCRA, CERCLA, and the Toxic Substances Control Act (TSCA). The SDWA, as amended in 1986, required EPA to publish maximum contaminant level goals (MCLGs) for contaminants which may have adverse health effects and are anticipated to occur in public water systems. MCLGs are nonenforceable goals and represent levels in which no anticipated adverse health effects would occur.

For noncarcinogens, an MCLG is determined by a drinking water equivalent level (DWEL). DWEL is calculated by multiplying the reference dose (RfD) by an assumed adult body weight of 70 kg, and then dividing by an average daily water consumption of 2 l per day. The total daily exposure to a substance is assumed from drinking-water exposure. A portion of the total daily exposure contributed by drinking water, normally 20%, is called the relative source contribution (RSC). MCLG is calculated by multiplying the DWEL by the RSC.

For carcinogens, contaminants are grouped into three different categories: MCLGs for Category I, which consists of carcinogen groups A and B, are established by EPA as zero; Category II consists of carcinogen group C; and Category III includes groups D and E.

The SDWA also mandated EPA to promulgate National Primary Drinking Water Regulations (NPDWR) which established standards in terms of a maximum contaminant level (MCL), or a specified treatment technique. An MCL must be determined as close to the MCLG as possible. To assure that the supply of drinking water complies with MCLs, the NPDWR include monitoring, analytical, and quality assurance provisions.

The MCL for each contaminant is established based on a performance evaluation of various technologies for removing the contaminant, the ability of laboratories to accurately and consistently measure the concentration of the contaminant with available analytical methods, and the costs of utilizing the technologies. The MCLs for noncarcinogens are normally set at MCLGs. The MCLs for carcinogens are higher than the respective MCLGs due to the limitation of analytical detection. These MCLs are determined to be adequately protective of human health and are evaluated with a goal of a risk range of 10^{-4} to 10^{-6} excess cancer risks during a lifetime exposure. One exception to this range is arsenic.

40 CFR Part 143 contains EPA's National Secondary Drinking Water Regulations (NSDWR) which provide guidance to control contaminants in drinking water that degrade aesthetic qualities. At higher concentrations, the listed contaminants under the NSDWR may affect human health as well. Table 10.3 displays a list of EPA's MCLGs and MCLs as found in 40 CFR 141.11, 141.12, 141.50, 141.51, 141.61, and 141.62.

The notable regulatory level under the RCRA program is the toxicity characteristic provision which establishes the regulatory levels for some con-stituents. As briefly described in Chapter 4, the TCLP test is designed to determine the mobility of organic and inorganic analytes present in liquid, solid, and multi-phasic wastes. The test simulates the leaching action that occurs in the subsurface. By comparing the test result, i.e., the concentration of the liquid extract obtained from leaching procedures, with the regulatory level of the constituent, it can be determined whether the contaminant is hazardous due to its toxicity. If the numeric test result is lower than the regulatory level, the constituent is not a hazardous waste by RCRA definition and, more importantly, it will not likely leach in the subsurface.

EPA toxicity characteristics regulatory levels are specified in 40 CFR 261.24. These levels are listed in Table 4.3 of Chapter 4. The levels were derived using health-based concentration limits, and dilution and attenuation factors. The generic dilution and attenuation factor of 100 is applied, meaning that the regulatory concentration levels for the listed constituents are 100 times the health-based concentration levels for these contaminants.

The health-based concentration limits were chronic toxicity reference levels taken from one of three sources:

- MCL as specified in the SDWA regulations
- oral RSDs for carcinogenic compounds using a risk level of one in 100,000
- RfDs for noncarcinogenic constituents

Under current RCRA regulations, the remediation of sites contaminated with metallic residue derived from the burning or processing of hazardous waste in boilers and industrial furnaces (BIFs) must be to a level not to exceed the TCLP (40 CFR 266.112). For nonmetal residue, the remediation must not exceed the health-based levels established in 40 CFR 266 Appendix VII. These levels are shown in Table 10.4.

The land disposal restriction (LDR) program under RCRA prohibits the land disposal of untreated hazardous wastes. Among other provisions, the program requires that before the waste is disposed of on land, it must be treated using specified technology such as stabilization, or an allowable constituent concentration such as no more than 0.5 mg/l. In the existing regulatory scheme,

Table 10.3 EPA Drinking Water Standard
(40 CFR Part 141)

Constituent	Concentrations in groundwater (mg/l)	
	MCL	MCLG
Acrylamide		0
Alachlor	0.002	0
Aldicarb	0.003	0.001
Aldicarb sulfoxide	0.004	0.001
Aldicarb sulfone	0.002	0.001
Antimony	0.06	0.006
Arsenic	0.05	
Asbestos longer than 10 μm	7 million fibers/l	7 million fibers/l
Atrazine	0.003	0.003
Barium	2	2
Benzene	0.005	0
Benzo(a)pyrene	0.0002	0
Beryllium	0.04	0.04
Cadmium	0.005	0.005
Carbofuran	0.04	0.04
Carbon tetrachloride	0.005	0
Chlordane	0.002	0
Chromium	0.1	0.1
Copper		1.3
Cyanide (as free cyanide)	0.2	0.2
2,4-D	0.07	0.07
Dalapon	0.2	0.2
Dibromochloropropane	0.0002	0
o-Dichlorobenzene	0.6	0.6
para-Dichlorobenzene	0.075	0.075
1,2-Dichloroethane	0.005	0
1,1-Dichloroethylene	0.007	0.007
cis-1,2-Dichloroethylene	0.07	0.07
trans-1,2-Dichloroethylene	0.1	0.1
Dichloromethane (methylene chloride)	0.005	0
1,2-Dichloropropane	0.005	0
Di(2-ethylhexyl)adipate	0.4	0.4
Di(2-ethylhexyl)phthalate	0.006	0
Dinoseb	0.007	0.007
Diquat	0.02	0.02
Endothall	0.1	0.1
Endrin	0.002	0.002
Epichlorohydrin		0
Ethylbenzene	0.7	0.7
Ethylene dibromide	0.00005	0
Fluoride	4	4

Table 10.3 EPA Drinking Water Standard
(40 CFR Part 141) (Continued)

| Constituent | Concentrations in groundwater (mg/l) | |
	MCL	MCLG
Glyphosate	0.7	0.7
Heptachlor	0.0004	0
Heptachlor epoxide	0.0002	0
Hexachlorobenzene	0.001	0
Hexachlorocyclopentadiene	0.05	0.05
Lead	0.05	0
Lindane	0.0002	0.0002
Mercury	0.002	0.002
Methoxychlor	0.04	0.04
Monochlorobenzene	0.1	0.1
Nickel	0.1	0.1
Nitrate (as N)	10	10
Nitrite (as N)	1	1
Total nitrate and nitrite (as N)	10	10
Oxamyl (vydate)	0.2	0.2
PCB	0.0005	0
Pentachlorophenol	0.001	0
Picloram	0.5	0.5
Selenium	0.05	0.05
Simazine	0.004	0.004
Styrene	0.1	0.1
2,3,7,8-TCDD (dioxin)	3×10^{-8}	0
Tetrachloroethylene	0.005	0
Thallium	0.02	0.0005
Toluene	1	1
Toxaphene	0.003	0
2,4,5-TP	0.05	0.05
1,2,4-Trichlorobenzene	0.07	0.07
1,1,1-Trichloroethane	0.2	0.2
1,1,2-Trichloroethane	0.005	0.003
Trichloroethylene	0.005	0
Vinyl chloride	0.002	0
Xylenes (total)	10	10

a constituent regulated under the standard for one RCRA hazardous waste may also be a constituent under the treatment standard for another RCRA waste. In 40 CFR 268.41, for example, the LDR standard for cadmium which is one of the F006 constituents, is 0.066 mg/l. The standards for cadmium as the K061 constituent is 0.41 mg/l and as D006 is 1.0 m/l.

Because of this problem, in 1993 EPA proposed a rule package that introduces the concept of a universal treatment standard which applies to

Table 10.4 Residue Concentration Limits for Nonmetals

Constituent	Concentration limits for residues (mg/kg)
Acetonitrile	0.2
Acetophenone	4.0
Acrolein	0.5
Acrylamide	0.0002
Acrylonitrile	0.0007
Aldrin	0.00002
Alyl alcohol	0.2
Aluminum phosphide	0.01
Aniline	0.06
Barium cyanide	1.0
Benz(a)anthracene	0.0001
Benzene	0.005
Benzidine	0.000001
Bis(2-chloroethyl)ether	0.0003
Bis(chloromethyl)ether	0.000002
Bis(2-ethylhexyl)phthalate	30
Bromoform	0.7
Calcium cyanide	0.000001
Carbon disulfide	4.0
Carbon tetrachloride	0.005
Chlordane	0.0003
Chlorobenzene	1.0
Chloroform	0.06
Copper cyanide	0.2
Cresols (cresylic acid)	2.0
Cyanogen	1.0
DDT	0.001
Dibenz(a,h)anthracene	0.000007
1,2-Dibromo-3-chloropropane	0.00002
p-Dichlorobenzene	0.075
Dichlorodifluoromethane	7.0
1,1-Dichloroethylene	0.005
2,4-Dichlorophenol	0.1
1,3-Dichloropropene	0.001
Dieldrin	0.00002
Diethylphthalate	30
Diethylstilbesterol	0.0000007
Dimethoate	0.03
2,4-Dinitrotoluene	0.0005
Diphenylamine	0.9
1,2-Diphenylhydrazine	0.0005
Endosulfan	0.002
Endrin	0.0002
Epichlorohydrin	0.04

Table 10.4 Residue Concentration Limits for Nonmetals (Continued)

Constituent	Concentration limits for residues (mg/kg)
Ethylene dibromide	0.0000007
Ethylene oxide	0.0003
Fluorine	4.0
Formic acid	70
Heptachlor	0.00008
Heptachlor epoxide	0.00004
Hexachlorobenzene	0.0002
Hexachlorobutadiene	0.005
Hexachlorocyclopentadiene	0.2
Hexachlorodibenzo-*p*-dioxins	0.00000006
Hexachloroethane	0.03
Hydrazine	0.0001
Hydrogen cyanide	0.00007
Hydrogen sulfide	0.000001
Isobutyl alcohol	10
Methomyl	1.0
Methoxychlor	0.1
3-Methylcholanthrene	0.00004
4,4-Methylenebis (2-chloroaniline)	0.002
Methylene chloride	0.05
Methyl ethyl ketone (MEK)	2.0
Methyl hydrazine	0.0003
Methyl parathion	0.02
Naphthalene	10
Nickel cyanide	0.7
Nitric oxide	4.0
Nitrobenzene	0.02
N-Nitrosodi-*N*-butylamine	0.00006
N-Nitrosodiethylamine	0.000002
N-Nitroso-*N*-methylurea	0.0000001
N-Nitrosopyrrolidine	0.0002
Pentachlorobenzene	0.03
Pentachloronitrobenzene (PCNB)	0.1
Pentachlorophenol	1.0
Phenol	1.0
Phenylmercury acetate	0.003
Phosphine	0.01
Polchlorinated biphenyls NOS	0.00005
Potassium cyanide	2.0
Potassium silver cyanide	7.0
Pronamide	3.0
Pyridine	0.04
Reserpine	0.00003
Selenourea	0.2

Table 10.4 Residue Concentration Limits for Nonmetals (Continued)

Constituent	Concentration limits for residues (mg/kg)
Silver cyanide	4.0
Sodium cyanide	1.0
Strychnine	0.01
1,2,4,5-Tetrachlorobenzene	0.01
1,1,2,2-Tetrachloroethane	0.002
Tetrachloroethylene	0.7
2,3,4,6-Tetrachlorophenol	0.01
Tetraethyl lead	0.000004
Thiourea	0.0002
Toluene	10
Toxaphene	0.005
1,1,2-Trichloroethane	0.006
Trichloroethylene	0.005
Trichloromonofluoromethane	10
2,4,5-Trichlorophenol	4.0
2,4,6-Trichlorophenol	4.0
Vanadium pentoxide	0.7
Vinyl chloride	0.002

individual constituents regardless of the hazardous waste in which the constituent is present.[151] The applicability of the standards is differentiated for non-wastewater categories such as contaminated soil and for wastewater categories, e.g., contaminated groundwater. Table 10.5 shows the proposed universal treatment standards.

In 1990, under EPA's proposed corrective action rule, known as the Subpart S proposal, the agency introduced corrective action standards called action levels for constituents in contaminated soil and groundwater. For carcinogens, the action level represents a concentration associated with an excess upper-bound lifetime cancer risk of one in 1,000,000 due to continuous constant lifetime exposure, and considers the overall weight of evidence for carcinogeneity.[152] The action level for systemic contaminants represents a concentration to which the human population could be exposed daily without considerable risk of damaging effects during a lifetime exposure. Table 10.6 shows a list of constituents showing action level source data.

The cleanup standards under CERCLA require that remediation actions comply with applicable or relevant and appropriate requirements (ARARs) (see discussion of ARARs in Chapter 3). For a site in which the groundwater is used as the source for drinking water, for example, the MCLs established under the SDWA represent the ARAR for that site.

On September 28, 1993, EPA drafted a new initiative called the soil screening levels (SSLs), which was formerly known as soil trigger levels. An SSL represents a quantity of contaminant in the soil that indicates a level of

Table 10.5 Universal Treatment Standards, 58 FR 48092

Constituent	Nonwastewater concentration total composition (mg/kg)	Wastewater concentration total composition (mg/l)
Acenaphthalene	3.4	0.059
Acenaphthene	3.4	0.059
Acetone	160	0.28
Acetonitrile		0.17
Acetophenone	9.7	0.010
2-Acetylaminofluorene	140	0.059
Acrolein		0.29
Acrylonitrile	84	0.24
Aldrin	0.066	0.021
4-Aminobiphenyl		0.13
Aniline	14	0.81
Anthracene	3.4	0.059
Antimony	2.1	1.9
Aramite		0.36
Aroclor 1016	0.92	0.013
Aroclor 1221	0.92	0.014
Aroclor 1232	0.92	0.013
Aroclor 1242	0.92	0.017
Aroclor 1248	0.92	0.013
Aroclor 1254	1.8	0.014
Aroclor 1260	1.8	0.014
Arsenic	5.0	1.4
Barium	7.6	1.2
Benzal chloride	6.0	0.055
Benzene	10	0.14
Benz(a)anthracene	3.4	0.059
Benzo(a)pyrene	3.4	0.061
Benzo(b)fluorantene	6.8	0.11
Benzo(k)fluorantene	6.8	0.11
Benzo(g,h,i)perylene	1.8	0.0055
Beryllium	0.014	0.82
α-BHC	0.066	0.00014
β-BHC	0.066	0.00014
δ-BHC	0.066	0.023
γ-BHC	0.066	0.0017
Bromodichloromethane	15	0.35
Bromomethane	15	0.11
4-Bromophenyl phenyl ether	15	0.055
n-Butanol	2.6	
n-Butyl alcohol		5.6
Butylbenzylphthalate	28	0.017
2-sec-Butyl-4,6-dinitrophenol	2.5	0.066

Table 10.5 Universal Treatment Standards, 58 FR 48092 (Continued)

Constituent	Nonwastewater concentration total composition (mg/kg)	Wastewater concentration total composition (mg/l)
Cadmium	0.19	0.20
Carbon disulfide	4.81	0.014
Carbon tetrachloride	6.0	0.057
Chlordane	0.26	0.0033
p-Chloroaniline	16	0.46
Chlorobenzene	6.0	0.057
Chlorobenzilate		0.10
2-Chloro-1,3-butadiene		0.057
Chlorodibromomethane	15	0.057
Chloroethane	6.0	0.27
bis-(2-Chloroethoxy)methane	7.2	0.036
bis-(2-Chloroethyel)ether	6.0	0.033
bis-(2-Chloroisopropyl)ether	7.2	0.055
2-Chloroethylvinylether		0.062
p-Chloro-m-cresol	14	0.018
Chloroform	6.0	0.046
Chloromethane	30	0.19
2-Chloronaphthalene	5.6	0.055
2-Chlorophenol	5.7	0.044
3-Chloropropene	30	0.036
Chromium (total)	0.33	0.37
Chrysene	3.4	0.059
Cresol(m- and p-)	3.2	0.77
o-Cresol	5.6	0.11
Cyclohexanone	0.75	0.36
Dibenz(a,e)pyrene		0.061
Dibenz(a,h)-anthracene	8.2	0.055
1,2-Dibromo-3-chloropropane	15	0.11
1,2-Dibromoethane	15	0.028
Dibromomethane	15	0.11
tris-(2,3-Dibromopropyl) phosphate		0.11
m-Dichlorobenzene	6.0	0.036
o-Dichlorobenzene	6.0	0.088
p-Dichlorobenzene	6.0	0.090
1,1-Dichloroethane	6.0	0.059
1,2-Dichloroethane	6.0	0.21
2,4-Dichlorophenol	14	0.044
2,6-Dichlorophenol	14	0.044
2,4-Dichlorophenoxyacetic acid	10	0.72
o,p'-DDD	0.087	0.023
p,p'-DDD	0.087	0.023
o,p'-DDE	0.087	0.031

Table 10.5 Universal Treatment Standards, 58 FR 48092 (Continued)

Constituent	Nonwastewater concentration total composition (mg/kg)	Wastewater concentration total composition (mg/l)
p,p′-DDE	0.087	0.031
o,p′-DDT	0.087	0.0039
p,p′-DDT	0.087	0.0039
Dichlorodifluoromethane	7.2	0.23
1,1-Dichloroethylene	6.0	0.025
trans-1,2-Dichloroethylene	30	0.054
1,2-Dichloropropane	18	0.85
cis-1,3-Dichloropropene	18	0.036
trans-1,3-Dichloropropene	18	0.036
Dieldrin	0.13	0.017
Diethyl phthalate	28	0.20
p-Dimethylaminoazobenzene		0.13
2,4-Dimethyl phenol	14	0.036
Dimethyl phthalate	28	0.047
1,4-Dinitrobenzene	2.3	0.32
4,6-Dinitrocresol	160	0.28
2,4-Dinitrophenol	160	0.12
2,4-Dinitrotoluene	140	0.32
2,6-Dinitrotoluene	28	0.55
Di-*n*-butyl phthalate	28	0.057
Di-*n*-octyl phthalate	28	0.017
Di-*n*-propylnitrosoamine	14	0.40
1,4-Dioxane	170	0.12
Diphenylamine	13	0.92
1,2-Diphenyl hydrazine		0.087
Diphenylnitrosamine	13	0.92
Disulfoton	6.2	0.017
Endosulfan I	0.066	0.023
Endosulfan II	0.13	0.029
Endosulfan sulfate	0.13	0.029
Endrin	0.13	0.0028
Endrin aldehyde	0.13	0.025
Ethyl acetate	33	0.34
Ethyl benzene	10	0.057
Ethyl ether	160	0.12
bis-(2-Ethylhexyl)phthalate		0.28
Ethyl methacrylate	160	0.14
Ethylene oxide		0.12
Famphur	15	0.017
Fluoranthene	3.4	0.068
Fluorene	3.4	0.059
Heptachlor	0.066	0.0012

Table 10.5 Universal Treatment Standards, 58 FR 48092 (Continued)

Constituent	Nonwastewater concentration total composition (mg/kg)	Wastewater concentration total composition (mg/l)
Heptachlor epoxide	0.066	0.016
Hexachlorobenzene	10	0.055
Hexachlorobutadiene	5.6	0.055
Hexachlorocyclopentadiene	2.4	
Hexachlorodibenzo-furans	0.001	0.000063
Hexachlorodibenzo-p-dioxins	0.001	0.000063
Hexachloroethane	30	0.055
Hexachloropropene	30	0.035
Ideno(1,2,3-cd)pyrene	3.4	0.0055
Iodomethane	65	0.19
Isobutanol	170	
Isobutyl alcohol		5.6
Isodrin	0.066	0.021
Isosafrole	2.6	0.081
Kepone	0.13	0.0011
Lead	0.37	0.28
Mercury	0.009	0.15
Methacrylonitrile	84	0.24
Methanol	0.75	5.6
Methapyrilene	1.5	0.081
Methoxychlor	0.18	0.25
3-Methylchloanthrene	15	0.0055
4,4-Methylene-bis-(2-chloro aniline)	30	0.50
Methylene chloride	30	0.089
Methyl ethyl ketone	36	0.28
Methyl isobutyl ketone	33	0.14
Methyl metacrylate	160	0.14
Methyl methansulfonate		0.018
Methyl parathion	4.6	0.014
Naphthalene	5.6	0.059
2-Naphthylamine		0.52
Nickel	5.0	0.55
o-Nitroaniline	14	
p-Nitroaniline	28	0.028
Nitrobenzene	14	0.068
5-Nitro-o-toluidine	28	0.32
o-Nitrophenol	13	
p-Nitrophenol	29	0.12
N-Nitrosodiethylamine	28	0.40
N-Nitrododimethylamine		0.40
N-Nitroso-di-n-butylamine	17	0.40
N-Nitrosomethylethylamine	2.3	0.40

Table 10.5 Universal Treatment Standards, 58 FR 48092 (Continued)

Constituent	Nonwastewater concentration total composition (mg/kg)	Wastewater concentration total composition (mg/l)
N-Nitrosomorpholine	2.3	0.40
N-Nitrosopipendine	35	0.013
N-Nitrosopyrrolidine	35	0.013
Parathion	4.6	0.014
Pentachlorobenzene	10	0.055
Pentachlorodibenzo-furans	0.001	0.000035
Pentachlorodibenzo-p-dioxins	0.001	0.000063
Pentachloroethane	6	
Pentachloronitrobenzene	4.8	0.055
Pentachlorophenol	7.4	0.089
Phenacetin	16	0.081
Phenanthrene	5.6	0.059
Phenol	6.2	0.039
Phorate	4.6	0.021
Phthalic anhydride	28	0.055
Propanenitrile	360	0.24
Pronamide	1.5	0.093
Pyrene	8.2	0.067
Pyridine	16	0.014
Safrole	22	0.081
Selenium	0.16	0.82
Silver	0.30	0.29
Silvex (2,4,5-TP)	7.9	0.72
2,4,5-T	7.9	0.72
1,2,4,5-Tetrachlorobenzene	14	0.055
Tetrachlorodibenzo-furans	0.001	0.000063
Tetrachlorodibenzo-p-dioxins	0.001	0.000063
1,1,1,2-Tetrachloroethane	6.0	0.057
1,1,2,2-Tetrachloroethane	6.0	0.057
Tetrachloroethene		0.056
Tetrachloroethylene	6.0	
2,3,4,6-Tetrachlorophenol	7.4	0.030
Thallium	0.078	1.4
Toluene	10	0.080
Toxaphene	2.6	0.0095
Tribromomethane		0.63
1,2,4-Trichlorobenzene	19	0.055
1,1,1-Trichloroethane	6.0	0.054
1,1,2-Trichloroethane	6.0	0.054
Trichloroethene		0.054
Trichloroethylene	6.0	
Trichloromonofluoromethane		0.020

Table 10.5 Universal Treatment Standards, 58 FR 48092 (Continued)

Constituent	Nonwastewater concentration total composition (mg/kg)	Wastewater concentration total composition (mg/l)
2,4,5-Trichlorophenol	7.4	0.18
2,4,6-Trichlorophenol	7.4	0.035
1,2,3-Trichloropropane	30	0.85
1,1,2-Trichloro-1,2,2-trifluoroethane	30	0.057
Vanadium	0.23	0.042
Vinyl chloride	6.0	0.27
Xylene(s)	30	0.32
Zinc	5.3	1.0
Total PCBs	10	0.1

concentration above which there is a need to perform a site-specific study.[153] The need for further study, however, does not mean that the contamination level of the site automatically triggers a response action. If the contaminant concentration falls below the SSL and there are no major ecological receptors of concern, no further study is required under the CERCLA program.

The proposed SSLs are established using residential land-use exposure assumptions. The pathways of exposure to the contaminants include soil ingestion, inhalation of volatiles and fugitive dusts, and ingestion of contaminated groundwater caused by migration of contaminants through soil to an underlying potable aquifer. SSLs are calculated based on a 10^{-6} risk for carcinogens and a hazard quotient of 1 for noncarcinogens. To protect groundwater, SSLs are based on nonzero MCLGs and if the numeric MCLGs are not available, MCLs are used. If the values of MCLs are not available, the risk-based targets are used.

Among the substances regulated under TSCA, a major concern is polychlorinated biphenyl (PCB) because of its widespread use and adverse health effects. Under the statute, the intentional disposal or accidental release of 50 ppmw or more is considered improper and requires a site cleanup. The cleanup standards, however, vary among EPA regions. For some EPA regional offices the cleanup level is 50 ppmw, while other regions require cleanups to background levels or the limit of detection of PCB.

4. HUMAN HEALTH RISK STANDARD

The human health cleanup standards are based on levels of acceptable risks to human health. Although the risk is directly associated with human health, it is often established by giving substantial consideration to the ecological risk factor. As discussed in Chapter 9, the acceptability of risk is a management decision which is influenced by complex socio-political aspirations.

Table 10.6 List of Constituents Showing Action Level Source Data, 55 FR 30798

Constituent	Noncarcinogenic effects		Carcinogenic effects	
	Oral RfD (mg/kg/d)	Inhalation RfD (mg/kg/d)	Oral SCF (inverse mg/kg/d)	Inhalation SCF (inverse mg/kg/d)
Acetone	0.1			
Acetonitrile	0.006			
Acetophenone	0.1	0.00005		
Acrylamide	0.0002		4.5	4.5
Acrylonitrile			0.54	0.24
Aldicarb	0.0013			
Aldrin	0.00003		17	17
Allyl alcohol	0.005			
Aluminum phosphide	0.0004			
Aniline			0.0057	
Antimony	0.0004			
Arsenic	0.001			50
Asbestos				0.23
Barium cyanide	0.07			
Barium ionic	0.05	0.0001		
Benzidine	0.003		230	230
Beryllium	0.005		4.3	8.4
bis(2-ethylhexyl) phthalate	0.02		0.014	
bis(chloroethyl)ether			1.1	1.1
Bromodichloro-methane	0.02		1.3	
Bromoform	0.02			
Bromomethane	0.0014	0.008		
Butyl benzyl phthalate	0.2			
Cadmium	0.0005			6.1
Calcium cyanide	0.04			
Carbon disulfide	0.1			
Carbon tetrachloride	0.0007		0.13	0.13
Chloral	0.002			
Chlordane	0.00006		1.3	1.3
Chlorine cyanide	0.05			
Chlorobenzene	0.02	0.005		
Chloroform	0.01		0.0061	0.081
2-Chlorophenol	0.005			
Chromium (VI)	0.005			41
Copper cyanide	0.005			
m-Cresol	0.05			
o-Cresol	0.05			
p-Cresol	0.05			
Cyanide	0.02			

Table 10.6 List of Constituents Showing Action Level Source Data, 55 FR 30798 (Continued)

Constituent	Noncarcinogenic effects		Carcinogenic effects	
	Oral RfD (mg/kg/d)	Inhalation RfD (mg/kg/d)	Oral SCF (inverse mg/kg/d)	Inhalation SCF (inverse mg/kg/d)
Cyanogen	0.04			
Cyanogen bromide	0.09			
DDD			0.24	
DDE			0.34	
DDT	0.0005		0.34	0.34
Dibutylphthalate	0.1			
Dibutylnitrosamine			5.4	5.4
3,3′-Dichloro-benzidine			0.45	
Dichlorodifluoro-methane	0.2	0.05		
1,2-Dichloroethane			0.091	0.091
1,1-Dichloroethylene	0.009		0.6	1.2
2,4-Dichlorophenol	0.003			
2,4-Dichlorophen-oxyacetic acid	0.01			
1,3-Dichloropropene	0.0003			
Dieldrin	0.00005		16	16
Diethylphthalate	0.8			
Diethylnitrosamine			150	150
Dimethoate	0.02			
Dimethylnitrosamine			51	51
m-Dinitrobenzene	0.0001			
2,4-Dinitrophenol	0.002			
2,3-Dinitrotoluene (and 2,6-,mixture)			0.68	
1,4-Dioxane			0.011	
Diphenylamine	0.025			
1,2-Diphenyl-hydrazine			0.8	0.8
Disulfoton	0.00004			
Endosulfan	0.00005			
Endothall	0.02			
Endrin	0.0003			
Epichlorohydrin	0.002		0.0099	0.0042
Ethylbenzene	0.1			
Ethylene dibromide			85	0.76
Formaldehyde				0.045
Formic acid	2.0			
Glycidealdehyde	0.0004			
Heptachlor	0.0005		4.5	4.5
Heptachlor epoxide	0.000013		9.1	9.1

**Table 10.6 List of Constituents Showing Action Level
Source Data, 55 FR 30798 (Continued)**

	Noncarcinogenic effects		Carcinogenic effects	
Constituent	Oral RfD (mg/kg/d)	Inhalation RfD (mg/kg/d)	Oral SCF (inverse mg/kg/d)	Inhalation SCF (inverse mg/kg/d)
Hexachlordibenzo-*p*-dioxin			6200	6200
Hexachlorobutadiene	0.002		0.078	0.078
α-Hexachlorocyclo-hexane			6.3	6.3
β-Hexachlorocyclo-hexane			1.8	1.8
Hexachlorocyclopenta-diene	0.007	0.00002		
Hexachloroethane	0.001		0.014	0.014
Hexachlorophene	0.0003			
Hydrazine			3.0	17
Hydrogen cyanide	0.02			
Hydrogen sulfide	0.003			
Isobutyl alcohol	0.3			
Isophorone	0.2		0.0041	
Lindane	0.0003		1.3	
m-Phenylenediamine	0.006			
Maleic anhydride	0.1			
Maleic hydrazine	0.5			
Mercury (inorganic)	0.0003			
Metacrylonitrile	0.0001	0.0002		
Methomyl	0.025			
Methyl ethyl ketone	0.05	0.09		
Methyl isobutyl ketone	0.05	0.02		
Methyl parathion	0.00025			
Methylene chloride	0.06		0.0075	0.014
N-Nitroso-di-*n*-butylamine			5.4	5.4
N-Nitroso-*n*-methyl ethylamine			22	
N-Nitrosodi-*n*-propylamine			7.0	
N-Nitrosodiethano-lamine			2.8	
N-Nitrosodienylamine			0.0049	
N-Nitrosopyrrolidine			2.1	2.1
Nickel	0.02			
Nickel refinery dust				0.64
Nitric oxide	0.1			

**Table 10.6 List of Constituents Showing Action Level
Source Data, 55 FR 30798 (Continued)**

Constituent	Noncarcinogenic effects		Carcinogenic effects	
	Oral RfD (mg/kg/d)	Inhalation RfD (mg/kg/d)	Oral SCF (inverse mg/kg/d)	Inhalation SCF (inverse mg/kg/d)
Nitrobenzene	0.0005	0.0006		
Nitrogen dioxide	1.0			
Osmium tetroxide	0.00001			
Parathion	0.006			
Pentachlorobenzene	0.0008			
Pentachloronitro-benzene	0.003			0.25
Pentachlorophenol	0.03			
Phenol	0.6			
Phenyl mercuric acetate	0.00006			
Phosphine	0.0003			
Phthalic anhydride	2.0			
PCB			7.7	
Potassium cyanide	0.05			
Potassium silver cyanide	0.2			
Pronamide	0.075			
Pyridine	0.001			
Selenious acid	0.003			
Selenoures	0.005			
Silver	0.003			
Silver cyanide	0.1			
Sodium cyanide	0.04			
Strychnine	0.0003			
Styrene	0.2			
1,1,1,2-Tetrachlorethane	0.03		0.026	0.026
1,2,4,5-Tetrachloro-benzene	0.0003			
1,1,2,2-Tetrachloro-ethane	0.2			0.2
Tetrachloroethylene	0.01		0.051	0.0033
2,3,4,6-Tetrachloro-phenol	0.03			
Tetraethyl lead	0.0000001			
Tetraethyldithiopyro-phosphate	0.0005			
Thallic oxide	0.00007			
Thallium acetate	0.00009			
Thallium carbonate	0.00008			
Thallium chloride	0.00008			
Thallium nitrate	0.00009			
Thallium sulfate	0.00008			

Table 10.6 List of Constituents Showing Action Level
Source Data, 55 FR 30798 (Continued)

Constituent	Noncarcinogenic effects		Carcinogenic effects	
	Oral RfD (mg/kg/d)	Inhalation RfD (mg/kg/d)	Oral SCF (inverse mg/kg/d)	Inhalation SCF (inverse mg/kg/d)
Thiosemicarbazide	0.006			
Thiram	0.005			
Toluene	0.3	2.0		
Toxaphene			1.1	1.1
1,2,4-Trichloro-benzene	0.02	0.003		
1,1,1-Trichloroethane	0.09	0.3		
1,1,2-Trichloroethane	0.004		0.057	0.057
Trichloroethylene			0.011	
Trichloromono-fluoromethane	0.3	0.2		
2,4,5-Trichlorophenol	0.1			
2,4,6-Trichlorophenol			0.02	0.02
2,4,5-Trichloro-phenoxyacetic acid	0.02			
1,2,3-Trichloropropane	0.006			
Vanadium pentoxide	0.009			
Xylenes	2.0	0.3		
Zinc cyanide	0.05			
Zinc phosphide	0.0003			

There are two types of health-based cleanup standards: the generic type and the site-specific health risk standards. Both standards are calculated from the risk assessment approach. The two components of the methodology are the exposure assessment and the toxicity evaluation.

The degree of human exposure to a contaminant can be expressed in terms of chemical intake. The exposure scenario affects the chemical intake. At a given site, there may be several exposure scenarios that are applicable. The two most common scenarios are residential and industrial scenarios. Within each pertinent scenario, exposure pathways can be identified. The pathways indicate, for example, the sources of contamination, the routes, and the mechanisms of release. The amount of chemical intake in an environmental medium is calculated from the contaminant concentrations in that medium over the exposure period, the contact rate, frequency of exposure, duration of exposure, body weight, and averaging time.

The toxicity of a contaminant is a function of its carcinogeneity and systemic toxicity. Carcinogeneity is indicated by the value of the cancer slope factor (CSF) and systemic toxicity is expressed in terms of the RfD. For each exposure scenario, the cancer human health risk is calculated as the chemical intake times the CSF and the noncancer risk is equal to the chemical intake multiplied by the RfD. In calculating the cancer risk, the slope factors between one exposure route (e.g., inhalation) and another route (e.g., ingestion) for the same chemical are not necessarily the same. The same statement holds true for the reference dose. For the same chemical, the values of the RfD between one exposure route and another are not necessarily the same.

In the generic human health risk standard the parameters of the risk equation are defaulted, using standardized assumptions that are commonly applicable to contaminated sites. The health-based risk is calculated strictly on the basis of toxicology. The calculation does not consider, for example, the limits of detection. Table 10.7 provides an example of the health-based risk standard used in Arizona's 1995 soil remediation interim rule. The standards, called the health-based guidance levels (HBGLs), are established for residential and nonresidential land uses. They are calculated only for ingestion of contaminants in soil.[154] For the upcoming permanent rule, the HBGLs will also take into consideration exposure from inhalation of contaminants.

The use of default parameters obviously does not reflect the unique characteristics of a site. A site-specific risk assessment can be utilized as an alternative to "cookbook" calculations based on default exposure values that result in stringent cleanup levels.[155] The use of a site-specific assessment is advantageous because it manipulates the accuracy of estimating the fate and transport behavior of the contaminants resulting in a remediation effort that sufficiently maintains an adequate level of protection.[156]

5. TECHNOLOGY-BASED CLEANUP

The technology-based standards are the achievable levels of cleanup from the utilization of available technologies in destroying or reducing contamination. The effectiveness of a technology to remediate a site is largely determined by the type of contaminant and the type of contaminated media. The most commonly used technologies for soil and groundwater remediation are

Type of contaminant	Media	Technologies
Metals	Soil	Solidification/stabilization
Organics	Groundwater	Carbon adsorption, oxidation
	Soil, groundwater	Biological treatment, chemical extraction, incineration.
Organics and metals	Soil	Soil washing, vitrification
	Soil, groundwater	Physical treatment, chemical treatment
Volatile organics	Soil	Soil vapor extraction, thermal desorption
	Groundwater	Air stripping

The next chapter describes in detail various technologies for site remediation. The technology screening process is examined in Chapter 12.

**Table 10.7 Arizona's Human Health-Based Guidance Levels (HBGLs)
for Ingestion of Contaminants in Soil (June 1995 Update)**

	Chemical	CAS number	Cancer group	Soil ingestion Residential oral HBGL (mg/kg)	Nonresidential oral HBGL (mg/kg)
A					
1.	Acenaphthene	83-32-9	ND	7000	24500
2.	*Acenaphthy-lene (PAH)	208-96-8	D	7000	24500
3.	*Acephate	30560-19-1	C	160	560
4.	Acetochlor	34256-82-1	NA	2300	8050
5.	*Acetone	67-64-1	D	12000	42000
6.	Acetonitrile	75-05-8	ND	700	2450
7.	Acetophenone	98-86-2	D	12000	42000
8.	Acifluorfen	62476-59-9	ND	1500	5250
9.	*Acrolein	107-02-8	C	2300	8050
10.	*Acrylamide	79-06-1	B2	0.3	1.3
11.	Acrylic acid	79-10-7	NA	58000	203000
12.	*Acrylonitrile	107-13-1	Bl	2.5	10.5
13.	*Alachlor	15972-60-8	B2	17	71
14.	Alar	1596-84-5	NA	18000	63000
15.	*Aldicarb	116-06-3	D	120	420
16.	*Aldicarb sulfone	1646-88-4	D	120	420
17.	*Aldicarb sulfoxide	1646-87-3	D	150	525
18.	*Aldrin	309-00-2	B2	0.08	0.34
19.	Allyl alcohol	107-18-6	NA	580	2030
20.	Allyl chloride	107-05-1	C	5800	20300
21.	Aluminum phosphide	20859-73-8	NA	47	165
22.	Amdro	67485-29-4	ND	35	123
23.	*Ametryn	834-12-8	D	1100	3850
24.	Aminopyridine	504-24-5	D	2.3	8.0
25.	Amitraz	33089-61-1	ND	290	1015
26.	Ammonia (NH$_3$)	7664-41-7	D	120000	420000
27.	*Ammonium sulfamate	7773-06-0	D	23000	80500
28.	Aniline	62-53-3	B2	240	1008
29.	*Athracene (PAH)	120-12-7	D	35000	122500
30.	*Antimony (Sb)	7440-36-0	D	47	165
31.	Apollo	74115-24-5	C	150	525
32.	Aramite	140-57-8	B2	54	227
33.	*Arsenic, inorganic (As)	7440-38-2	A	0.91	3.82
34.	Assure	76578-14-8	D	1100	3850
35.	*Asulam	3337-71-1	D	5800	20300

Table 10.7 Arizona's Human Health-Based Guidance Levels (HBGLs) for Ingestion of Contaminants in Soil (June 1995 Update) (Continued)

	Chemical	CAS number	Cancer group	Residential oral HBGL (mg/kg)	Nonresidential oral HBGL (mg/kg)
36.	*Atrazine	1912-24-9	C	6.1	21.4
37.	Avermectin b₁	65195-55-3	ND	47	165
38.	*Azinphos-methyl	86-50-0	E	290	1015
39.	Azobenzene	103-33-3	B2	12	50
B					
40.	*Barium (Ba)	7440-39-3	D	8200 ##	28700 ##
41.	Barium cyanide	542-62-1	ND	12000	42000
42.	Baythroid	68359-37-5	ND	2900	10150
43.	Benefin	1861-40-1	NA	35000	122500
44.	*Benomyl	17804-35-2	D	5800	20300
45.	Bentazon	25057-89-0	D	290	1015
46.	Benzaldehyde	100-52-7	NA	12000	42000
47.	*Benz(a)anthracene (PAH)	56-55-3	B2	1.1	4.6
48.	*Benzene (BNZ)	71-43-2	A	47	197
49.	*Benzidine	92-87-5	A	0.006	0.025
50.	*Benzo(a)pyrene (PAH) (BaP)	50-32-8	B2	0.19	0.80
51.	*Benzo(b)fluoranthene (PAH)	205-99-2	B2	1.1	4.6
52.	*Benzo(k)fluoranthene (PAH)	207-08-9	B2	1.1	4.6
53.	Benzoic acid	65-85-0	D	470000	1645000
54.	Benzotri-chloride	98-07-7	B2	0.1	0.4
55.	*Benzyl alcohol	100-51-6	ND	35000	122500
56.	Benzyl chloride	100-44-7	B2	8	34
57.	*Beryllium (Be)	7440-41-7	B2	0.32	1.34
58.	Bidrin	141-66-2	NA	12	42
59.	Biphenthrin	82657-04-3	NA	1800	6300
60.	1,1-Biphenyl	92-52-4	D	5800	20300
61.	*bis(2-chloroethyl) ether (BCEE)	111-44-4	B2	1.2	5.0
62.	*bis(2-chloroisopropyl) ether	39638-32-9	ND	19	67
63.	bis(chloromethyl) ether (CME)	542-88-1	A	0.006	0.025
64.	Bisphenol A	80-05-7	NA	5800	20300
65.	*Boron and borates only (B)	7440-42-8	D	11000	38500
66.	*Bromacil	314-40-9	C	1500	5250

Table 10.7 Arizona's Human Health-Based Guidance Levels (HBGLs) for Ingestion of Contaminants in Soil (June 1995 Update) (Continued)

			Soil ingestion	
			Residential	Nonresidential
	CAS	Cancer	oral HBGL	oral HBGL
Chemical	number	group	(mg/kg)	(mg/kg)
67. *Bromodi-chloroethane (THM) (BDCM)	75-27-4	B2	22	92
68. *Bromoform (THM) (BRFM)	75-25-2	B2	170	714
69. *Bromo-methane (BMM)	74-83-9	D	160	560
70. *Bromoxynil	1689-84-5	D	2300	8050
71. Bromoxynil octanoate	1689-99-2	NA	2300	8050
72. *N*-butanol	71-36-3	D	12000	42000
73. *Butyl benzyl phthalate	85-68-7	C	2300	8050
74. *Butylate	2008-41-5	D	5800	20300
75. Butylphthalyl butylglycolate	85-70-1	NA	120000	420000
C				
76. Cacodylic acid	75-60-5	D	350	1225
77. *Cadmium (Cd)	7440-43-9	B1	58	244
78. Calcium cyanide	592-01-8	ND	4700	16450
79. Caprolactam	105-60-2	NA	58000	203000
80. Captafol	2425-06-1	ND	160	560
81. *Captan	133-06-2	D	390	1365
82. *Carbaryl	63-25-2	D	12000	42000
83. *Carbofuran	1563-66-2	E	580	2030
84. *Carbon disulfide	75-15-0	D	12000	42000
85. *Carbon tetrachloride (CCL4)	56-23-5	B2	10	42
86. Carbosulfan	55285-14-8	ND	1200	4200
87. *Carboxin	5234-68-4	D	12000	42000
88. Chloral	75-87-6	NA	230	805
89. *Chloramben	133-90-4	D	1800	6300
90. *Chlordane	57-74-9	B2	1	4
91. *Chlordime-form	6164-98-3	B2	1.2	5.0
92. Chlorimuron-ethyl	90982-32-4	NA	2300	8050
93. Chlorine cyanide	506-77-4	ND	5800	20300
94. *p*-Chloroaniline	106-47-8	NA	470	1645

Table 10.7 Arizona's Human Health-Based Guidance Levels (HBGLs) for Ingestion of Contaminants in Soil (June 1995 Update) (Continued)

	Chemical	CAS number	Cancer group	Soil ingestion Residential oral HBGL (mg/kg)	Nonresidential oral HBGL (mg/kg)
95.	*Chlorobenzene(monochloro-benzene) (MCB)	108-90-7	D	2300	8050
96.	Chlorobenzilate	510-15-6	B2	5	21
97.	1-Chlorobutane	109-69-3	D	47000	164500
98.	*Chloroform (THM) (CLFM)	67-66-3	B2	220	924
99.	*Chloromethane (CM)	74-87-3	C	100	350
100.	β-Chloro-naphthalene	91-58-7	NA	9400	32900
101.	*2-Chloro-phenol	95-57-8	D	580	2030
102.	*Chlorothalonil	1897-45-6	B2	120	504
103.	*o-Chloro-toluene	95-49-8	D	2300	8050
104.	Chlorpropham	101-21-3	NA	23000	80500
105.	*Chlorpyrifos	2921-88-2	D	350	1225
106.	Chlorpyrifos-methyl	5598-13-0	NA	1200	4200
107.	*Chlorsulfuron	64902-72-3	D	5800	20300
108.	Chromium (III)	16065-83-1	NA	120000	420000
109.	Chromium (VI)	18540-29-9	A	580	2436
110.	Chromium (VI) (CrVI)	7440-47-3	A	580	2436
111.	*Chromium (Total) (Cr)	NA	D	1700 ##	5950 ##
112.	*Chrysene (PAH)	218-01-9	B2	110	462
113.	*Copper (Cu)	7440-50-8	D	4300 ##	15050 ##
114.	Copper cyanide	544-92-3	ND	580	2030
115.	*Cresols (total)	NA	D	5800	20300
116.	Crotonaldehyde	123-73-9	C	0.72	2.52
117.	Cumene	98-82-8	NA	4700	16450
118.	*Cyanazine	21725-46-2	D	1.6	5.6
119.	*Cyanide (Cn)	57-12-5	D	2300	8050
120.	Cyanogen	460-19-5	ND	4700	16450
121.	Cyanogen bromide	506-68-3	ND	11000	38500
122.	Cyclohexanone	108-94-1	NA	580000	2030000
123.	Cyclohexylamine	108-91-8	NA	23000	80500
124.	Cyhalothrin (Karate)	68085-85-8	ND	580	2030

Table 10.7 Arizona's Human Health-Based Guidance Levels (HBGLs) for Ingestion of Contaminants in Soil (June 1995 Update) (Continued)

	Chemical	CAS number	Cancer group	Soil ingestion Residential oral HBGL (mg/kg)	Nonresidential oral HBGL (mg/kg)
125.	Cypermethrin	52315-07-8	ND	1200	4200
126.	*Cyromazine	66215-27-8	D	880	3080
D					
127.	*2,4-D (2,4-dichlorophen-oxyace)	94-75-7	D	1200	4200
128.	*Dalapon	75-99-0	D	3500	12250
129.	Danitol	39515-41-8	ND	2900	10150
130.	*DCPA (dimethyl tetrachloroter)	1861-32-1	D	1200	4200
131.	*DDD (p,p'-dichlorodi-phenyldic) (DDD)	72-54-8	B2	5.7	23.9
132.	*DDE (p,p'-dichlorodi-phenyldic) (DDE)	72-55-9	B2	4	17
133.	*DDT (p,p'-dichlorodi-phenyltri) (DDT)	50-29-3	B2	4	17
134.	*DDT/DDD/DDE (total) (DDT)	NA	B2	4.0	17
135.	Decabromo-diphenyl ether	1163-19-5	C	1200	4200
136.	Demeton	8065-48-3	NA	4.7	16.5
137.	2,4-Diamino-toluene	95-80-7	MA	0.43	1.51
138.	*Diazinon	333-41-5	E	110	385
139.	*Dibenz(a,h) anthracene (PAH)	53-70-3	B2	0.11	0.46
140.	1,4-Dibromo-benzene	106-37-6	NA	1200	4200
141.	*Dibromo-chloromethane (THM) (DBCM)	124-48-1	C	16	56
142.	*1,2-Dibromo-3-chloro-propane (DBCP)	96-12-8	B2	0.97	4.07

Table 10.7 Arizona's Human Health-Based Guidance Levels (HBGLs) for Ingestion of Contaminants in Soil (June 1995 Update) (Continued)

	Chemical	CAS number	Cancer group	Residential oral HBGL (mg/kg)	Nonresidential oral HBGL (mg/kg)
				Soil ingestion	
143.	*Dibutyl phthalate	84-74-2	D	12000	42000
144.	*Dicamba	1918-00-9	D	3500	12250
145.	*Dichlobenil	1194-65-6	D	58	203
146.	*1,2-Dichloro-benzene (DCB2)	95-50-1	D	11000	38500
147.	*1,3-Dichloro-benzene (DCB3)	541-73-1	D	10000	35000
148.	*1,4-Dichloro-benzene (DCB4)	106-46-7	C	57	200
149.	*3,3'-Dichloro-benzidine	91-94-1	B2	3	13
150.	*Dichlorodi-fluoromethane (DCDFM)	75-71-8	D	23000	80500
151.	1,1-Dichloro-ethane (DCA)	75-34-3	C	1200	4200
152.	*1,2-Dichloro-ethane (DCA2)	107-06-2	B2	15	63
153.	*1,1-Dichloro-ethylene (DCE)	75-35-4	C	2.3	8.0
154.	1,2-Dichloro-ethylene (DCE2)	540-59-0	D	2300	8050
155.	1,2-Dichloro-ethylene (TOTAL)	NA	D	1200	4200
156.	*cis-1,2-Dichloro-ethylene	156-59-2	D	1200	4200
157.	*trans-1,2-Dichloro-ethylene	156-60-5	D	2300	8050
158.	*Dichloro-methane (DCM)	75-09-2	B2	180	756
159.	4-(2,4-Dichloro-phenoxy)butyric acid	94-82-6	NA	940	3290
160.	*2,4-Dichloro-phenol	120-83-2	D	350	1225
161.	*1,2-Dichloro-propane (DCP2)	78-87-5	B2	20	84
162.	2,3-Dichloro-propanol	616-23-9	ND	350	1225

Table 10.7 Arizona's Human Health-Based Guidance Levels (HBGLs) for Ingestion of Contaminants in Soil (June 1995 Update) (Continued)

			Soil ingestion	
Chemical	CAS number	Cancer group	Residential oral HBGL (mg/kg)	Nonresidential oral HBGL (mg/kg)
163. *1,3-Dichloro-propene	542-75-6	B2	7.6	31.9
164. Dichlorvos	62-73-7	B2	4.7	19.7
165. *Dicloran	99-30-9	E	2900	10150
166. *Dicofol	115-32-2	C	3.1*	13.0*
167. *Dieldrin	60-57-1	B2	0.09	0.38
168. *Diethyl phthalate	84-66-2	D	94000	329000
169. *Di(2-ethyl-hexyl) adipate	103-23-1	C	1100	3850
170. *Di(2-ethyl-hexyl) phthalate (DEHP)	117-81-7	B2	97	407
171. *Difenzoquat	43222-48-6	D	9400	32900
172. Diflubenzuron	35367-38-5	ND	2300	8050
173. *Diisopropyl methylphos-phonate (DIMP)	1445-75-6	D	9400	32900
174. Dimethipin	55290-64-7	C	230	805
175. *Dimethoate	60-51-5	D	23	81
176. Dimethyl phthalate	131-11-3	D	1200000	4200000
177. Dimethyl sulfate	77-78-1	B2	0.04	0.17
178. Dimethyl terephthalate	120-61-6	NA	12000	42000
179. *N,N*-Dimethy-laniline	121-69-7	NA	230	805
180. 1,2-Dimethyl-benzene (Xylene-*o*)	95-47-6	ND	230000	805000
181. 1,3-Dimethyl-benzene (Xylene-*m*)	108-38-3	ND	230000	805000
182. 1,4-Dimethyl-benzene (Xylene-*p*)	106-42-3	ND	230000	805000
183. 3,3-Dimethyl-benzidine	119-93-7	NA	0.15	0.53
184. *N,N*-Dimethyl-formamide	68-12-2	ND	12000	42000
185. 1,1-Dimethyl-hydrazine	57-14-7	NA	0.52	1.82

Table 10.7 Arizona's Human Health-Based Guidance Levels (HBGLs) for Ingestion of Contaminants in Soil (June 1995 Update) (Continued)

	Chemical	CAS number	Cancer group	Residential oral HBGL (mg/kg)	Nonresidential oral HBGL (mg/kg)
186.	2,4-Dimethyl-phenol	105-67-9	NA	2300	8050
187.	2,6-Dimethyl-phenol	576-26-1	ND	70	245
188.	3,4-Dimethyl-phenol	95-65-8	NA	120	420
189.	o-Dinitro-benzene	528-29-0	D	47	165
190.	m-Dinitro-benzene	99-65-0	D	12	42
191.	4,6-Dinitro-o-cyclohexyl phenol	131-89-5	NA	230	805
192.	*2,4-Dinitro-phenol	51-28-5	ND	230	805
193.	*2,4-Dinitro-toluene	121-14-2	B2	2	8
194.	2,6-Dinitro-toluene	606-20-2	ND	120	420
195.	*Dinoseb	88-85-7	D	120	420
196.	Dioctylphthalate	117-84-0	ND	2300	8050
197.	*1,4-Dioxane	123-91-1	B2	120	504
198.	*Diphenamid	957-51-7	D	3500	12250
199.	Diphenylamine	122-39-4	NA	2900	10150
200.	*1,2-Diphenyl-hydrazine	122-66-7	B2	1.7	7.1
201.	*Diquat dibromide	85-00-7	D	260	910
202.	Direct black 38	1937-37-7	NA	0.16	0.56
203.	Direct blue 6	2602-46-2	NA	0.17	0.60
204.	Direct brown 95	16071-86-6	NA	0.15	0.53
205.	*Disulfoton	298-04-4	E	4.7	16.5
206.	Dithiane	505-29-3	D	1200	4200
207.	*Diuron	330-54-1	D	230	805
208.	Dodine	2439-10-3	ND	470	1645
209.	DPX-M6316 (thifensulfuron met)	79277-27-3	ND	1500	5250
E					
210.	*Endosulfan	115-29-7	D	700	2450
211.	Endosulfan i	959-98-8	D	5.8	20.3
212.	*Endothall	145-73-3	D	2300	8050

Table 10.7 Arizona's Human Health-Based Guidance Levels (HBGLs) for Ingestion of Contaminants in Soil (June 1995 Update) (Continued)

	Chemical	CAS number	Cancer group	Soil ingestion Residential oral HBGL (mg/kg)	Nonresidential oral HBGL (mg/kg)
213.	*Endrin	72-20-8	D	35	123
214.	*Epichlorohydrin	108-89-8	B2	140	588
215.	*Ethephon	16672-87-0	D	580	2030
216.	*EPTC (s-ethyl dipropylthiocar)	759-94-4	D	2900	10150
217.	Ethion	563-12-2	ND	58	203
218.	2-Ethoxy-ethanol	110-80-5	NA	47000	164500
219.	Ethyl acetate	141-78-6	NA	110000	385000
220.	Ethyl acrylate	140-88-5	NA	28	98
221.	Ethyl ether	60-29-7	ND	23000	80500
222.	Ethyl methacrylate	97-63-2	NA	11000	38500
223.	Ethyl p-nitraphenyl phenylphospho-rothioat	2104-64-5	NA	1.2	4.2
224.	*Ethylbenzene (ETB)	100-41-4	D	12000	42000
225.	Ethylene diamine	107-15-3	D	2300	8050
226.	*Ethylene dibromide (EDB)	106-93-4	B2	0.02	0.08
227.	*Ethylene glycol	107-21-1	D	230000	805000
228.	*Ethylene thiourea (ETU)	96-45-7	B2	12	50
229.	Ethylphthalyl ethylglycolate	84-72-0	NA	350000	1225000
230.	*N-Ethyltoluene sulfonamide	26914-52-3	ND	290	1015
231.	Express	101200-48-0	NA	940	3290
F					
232.	*Fenamiphos	22224-92-6	D	29	102
233.	*Fenarimol	60168-88-9	E	7600	26600
234.	*Fenvalerate	51630-58-1	ND	2900	10150
235.	*Fluometuron	2164-17-2	D	1500	5250
236.	*Fluoranthene (PAH)	206-44-0	D	4700	16450
237.	*Fluorene (PAH)	86-73-7	D	4700	16450
238.	*Fluoride (F)	7782-41-4	D	7000	24500
239.	*Fluridone	59756-60-4	D	9400	32900
240.	Flurprimidol	56425-91-3	ND	2300	8050

Table 10.7 Arizona's Human Health-Based Guidance Levels (HBGLs) for Ingestion of Contaminants in Soil (June 1995 Update) (Continued)

			Soil ingestion	
Chemical	CAS number	Cancer group	Residential oral HBGL (mg/kg)	Nonresidential oral HBGL (mg/kg)
241. Flutolanil	66332-96-5	ND	7000	24500
242. *Fluvalinate	69409-94-5	D	1200	4200
243. Folpet	133-07-3	B2	390	1638
244. Fomesafen	72178-02-0	C	7.2	25.2
245. *Fonofos	944-22-9	D	230	805
246. Formaldehyde	50-00-0	B1	23000	96600
247. *Formetanate hydrochloride	23422-53-9	E	180	630
248. Formic acid	64-18-6	ND	230000	805000
249. *Fosetyl-al	39148-24-8	C	35000	122500
250. Furan	110-00-9	NA	120	420
251. Furfural	98-01-1	NA	350	1225
252. Furmecyclox	60568-05-0	B2	45	189
G				
253. Glufosinate-ammonium	77182-82-2	NA	47	165
254. Glycidaldehyde	765-34-4	B2	47	197
255. *Glyphosate	1071-83-6	D	12000	42000
H				
256. Haloxyfop-methyl	69806-40-2	ND	5.8	20.3
257. *Heptachlor	76-44-8	B2	0.3	1.3
258. *Heptachlor epoxide	1024-57-3	B2	0.15	0.63
259. Hexabromo-benzene	87-82-1	NA	230	805
260. *Hexachloro-benzene	118-74-1	B2	0.85	3.57
261. *Hexachloro-butadiene	87-68-3	C	17	60
262. *α-Hexachloro-cyclohexane (α-HCH)	319-84-6	B2	0.22	0.92
263. *β-Hexachloro-cyclohexane (β-HCH)	319-85-7	C	0.76[**]	3.19[**]
264. Technical-hexachloro-cyclohexa	608-73-1	B2	0.76	3.19
265. *Hexachloro-cyclopenta-diene (HCCPD)	77-47-4	D	820	2870

Table 10.7 Arizona's Human Health-Based Guidance Levels (HBGLs) for Ingestion of Contaminants in Soil (June 1995 Update) (Continued)

			Soil ingestion	
Chemical	CAS number	Cancer group	Residential oral HBGL (mg/kg)	Nonresidential oral HBGL (mg/kg)
266. Hexachloro-dibenzo-*p*-dioxin, mixture	19408-74-3	B2	0.0002	0.0008
267. *Hexachloro-ethane	67-72-1	C	97	340
268. Hexachloro-phene	70-30-4	NA	35	123
269. *n-Hexane	110-54-3	D	7000	24500
270. *Hexazinone	51235-04-2	D	3900	13650
271. Hiobencarb	28249-77-6	ND	1200	4200
272. *HMX (octohydro-1,3,5,7-tetranitro-1,3,5,7-tetrazocine)	2691-41-0	D	5800	20300
273. Hydrazine	302-01-2	B2	0.45	1.89
274. Hydrogen cyanide	74-90-8	ND	2300	8050
275. Hydrogen sulfide	7783-06-4	NA	350	1225
276. Hydroquinone	123-31-9	NA	4700	16450
I				
277. *Imazalil	35554-44-0	D	1500	5250
278. *Imazaquin	81335-37-7	D	29000	101500
279. *Indenopyrene (PAH)	193-39-5	B2	1.1	4.6
280. Iprodione	36734-19-7	ND	4700	16450
281. Isobutyl alcohol	78-83-1	NA	35000	122500
282. *Isophorone	78-59-1	C	1400	4900
283. Isopropalin	33820-53-0	ND	1800	6300
284. Isopropyl methyl phosphonic acid	1832-54-8	D	12000	42000
285. Isoxaben	82558-50-7	C	5800	20300
L				
286. Lactofen	77501-63-4	NA	230	805
287. *Lead and compounds (inorganic) (Pb)	7439-92-1	B2	400 ##	1400 ##
288. *Lindane (gamma-hexachlorocycl) (HCH)	58-89-9	C	1	4
289. *Linuron	330-55-2	C	23	81

Table 10.7 Arizona's Human Health-Based Guidance Levels (HBGLs) for Ingestion of Contaminants in Soil (June 1995 Update) (Continued)

| | | | Soil ingestion | |
| | | | Residential | Nonresidential |
Chemical	CAS number	Cancer group	oral HBGL (mg/kg)	oral HBGL (mg/kg)
290. Londax	83055-99-6	NA	23000	80500
M				
291. *Malathion	121-75-5	D	2300	8050
292. Maleic anhydride	108-31-6	NA	12000	42000
293. *Maleic hydrazide	123-33-1	D	58000	203000
294. *Mancozeb	8018-01-7	ND	3500	12250
295. *Maneb	12427-38-2	D	580	2030
296. *Manganese (Mn)	7439-96-5	D	580	2030
297. *MCPA (2-methyl-4-chlorophenox (MCPA)	94-74-6	D	58	203
298. *Mepiquat chloride	24307-26-4	D	3500	12250
299. *Mercuxy (inorganic) (Hg)	7439-97-6	D	35	123
300. Merphos	150-50-5	NA	3.5	12.3
301. *Merphos oxide	78-48-8	NA	3.5	12.3
302. *Metalaxyl	57837-19-1	D	7000	24500
303. Methacrylo-nitrile	126-98-7	NA	12	42
304. *Metha-midophos	10265-92-6	D	5.8	20.3
305. Methanol	67-56-1	ND	58000	203000
306. Methidathion	950-37-8	C	120	420
307. *Methiocarb	2032-65-7	E	150	525
308. *Methomyl	16752-77-5	D	2900	10150
309. *Methoxychlor	72-43-5	D	580	2030
310. 2-Methoxy-ethanol	109-86-4	NA	120	420
311. Methyl acetate	79-20-9	NA	120000	420000
312. 2-(2-Methyl-4-chlorophenoxy) propionic acid	93-65-2	NA	120	420
313. 4-(2-Methyl-4-chlorophenoxy) butyric acid	94-81-5	NA	1200	4200
314. *Methyl ethyl ketone (MEK)	78-93-3	D	70000	245000
315. Methyl isobutyl ketone	108-10-1	NA	9400	32900
316. Methyl mercury	22967-92-6	C	12	42

Table 10.7 Arizona's Human Health-Based Guidance Levels (HBGLs)
for Ingestion of Contaminants in Soil (June 1995 Update) (Continued)

	Chemical	CAS number	Cancer group	Soil ingestion Residential oral HBGL (mg/kg)	Nonresidential oral HBGL (mg/kg)
317.	Methyl methacrylate	80-62-6	NA	9400	32900
318.	*Methyl parathion	298-00-0	D	29	102
319.	*Methyl *tert* butyl ether (MTBE)	1634-04-4	D	580	2030
320.	4,4'-Methylene dianiline	101-77-9	NA	5.4	18.9
321.	4,4'-Methylene *bis* (*N,N'*-dimethyl) aniline	101-61-1	B2	30	126
322.	2-Methyl-lactanitrile	75-86-5	NA	8200	28700
323.	2-Methylphenol (*o*-Cresol)	95-48-7	C	580	2030
324.	3-Methylphenol (*m*-Cresol)	108-39-4	C	580	2030
325.	4-Methylphenol	106-44-5	C	580	2030
326.	*Metolachlor	512-8-45-2	C	1800	6300
327.	*Metribuzin	21087-64-9	D	2900	10150
328.	*Metsulfuron-methyl	74223-64-6	D	29000	101500
329.	Mirex	2385-85-5	ND	0.76	2.66
330.	Molinate	2212-67-1	ND	230	805
331.	*Molybdenum	7439-98-7	D	580	2030
332.	Monochlor-amine	10599-90-3	D	12000	42000
333.	*Mono-crotophos	6923-22-4	E	5.3	18.6
334.	Monomethyl-hydrazine	60-34-4	NA	1.2	4.2
335.	*Monuron	150-68-5	ND	82	287
336.	*Msma (monosodium methanearson)	2163-80-6	A	840 ##	3528 ##
337. N	*Myclobutanil	88671-89-0	ND	2900	10150
338.	*Naled	300-76-5	D	230	805
339.	*Naphthalene (PAH)	91-20-3	D	4700	16450
340.	*Napropamide	15299-99-7	ND	12000	42000
341.	*Nickel, soluble salts (Ni)	7440-02-0	D	2300	8050
342.	*Nitrate (No.$_3$)	14797-55-8	D	190000	665000

Table 10.7 Arizona's Human Health-Based Guidance Levels (HBGLs) for Ingestion of Contaminants in Soil (June 1995 Update) (Continued)

	Chemical	CAS number	Cancer group	Soil ingestion Residential oral HBGL (mg/kg)	Nonresidential oral HBGL (mg/kg)
343.	*Nitrate/nitrite (total)	NA	D	190000	665000
344.	Nitric oxide	10102-43-9	NA	12000	42000
345.	*Nitrite	14797-65-0	D	12000	42000
346.	2-Nitroaniline	88-74-4	NA	7	25
347.	*Nitrobenzene	98-95-3	D	58	203
348.	Nitrogen dioxide	10102-44-0	NA	120000	420000
349.	*Nitroguanidine	556-88-7	D	12000	42000
350.	N-Nitroso-di-n-butylamine	924-16-3	B2	0.25	1.05
351.	*N-Nitroso-di-n-propylamine	621-64-7	B2	0.19	0.80
352.	N-Nitroso-diethylamine	55-18-5	B2	0.009	0.038
353.	*N-Nitroso-dimethylamine	62-75-9	B2	0.03	0.13
354.	*N-Nitroso-diphenylamine	86-30-6	B2	280	1176
355.	N-Nitroso-N-ethylurea	759-73-9	B2	0.01	0.04
356.	N-Nitroso-N-methyl-ethylamine	10595-95-6	B2	0.06	0.25
357.	N-Nitro-sodiethano-lamine	1116-54-7	B2	0.49	2.06
358.	*N-Nitro-sopyrrolidine	930-55-2	B2	0.65	2.73
359.	*Norflurazon	27314-13-2	D	4700	16450
360.	NuStar	85509-19-9	NA	82	287
O					
361.	Octabromo-diphenyl ether	32536-52-0	D	350	1225
362.	Octamethyl-pyrophos-phoramide	152-16-9	NA	230	805
363.	*Oryzalin	19044-88-3	C	580	2030
364.	Oxadiazon	19666-30-9	NA	580	2030
365.	*Oxamyl	23135-22-0	E	2900	10150
366.	*Oxydemeton-methyl	301-12-2	D	58	203
367.	Oxyfluorfen	42874-03-3	ND	350	1225

Table 10.7 Arizona's Human Health-Based Guidance Levels (HBGLs) for Ingestion of Contaminants in Soil (June 1995 Update) (Continued)

			Soil ingestion	
Chemical	CAS number	Cancer group	Residential oral HBGL (mg/kg)	Nonresidential oral HBGL (mg/kg)
P				
368. Paclobutrazol	76738-62-0	NA	1500	5250
369. *Paraquat	1910-42-5	C	53	186
370. *Parathion	56-38-2	C	70	245
371. *Pendimethalin	40487-42-1	D	4700	16450
372. Pentabromodi-phenyl ether	32534-81-9	D	230	805
373. *Pentachloro-benzene	608-93-5	D	94	329
374. Pentachloro-nitrobenzene	82-68-8	NA	5.2	18.2
375. *Pentachloro-phenol	87-86-5	B2	11	46
376. *Permethrin	52645-53-1	D	5800	20300
377. Phenmedipham	13684-63-4	NA	29000	101500
378. *Phenol	108-95-2	D	70000	245000
379. *m*-Phenylene-diamine	108-45-2	NA	700	2450
380. Phenylmercuric acetate	62-38-4	ND	9.4	32.9
381. *Phorate	298-02-2	E	23	81
382. *Phosmet	732-11-6	D	2300	8050
383. *Phosphamidon	13171-21-6	D	20	70
384. Phosphine	7803-51-2	D	35	123
385. Phthalic anhydride	85-44-9	NA	230000	805000
386. *Picloram	1918-02-1	D	8200	28700
387. Pirimiphos-methyl	29232-93-7	ND	1200	4200
388. *Polychlorinated biphenyls (PCBs)	1336-36-3	B2	0.18	0.76
389. Polychlorinated biphenyl — ar	12674-11-2	ND	8.2	28.7
390. Potassium cyanide	151-50-8	NA	5800	20300
391. Potassium silver cyanide	506-61-6	ND	23000	80500
392. Prochloraz	67747-09-5	C	9.1	31.9
393. *Profenofos	41198-08-7	D	5.8	20.3
394. *Prafluralin	26399-36-0	ND	700	2450
395. *Prometon	1610-18-0	D	1800	6300

**Table 10.7 Arizona's Human Health-Based Guidance Levels (HBGLs)
for Ingestion of Contaminants in Soil (June 1995 Update) (Continued)**

			Soil ingestion	
			Residential	Nonresidential
	CAS	Cancer	oral HBGL	oral HBGL
Chemical	number	group	(mg/kg)	(mg/kg)
396. *Prometryn	7287-19-6	D	470	1645
397. *Pronamide	23950-58-5	C	880	3080
398. *Propachlor	1918-16-7	D	1500	5250
399. Propanil	709-98-8	ND	580	2030
400. *Propargite	2312-35-8	ND	2300	8050
401. Propargyl alcohol	107-19-7	NA	230	805
402. *Propazine	139-40-2	C	230	805
403. *Propham	122-42-9	D	2300	8050
404. *Propiconazole	60207-90-1	D	1500	5250
405. *Propoxur	114-26-1	C	47	165
406. Propylene glycol	57-55-6	ND	2300000	8050000
407. Propylene glycol mono-ethyl ether	52125-53-8	ND	82000	287000
408. Propylene glycol mono-methyl ether	107-98-2	NA	82000	287000
409. Propylene oxide	75-56-9	B2	5.7	23.9
410. Pursuit	81335-77-5	NA	29000	101500
411. *Pyrene (PAH)	129-00-0	D	3500	12250
412. Pyridine	110-86-1	NA	120	420
Q				
413. Quinalphos	13593-03-8	NA	58	203
414. Quinoline	91-22-5	NA	0.11	0.39
R				
415. *RDX (hexahydro-1,3,5-trinitro)	121-82-4	C	12	42
416. Resmethrin	10453-86-8	NA	3500	12250
417. Ronnel	299-84-3	NA	5800	20300
418. Rotenone	83-79-4	NA	470	1645
S				
419. Savey	78587-05-0	NA	2900	10150
420. Selenious acid	7783-00-8	D	580	2030
421. *Selenium and compounds (Se)	7782-49-2	D	580 ##	2030 ##
422. Selenourea	630-10-4	ND	580	2030
423. *Sethoxydim	74051-80-2	D	11000	38500
424. *Silver (Ag)	7440-22-4	D	580	2030
425. Silver cyanide	506-64-9	ND	12000	42000
426. *Simazine	122-34-9	C	11	39
427. Sodium azide	26628-22-8	ND	470	1645
428. Sodium cyanide	143-33-9	NA	4700	16450

**Table 10.7 Arizona's Human Health-Based Guidance Levels (HBGLs)
for Ingestion of Contaminants in Soil (June 1995 Update) (Continued)**

	Chemical	CAS number	Cancer group	Soil ingestion Residential oral HBGL (mg/kg)	Nonresidential oral HBGL (mg/kg)
429.	Sodium diethyldithio-carbamate	148-18-5	NA	5	18
430.	Sodium fluoroacetate	62-74-8	ND	2.3	8.0
431.	*Strontium	7440-24-6	D	70000	245000
432.	Strychnine	57-24-9	ND	35	123
433.	*Styrene	100-42-5	C	2300	8050
434.	*Sulfate (SO$_4$)	14808-79-8	D	6700000 ##	23450000 ##
435.	*Sulprofos	35400-43-2	E	290	1015
T					
436.	*2,3,7,8-TCDD (TCDD)	1746-01-6	B2	0.000009	0.000038
437.	*Tebuthiuron	34014-18-1	D	8200	28700
438.	*Terbacil	5902-51-2	E	1500	5250
439.	*Terbufos	13071-79-9	D	2.9	10.2
440.	*Terbutryn	886-50-0	ND	120	420
441.	*1,2,4,5-Tetrachloro-benzene	95-94-3	D	35	123
442.	*1,1,1,2-Tetrachloro-ethane	630-20-6	C	52	182
443.	*1,1,2,2-Tetrachloro-ethane (TET)	79-34-5	C	6.8 ##	28.6 ##
444.	*Tetrachloro-ethylene (PCE)	127-18-4	B2	27	113
445.	2,3,4,6-Tetrachloro-phenol	58-90-2	ND	3500	12250
446.	Tetrachloro-vinphos	961-11-5	NA	57	200
447.	*Tetraethyl lead	78-00-2	D	0.01	0.04
448.	Tetraethyl-dithiopyro-phosphate	3689-24-5	ND	58	203
449.	Thallic-oxide	1314-32-5	D	8.2	28.7
450.	*Thallium (Tl)	7440-28-0	ND	8.2	28.7
451.	Thallium acetate	563-68-8	D	11	39
452.	Thallium carbonate	6533-73-9	D	9.4	32.9
453.	Thallium chloride	7791-12-0	D	9.4	32.9

**Table 10.7 Arizona's Human Health-Based Guidance Levels (HBGLs)
for Ingestion of Contaminants in Soil (June 1995 Update) (Continued)**

			Soil ingestion	
Chemical	CAS number	Cancer group	Residential oral HBGL (mg/kg)	Nonresidential oral HBGL (mg/kg)
454. Thallium nitrate	10102-45-1	D	11	39
455. Thallium selenite	12039-52-0	D	11	39
456. Thallium sulfate	7446-18-6	D	9.4	32.9
457. Thiofanox	39196-18-4	NA	35	123
458. *Thiophanate-methyl	23564-05-8	D	9400	32900
459. Thiophenol	108-98-5	NA	1.2	4.2
460. *Thiram	137-26-8	D	580	2030
461. Tin (Sn)	NA	ND	70000	245000
462. *Toluene (TOL)	108-88-3	D	23000	80500
463. Total petroleum hydrocarbons	NA	ND	7000^{su}	24500^{su}
464. *Toxaphene	8001-35-2	B2	1.2	5.0
465. Tralomethrin	66841-25-6	ND	880	3080
466. *Triadimefon	43121-43-3	D	3500	12250
467. Triallate	2303-17-5	ND	1500	5250
468. Triasulfuron	82097-50-5	NA	1200	4200
469. 1,2,4-Tribromo-benzene	615-54-3	ND	580	2030
470. Tributyltin oxide	56-35-9	ND	3.5	12.3
471. *Trichlorfon	52-68-6	C	150	525
472. *1,2,4-Trichloro-benzene	120-82-1	D	1200	4200
473. *1,1,1-Trichloro-ethane (TCA)	71-55-6	D	11000	38500
474. *1,1,2-Trichloro-ethane (TCA2)	79-00-5	C	24	84
475. *Trichloro-ethylene (TCE)	79-01-6	B2	120	504
476. *Trichloro-fluoromethane (TCFM)	75-69-4	D	35000	122500
477. *2,4,5-Trichloro-phenol	95-95-4	D	12000	42000
478. *2,4,6-Trichloro-phenol	88-06-2	B2	120	504
479. *2,4,5-T (2,4,5-trichloropheno propionic acid)	93-76-5	D	1200	4200
480. *2,4 5-TP (2,4,5-Trichloro-phenoxy)	93-72-1	D	940	3290

Table 10.7 Arizona's Human Health-Based Guidance Levels (HBGLs) for Ingestion of Contaminants in Soil (June 1995 Update) (Continued)

	Chemical	CAS number	Cancer group	Soil ingestion Residential oral HBGL (mg/kg)	Nonresidential oral HBGL (mg/kg)
481.	1,1,2-Trichloro-propane	598-77-6	ND	580	2030
482.	*1,2,3-Trichloro-propane	96-18-4	D	0.19	0.67
483.	*Trichloro-trifluoroethane (F113)	76-13-1	D	3500000	12250000
484.	*Triclopyr	55335-06-3	E	290	1015
485.	Tridiphane	58138-08-2	ND	350	1225
486.	*Trifluralin	1582-09-8	C	180	630
487.	*Triforine	26644-46-2	D	2900	10150
488.	1,3,5-Trinitro-benzene	99-35-4	NA	5.8	20.3
489.	*2,4,6-Tri-nitrotoluene (TNT)	118-96-7	C	45	158
U					
490.	Uranium (U)	7440-61-1	A	350 ##	1225 ##
V					
491.	*Vanadium (V)	7440-62-2	D	820	2870
492.	Vanadium pentoxide	1314-62-1	NA	1100	3850
493.	*Vernolate	1929-77-7	ND	120	420
494.	*Vinclazolin	50471-44-8	D	2900	10150
495.	Vinyl acetate	108-05-4	NA	120000	420000
496.	*Vinyl chloride (VC)	75-01-4	A	0.72	3.02
W					
497.	Warfarin	81-81-2	NA	35	123
498.	White phosphorus	7723-14-0	D	2.3	8.0
X					
499.	*Xylenes (total) (XYL)	1330-20-7	D	230000	805000
Z					
500.	*Zinc and compounds (Zn)	7440-66-6	D	35000	122500
501.	Zinc cyanide	557-21-1	ND	5800	20300
502.	Zinc phosphide	1314-84-7	NA	35	123
503.	*Zineb	12122-67-7	D	5800	20300

Source: Arizona Department of Environmental Quality, Notice of Proposed Interim Rulemaking, Phoenix, AZ, 1995, Appendix A.

PART III
REFERENCES

1. U.S. Environmental Protection Agency, *Test Methods for Evaluating Solid Waste — SW 846*, Washington, D.C., 1989.
2. LaGrega, Michael D., Buckingham, Phillip L., and Evans, Jeffrey C., *Hazardous Waste Management*, McGraw-Hill, New York, 1994.
3. Metcalfe, H. Clark, Williams, John E., and Castka, Joseph F., *Modern Chemistry*, Holt, Rinehart and Winston, New York, 1970.
4. U.S. Department of Health and Human Services, *NIOSH Pocket Guide to Chemical Hazards*, Washington, D.C., 1994.
5. Sara, Martin N., *Standard Handbook for Solid and Hazardous Waste Facility Assessments*, Lewis Publishers, Boca Raton, FL, 1993.
6. Werner, Karl and Michalski, Andrew, DNAPL-Contaminated Groundwater, *Industrial Wastewater*, Vol. 3, No. 5, 1995, 47–53.
7. Vance, David B., Physical Properties of DNAPLs and Groundwater Contamination, *The National Environmental Journal*, Vol. 5, Issue 2, 1995, 23–24.
8. Nickerson, Charles A. and Nickerson, Ingeborg A., *Statistical Analysis for Decision Making*, Petrocelli Books, New York, 1978.
9. Berthouex, Paul Mac and Brown, Linfield C., *Statistics for Environmental Engineers*, Lewis Publishers, Boca Raton, FL, 1994.
10. Bajpai, A.C., Calus, I.M., and Fairley, J.A., Descriptive Statistical Techniques, in C.N. Hewitt, ed., *Methods of Environmental Data Analysis*, Elsevier, London, 1992, 1–35.
11. Watts, D.G., Why is Introductory Statistics Difficult to Learn? And What Can We Do to Make It Easier?, *American Statistician*, Vol. 45, No. 4, 1991, 290–291.
12. Knowles, Porter C., Geology and Groundwater Hydrology, in Kenneth W. Ayers et al., eds., *Environmental Science and Technology Handbook*, Government Institutes Inc., Rockville, MD, 1994, 249–316.
13. Bishop, Margaret S., Lewis, Phillis G., and Bronaugh, Richmond L., *Earth Science*, Charles Merrill, Columbus, OH, 1972.
14. Gardner, Walter H., Physical Properties, in Sybil P. Parker and Robert A. Corbitt, eds., *Encyclopedia of Environmental Science and Engineering*, McGraw-Hill, New York, 1993, 556–562.

15. Domenico, Patrick A. and Schwartz, Franklin W., *Physical and Chemical Hydrogeology*, John Wiley & Sons, New York, 1990.
16. Harlan, R.L., Kolm, K.E., and Gutentag, E.D., *Water-Well Design and Construction*, Elsevier, Amsterdam, The Netherlands, 1989.
17. Blackman, William C., Jr., *Basic Hazardous Waste Management*, Lewis Publishers, Boca Raton, FL, 1994.
18. Barcelona, Michael J., Overview of the Sampling Process, in Lawrence H. Keith, ed., *Principles of Environmental Sampling*, American Chemical Society, Washington, D.C., 1988, 3–23.
19. Keith, Lawrence, *Environmental Sampling and Analysis, A Practical Guide*, Lewis Publishers, Boca Raton, FL, 1991.
20. Canadian Council of Ministers of the Environment, *Guidance Manual on Sampling, Analysis, and Data Management for Contaminated Sites, Volume I: Main Report*, Winnipeg, Manitoba, 1993.
21. U.S. Environmental Protection Agency, *Data Quality Objectives for Remedial Response Activities — Development Process*, Washington, D.C., 1987.
22. U.S. Environmental Protection Agency, *Characterizing Heterogeneous Wastes: Methods and Recommendations*, Las Vegas, 1992.
23. Shashidara, N.S., Quality Assurance for Hazardous Waste Projects, *Civil Engineering*, Vol. 64, No. 12, 1994, 66–68.
24. Taylor, J.K., What is Quality Assurance? in J.K. Taylor and T.W. Stanley, eds., *Quality Assurance for Environmental Measurements*, American Society for Testing and Materials, Philadelphia, 1985, 5–11.
25. Maney, J.P. and Dallas, A., The Importance of Measurement Integrity, *Environmental Laboratory*, Vol. 3, No. 5, 1991, 20–25, 52.
26. McCoy and Associates, Soil Sampling and Analysis — Practices and Pitfalls, *The Hazardous Waste Consultant*, Vol. 10, Issue 6, 1992, 4.1–4.26.
27. Gilbert, Richard O., *Statistical Methods for Environmental Pollution Monitoring*, van Nostrand Reinhold, New York, 1987.
28. Bauer, Rich et al., *RCRA Corrective Action Data Review Guidance Manual*, U.S. EPA Region 9, San Francisco, 1995.
29. Keith, Lawrence H., *Quality Control and GC Advisor*, Lewis Publishers, Boca Raton, FL, 1991.
30. ASTM, *ASTM Standards on Environmental Sampling*, Philadelphia, 1995.
31. Canadian Council of Ministers of the Environment, *Guidance Manual on Sampling, Analysis, and Data Management for Contaminated Sites — Volume II: Analytical Method Summaries*, Winnipeg, Manitoba, 1993.
32. Boulding, J. Russell, *Subsurface Characterization and Monitoring Techniques, A Desk Reference Guide — Volume 1: Solids and Groundwater, Appendices A and B*, U.S. Environmental Protection Agency, Washington, D.C., 1993.
33. Brown, K.W., Breckenridge, R.P., and Rope, R.C., Soil Sampling Reference Field Methods, in *U.S. Fish and Wildlife Service Lands Contaminant Monitoring Operations Manual*, Idaho National Engineering Laboratory, Idaho Falls, Appendix J.
34. Kearl, P.M., Korte, N.E., and Cronk, T.A., Suggested Modifications to Groundwater Sampling Procedures Based on Observations from the Colloidal Boroscope, *Groundwater Monitoring Review*, Vol. 12, No. 2, 1992, 155–161.

35. LaGrega, Michael D., Buckingham, Phillip L., and Evans, Jeffrey C., *Hazardous Waste Management*, McGraw-Hill, New York, 1994.

36. Canadian Council of Ministers of the Environment, *Guidance Manual on Sampling, Analysis, and Data Management for Contaminated Sites — Volume 1: Main Report*, Winnipeg, Manitoba, 1993.

37. National Institute for Occupational Safety and Health, *NIOSH Pocket Guide to Chemical Hazards*, Washington, D.C., 1994.

38. McVeigh, Thomas P., Sampling and Monitoring of Remedial Action Sites, in Harry M. Freeman, ed., *Standard Handbook of Hazardous Waste Treatment and Disposal*, McGraw-Hill, New York, 1989, 12.2–12.34.

39. U.S. Environmental Protection Agency, *Guide to Management of Investigation Derived Waste*, Washington, D.C., 1992.

40. U.S. Environmental Protection Agency, *Management of Investigation Derived Wastes During Site Inspections*, Washington, D.C., 1991.

41. Saunders, Richard D., Minimizing Subsurface Investigation Derived Wastes, *Environmental Testing and Analysis*, January/February, 1995, 62–68.

42. Smith, Roy-Keith, Owens, Robert G., Jr., and Geier, Denise S., Ending the Laborious Search for Defensible Data, *Industrial Wastewater*, Vol. 3, No. 1, 1995, 51–56.

43. McCoy and Associates, Soil Sampling and Analysis — Practices and Pitfalls, in *The Hazardous Waste Consultants*, Vol. 10, No. 6, 1992, 4.1–4.26.

44. LaGrega, Michael D., Buckingham, Phillip L., and Evans, Jeffrey C., *Hazardous Waste Management*, McGraw-Hill, New York, 1994.

45. Sara, Martin N., *Standard Handbook for Solid and Hazardous Waste Facility Assessments*, Lewis Publishers, Boca Raton, FL, 1993.

46. Durant, Neal D. and Myers, Vernon B., EPA's Approach to Vadose Zone Monitoring at RCRA Facilities, in L.G. Wilson, Lorne G. Everett, and Stephen J. Cullen, eds., *Handbook of Vadose Zone Characterization and Monitoring*, Lewis Publishers, Boca Raton, FL, 1995, 9–22.

47. U.S. Environmental Protection Agency, *Geophysics: A Key Step in Site Characterization*, Las Vegas, 1991.

48. Olhoeft, G.R., Low Frequency Electrical Properties, *Geophysics*, Vol. 50, 1985, 2492.

49. U.S. Environmental Protection Agency, *Field Screening Methods for Radioactive Contamination*, Las Vegas, 1990.

50. Sara, Martin N., *Standard Handbook for Solid and Hazardous Waste Facility Assessments*, Lewis Publishers, Boca Raton, FL, 1993.

51. Greenhouse, J. and Gudjurgis, P., Geophysics, in *Subsurface Assessment Handbook for Contaminated Sites*, Canadian Council of Ministers of the Environment, Winnipeg, Manitoba, 1994, 47–116.

52. Olhoeft, Gary R., *Geophysics Advisor Expert System, Version 2.0*, U.S. Environmental Protection Agency, Las Vegas, 1992.

53. Rudolph, D., Hydrogeological Site Investigation, in *Subsurface Assessment Handbook for Contaminated Sites*, Canadian Council of Ministers of the Environment, Winnipeg, Manitoba, 1994, 119–174.

54. Rudolph, D., Hydrogeological Site Investigation, in *Subsurface Assessment Handbook for Contaminated Sites*, Canadian Council of Ministers of the Environment, Winnipeg, Manitoba, 1994, 119–174.

55. Barker, J., Chemical Hydrogeology, in *Subsurface Assessment Handbook for Contaminated Sites*, Canadian Council of Ministers of the Environment, Winnipeg, Manitoba, 1994, 177–208.

56. U.S. Environmental Protection Agency, *Field Portable X-Ray Fluorescence*, Las Vegas, 1990.

57. Marrin, D.L. and Thompson, G.M., Gaseous Behavior of TCE Overlying a Contaminated Aquifer, *Groundwater*, Vol. 25, No. 1, 1987, 21–27.

58. U.S. Environmental Protection Agency, *Soil Gas Measurement*, US EPA, Las Vegas, 1990.

59. Godoy, F.E. and Naleid, D.S., Optimizing the Use of Soil Gas Surveys, *Hazardous Materials Control*, Vol. 3, No. 5, 1990, 23–29.

60. Lucero, D.P., A Soil Gas Sampler Implant for Monitoring Dump Site Subsurface Hazardous Fluids, *Hazardous Materials Control*, Vol. 3, No. 5, 1990, 36–44.

61. Ullom, William L., Soil Gas Sampling, in L.G. Wilson, Lorne G. Everett, and Stephen J. Cullen, eds., *Handbook of Vadose Zone Characterization and Monitoring*, Lewis Publishers, Boca Raton, FL, 1995, 555–567.

62. U.S. Environmental Protection Agency, *High Resolution Mass Spectrometry*, Las Vegas, 1991.

63. Lewis, T.E. et al., *Soil Sampling and Analysis for Volatile Organic Compounds*, U.S. Environmental Protection Agency, Washington, D.C., 1991.

64. Smith, J.A. et al., Effect of Soil Moisture on the Sorption of Trichloroethene Vapor to Vadose Zone Soil at Picatinny Arsenal, New Jersey, *Environmental Science and Technology*, No. 24, 676–683.

65. Marks, B.J. and Singh, M., Comparison of Soil Gas, Soil, and Groundwater Contaminant Levels of Benzene and Toluene, *Hazardous Materials Control*, Vol. 3, No. 6, 1990, 40–45.

66. Barker, J., Chemical Hydrogeology, in *Subsurface Assessment Handbook for Contaminated Sites*, Canadian Council of Ministers of the Environment, Winnipeg, Manitoba, 1994, 177–208.

67. Goldberg, Steven P., Carey, Peter J., and Fitzsimmons, James H., Tracking Groundwater with Fluoride Tracers, *Waste Age*, Vol. 26, No. 10, 1995, 99–108.

68. Davis, S.N. et al., *Groundwater Tracers*, National Water Well Association, Worthington, OH, 1985.

69. Barcelona, Michael et al., *Contamination of Groundwater: Prevention, Assessment, Restoration*, Noyes Data Corporation, Park Ridge, NJ, 1990.

70. Cullen, Stephen J. et al., Is Our Groundwater Monitoring Strategy Illogical?, in L.G. Wilson, Lorne G. Everett, and Stephen J. Cullen, eds., *Handbook of Vadose Zone Characterization and Monitoring*, Lewis Publishers, Boca Raton, FL, 1995, 1–8.

71. Weiner, Eugene R., *Movement and Fate of Contaminants in Surface Water, Groundwater and Soils*, American Society of Civil Engineers, Chicago, 1995.

72. Barcelona, Michael, Wehrmann, Allen, Keely, Joseph F., and Pettyjohn, Wayne A., *Contamination of Groundwater — Prevention, Assessment, Restoration*, Noyes Data Corporation, Park Ridge, NJ, 1990.

73. Molson, J., Frind, E., and Sudicky, E., Groundwater Flow and Contaminated Transport Model, in *Subsurface Assessment Handbook for Contaminated Sites*, Canadian Council of Ministers of the Environment, Winnipeg, Manitoba, 1994, 211–254.

74. Brusseau, Mark L. and Wilson, L.G., Estimating the Transport and Fate of Contaminants in the Vadose Zone Based on Physical and Chemical Properties of the Vadose Zone and Chemicals of Interest, in L.G. Wilson, Lorne G. Everett, and Stephen J. Cullen, eds., *Handbook of Vadose Zone Characterization and Monitoring*, Lewis Publishers, Boca Raton, FL, 203–215.

75. Breckenridge, R.P., Williams, J.R., and Keck, J.F., *Characterizing Soils for Hazardous Waste Site Assessments*, U.S. Environmental Protection Agency, Washington, D.C., 1991.

76. Panian, T.F., *Unsaturated Flow Properties Data Catalog*, Vol. 2, Water Resources Center, University of Nevada, Las Vegas, 1987.

77. Campbell, G.S., *Soil Physics With Basic*, Elsevier, New York, 1985.

78. Nielsen, D.R. and Bouwma, J., eds., *Soil Spatial Variability*, Center for Agricultural Publishing and Documentation, Wageningen, The Netherlands, 1985.

79. Davis, J.C., *Statistics and Data Analysis in Geology*, John Wiley & Sons, New York, 1986.

80. Borgman, Leon E. and Quimby, William F., Sampling for Tests of Hypothesis When Data are Correlated in Space and Time, in Lawrence H. Keith, ed., *Principles of Environmental Sampling*, American Chemical Society, Washington, D.C., 1988, 25–43.

81. Van der Heijde, P.K.M. and Elnawawy, O.A., *Compilation of Groundwater Models*, U.S. Environmental Protection Agency, Washington, D.C., 1993.

82. Nielsen, D.R., Van Genuchten, M.T., and Biggar, J.W., Water Flow and Solute Transport Processes in the Unsaturated Zone, *Water Resources Research*, Vol. 22, 1986, 89–108.

83. Van der Heijde, P.K.M., *Identification and Compilation of Unsaturated/Vadose Zone Models*, U.S. Environmental Protection Agency, Ada, OK, 1993.

84. Kramer, John H. and Cullen, Stephen J., Review of Vadose Zone Flow and Transport Models, in L.G. Wilson, Lorne G. Everett, and Stephen J. Cullen, eds., *Handbook of Vadose Zone Characterization and Monitoring*, Lewis Publishers, Boca Raton, FL, 1995, 267–289.

85. Liu, C.W. and Narasimhan, T.N., Redox-Controlled Multiple Species Reactive Chemical Transport 1, Model Development, *Water Resources Research*, Vol. 25, 1989, 869–882.

86. Van Genuchten, M.T., Leij, F.J., and Yates, S.R., *The RETC Code of Quantifying the Hydraulic Functions of Unsaturated Soils*, U.S. Environmental Protection Agency, Ada, OK, 1991.

87. Fogg, Graham E., Nielsen, D.R., and Shibberu, D., Modeling Contaminant Transport in the Vadose Zone: Perspective on State of the Art, in L.G. Wilson, Lorne G. Everett, and Stephen J. Cullen, eds., *Handbook of Vadose Zone Characterization and Monitoring*, Lewis Publishers, Boca Raton, FL, 1995, 249–265.

88. Bair, E. Scott and Roadcap, George S., Comparison of Flow Models Used to Delineate Capture Zones of Wells: 1. Leaky-Confined Fractured-Carbonate Aquifer, *Groundwater*, Vol. 30, No. 2, 1992, 199–211.

89. Molson, J., Frind, E., and Sudicky, E., Groundwater Flow and Contaminated Transport Model, in *Subsurface Assessment Handbook for Contaminated Sites*, Canadian Council of Ministers of the Environment, Winnipeg, Manitoba, 1994, 211–254.

90. Bair, E. Scott, Springer, Abraham E., and Roadcap, George S., Delineation of Traveltime-Related Capture Areas of Wells Using Analytical Flow Models and Particle-Tracking Analysis, *Groundwater*, Vol. 29, No. 3, 1991, 387–397.

91. U.S. Environmental Protection Agency, *Guidance for Delineation of Wellhead Protection Areas*, Washington, D.C., 1987.

92. McDonald, M.G. and Harbaugh, A.W., *A Modular Three-Dimensional Finite Difference Groundwater Flow Model*, U.S. Geological Survey, Washington, D.C., 1988.

93. Pollock, D.W., *Documentation of Computer Programs to Compute and Display Pathlines Using Results from the U.S. Geological Survey Modular Three-Dimensional Finite-Difference Groundwater Flow Model*, U.S. Geological Survey, Washington, D.C., 1989.

94. Blandford, T.N. and Huyakorn, P.S., *An Integrated Semi-Analytical Model for Delineation of Wellhead Protection Areas* (draft), prepared by HydroGeoLogic Inc. for U.S. EPA, Washington, D.C., 1989.

95. Springer, Abraham E. and Bair, E. Scott, Comparison of Methods Used to Delineate Capture Zones of Wells: 2. Stratified-Drift Buried-Valley Aquifer, *Groundwater*, Vol. 30, No. 6, 1992, 908–917.

96. Kates, Robert W., *Natural Hazard in Human Ecological Perspective, Hypotheses and Model*, Clark University, Worcester, MA, 1970.

97. White, Gilbert F., Natural Hazard Research, in R.J. Chorley, ed., *Dimensions in Geography*, Methuen, London, 1973, 6–15.

98. Mitchell, Bruce, *Geography and Resource Analysis*, Longman, London, 1979.

99. Harris, R.C., Hohenemser, C., and Kates, R.W., Our Hazards Environment, *Environment*, Vol. 20, No. 7, 1978, 6–15.

100. Zeigler, D.J., Johnson, J.H., and Brunn, S.D., *Technological Hazards*, Association of American Geographers, Washington, D.C., 1983.

101. Kaplan, S. and Garrick, B.J., On the Quantitative Definition of Risk, *Risk Analysis*, Vol. 1, No. 1, 1981, 11–27.

102. Coates, Joseph F., Why Government Must Make a Mess of Technological Risk Management, in Christoph Hohenemser and Jeanne X. Kasperson, eds., *Risk in the Technological Society*, Westview, Boulder, CO, 1981, 21–34.

103. Lowrance, W.W., *Of Acceptable Risk*, Kaufman, Los Altos, CA, 1976.

104. Hadden, S.G., ed., *Risk Analysis, Institutions, and Public Policy*, Associated Faculty Press, Port Washington, 1984.

105. Rowe, W.D., *An Anatomy of Risk*, John Wiley & Sons, New York, 1977.

106. Dreith, Richard H., An Industry's Guidelines for Risk Assessment, in F.A. Long and Glenn E. Schweitzer, eds., *Risk Assessment at Hazardous Wastes Sites*, American Chemical Society, Washington, D.C., 1982, 45–53.

107. Derby, S.L. and Keeney, R.L., Risk Analysis, Understanding 'How Safe is Safe Enough', *Risk Analysis*, Vol. 1, No. 3, 1981, 217–224.

108. National Research Council, *Risk Assessment in the Federal Government: Managing the Process*, National Academy of Science Press, Washington, D.C., 1983.

109. U.S. Environmental Protection Agency, *Risk Assessment Guidance for Superfund: Volume I — Human Health Evaluation Manual (Part A), Interim Final*, Washington, D.C., 1989.

110. Asante-Duah, D. Kofi, *Hazardous Waste Risk Assessment*, Lewis Publishers, Boca Raton, FL, 1993.

111. U.S. Environmental Protection Agency, *Population Exposure to External Natural Radiation Background in the United States*, Washington, D.C., 1987.

112. Asante-Duah, D. Kofi, *Hazardous Waste Risk Assessment*, Lewis Publishers, Boca Raton, FL, 1993.

113. U.S. Environmental Protection Agency, *Risk Assessment Guidance for Superfund: Volume 1 — Human Health Evaluation Manual (Part A), Interim Final*, Washington, D.C., 1989.

114. U.S. Environmental Protection Agency, *Risk Assessment Guidance for Superfund: Volume 1 — Human Health Evaluation Manual (Part B)*, Washington, D.C., 1991.

115. Arizona Department of Environmental Quality, *Human Health-Based Guidance Level for the Ingestion of Contaminants in Drinking Water and Soil*, Phoenix, 1992.

116. Milloy, Steven J., *Science-Based Risk Assessment: A Piece of the Superfund Puzzle*, National Environmental Policy Institute, Washington, D.C., 1995.

117. U.S. Environmental Protection Agency, National Oil and Hazardous Substances Pollution Contingency Plan, Final Rule, 55 *Federal Register* 8666, March 8, 1990.

118. U.S. Environmental Protection Agency, *Risk Assessment Guidance for Superfund: Volume 1 — Human Health Evaluation Manual (Part B, Development of Risk-Based Preliminary Remediation Goals), Interim*, Washington, D.C., 1991.

119. U.S. Environmental Protection Agency, *Risk Assessment Guidance for Superfund: Volume 1 — Human Health Evaluation Manual (Part C, Risk Evaluation of Remedial Alternatives), Interim*, Washington, D.C., 1991.

120. Burmaster, David E., and Appling, Jeanne W., Introduction to Human Health Risk Assessment, With an Emphasis on Contaminated Properties, *Environment Reporter*, April 7, 1995, 2431–2440.

121. Seip, H.M. and Heiberg, A.B., eds., *Risk Management of Chemicals in the Environment*, NATO Committee on the Challenges of Modern Society, Vol. 12, Plenum Press, New York, 1989.

122. National Research Council, *Risk Assessment in the Federal Government: Managing the Process*, National Academy Press, Washington, D.C., 1983.

123. U.S. Environmental Protection Agency, *Risk Assessment and Management: Framework for Decision Making*, Washington, D.C., 1984.

124. National Governors' Association, *Risk in Environmental Decision Making, A State Perspective Working Paper*, Washington, D.C., 1994.

125. Carpenter, Richard A., Risk Assessment, *Impact Assessment*, Vol. 13, No. 2, 1995, 153–187.

126. Allen, F.W., Towards a Holistic Appreciation of Risk: The Challenge for Communicators and Policy Makers, *Science, Technology and Human Values*, Vol. 12, 1987, 138–143.

127. Slovic, P., Informing and Educating the Public About Risk, *Risk Analysis*, Vol. 6, 1986, 403–415.

128. Kasperson, R.E. et al., The Social Amplification of Risk: A Conceptual Framework, *Risk Analysis*, Vol. 8, 1988, 177–187.

129. U.S. Environmental Protection Agency, *Risk Assessment Guidance for Superfund: Volume 1 — Human Health Evaluation Manual (Part C, Risk Evaluation of Remedial Alternatives), Interim*, Washington, D.C., 1991.

130. U.S. General Accounting Office, *Superfund: Improved Reviews and Guidance Could Reduce Inconsistencies in Risk Assessments*, Washington, D.C., 1994.

131. Locke, Paul A., *Reorienting Risk Assessment*, Environmental Law Institute, Washington, D.C., 1994.

132. Beanlands, G.E. and Duinker, P.N., An Ecological Framework for Environmental Impact Assessment, *Journal of Environmental Management*, Vol. 18, 1983, 267–277.

133. Spellerberg, I.F., *Monitoring Ecological Change*, Cambridge University Press, Cambridge, UK, 1991.

134. Treweek, Joanna, Ecological Impact Assessment, *Impact Assessment*, Vol. 13, No. 3, 1995, 289–315.

135. U.S. Environmental Protection Agency, *Risk Assessment Guidance for Superfund: Volume II — Environmental Evaluation Manual, Interim Final*, Washington, D.C., 1989.

136. U.S. Environmental Protection Agency, *Framework for Ecological Risk Assessment*, Washington, D.C., 1992.

137. U.S. Environmental Protection Agency, *Ecological Risk Assessment Issue Papers*, Washington, D.C., 1994.

138. U.S. Environmental Protection Agency, *Peer Review Workshop Report on Ecological Risk Assessment Issue Papers*, Washington, D.C., 1994.

139. U.S. Environmental Protection Agency, *A Review of Ecological Assessment Case Studies from a Risk Assessment Perspective*, Washington, D.C., 1993.

140. U.S. Environmental Protection Agency, *A Review of Ecological Assessment Case Studies from a Risk Assessment Perspective, Volume II*, Washington, D.C., 1994.

141. National Research Council, *Ecological Knowledge and Environmental Problem Solving: Concepts and Case Studies*, National Academy Press, Washington, D.C., 1986.

142. Suter, G.W. II, Endpoints for Regional Ecological Risk Assessments, *Environmental Management*, Vol. 14, No. 1, 1990, 19–23.

143. Harwell, Mark et al., Ecological Significance, in *Ecological Risk Assessment Issue Papers*, Washington, D.C., 1994, 2.1–2.49.

144. Norton, Bryan G., Ascertaining Public Values Affecting Ecological Risk Assessment, in *Ecological Risk Assessment Issue Papers*, Washington, D.C., 10.1–10.38.

145. Sagoff, M., *The Economy of the Earth: Philosophy, Law and the Environment*, Cambridge University Press, Cambridge, UK, 1988.

146. Smith, Eric P. and Shugart, H.H., Uncertainty in Ecological Risk Assessment, in *Ecological Risk Assessment Issue Papers*, Washington, D.C., 1994, 8.1–8.53.

147. Buonicore, Anthony J., ed., *Cleanup Criteria for Contaminated Soil and Groundwater*, ASTM, Philadelphia, 1995.

148. Shacklette, H.T., Hamilton, J.C., Boerngen, J.G., and Bowles, J.M., *Elemental Composition of Surficial Materials in the Conterminous United States*, U.S. Government Printing Office, Washington, D.C., 1971.

149. Shacklette, H.T. and Boerngen, J.G., *Elemental Concentrations in Soils and Surficial Materials of the Conterminous U.S.*, U.S. Government Printing Office, Washington, D.C., 1984.

150. U.S. Environmental Protection Agency, *Statistical Methods for Evaluating Groundwater Monitoring Data from Hazardous Waste Facilities*, Washington, D.C., 1988.
151. 58 FR 48092, September 14, 1993.
152. 55 FR 30798, July 27, 1990.
153. U.S. Environmental Protection Agency, *Soil Screening Guidance*, draft, Washington, D.C., 1994.
154. Arizona Department of Health Services, Human Health-Based Guidance Levels (HBGLs) for Ingestion of Contaminants in Soil, in *Notice of Proposed Interim Rulemaking*, Phoenix, 1995, Appendix A.
155. Maritato, Mark C. et al., Risk-Based Cleanups Form Powerful Approach to Prioritizing, Restoring Hazardous Waste Sites, *Environmental Solutions*, Vol. 8, No. 1, 1995, 51–56.
156. Vance, David B., Risk, Costs and Benefits — The New Regulatory Paradigm, *The National Environmental Journal*, Vol. 5, No. 6, 1995, 26–27.

PART IV
REMEDIAL ACTION
PLANNING

The first chapter of this part provides a review of some selected commercial site remediation technologies. The chapter describes the characteristics of each technology, its operating principle, process, limitations, and applicability to address site remediation. Biological treatment, physico-chemical, and thermal treatment technologies are represented in the discussion as well as the containment, monitoring, and measurement technologies. In this chapter, technologies for treating soil contamination and groundwater remediation technologies are discussed separately.

Chapter 12 describes the process of selecting a technology appropriate to a contaminated site. Given the variety of available alternative technologies, planners must reduce the number of options and identify the most appropriate technologies for the site. The framework outlined in this chapter indicates that the systematic selection process is largely determined by a site's soil and groundwater characteristics. How these characteristics affect the applicability of a technology to a site is also discussed in Chapter 12.

The last chapter of this part (Chapter 13) addresses the public participation aspect of site remediation. The importance of public participation to the success or failure of a remediation project is very obvious. The role of public participation in site remediation is to promote active communication between communities affected by contaminant releases and those responsible for response actions. The promotion is a prerequisite for a successful remedial action. This chapter addresses a public participation program in terms of its principles, strategy, structure, techniques, and pitfalls. As part of the discussion, risk communication is also briefly reviewed including the basic reasons why some risk communication programs fail.

11   REMEDIATION TECHNOLOGY

Understanding technological alternatives that are available for remediating a contaminated site is critical to site remediation planning. Planners seeking appropriate technologies encounter a wide array of treatment options; however, project goals usually dictate selecting technologies that permanently destroy or immobilize toxic contaminants, resulting in reductions of their toxicity, mobility, or quantity.

Available commercial site remediation technologies can be categorized in a variety of ways. One of the methods is to classify the technologies according to the way they treat contaminants. Based on this separation, remediation technologies may fall into either destruction technologies such as incineration, or technologies that reduce the solubility or leachability of metals.[1] An example of the second category is the stabilization of metals. This technology basically requires the addition of stabilizing agents that chemically bind the metals to the solid matrix.

Site remediation technologies can also be categorized into three major groupings based on their treatment methods. This standard categorization differentiates among the biological, physico-chemical, and thermal treatment technologies. The ordering of the categories represents the roughly ascending order of cost, which means that biological treatment is relatively less expensive than physico-chemical treatment technologies, while thermal treatment represents the most expensive category among the three categories.[2] An analysis by Mills[3] has found, for example, that the cost of soil bioremediation can be 30 to 60% less than thermal desorption methods.

In addition to the above grouping, site remediation technologies can also be classified into *in situ* and *ex situ* treatments, depending on the placement of the technology. *In situ* treatment is the method of treating soil and groundwater contamination where they are found without excavating the overlying soil; all of the treatment takes place in the subsurface. *Ex situ* treatment, on the other hand, refers to site remediation that occurs not in its original place; the remediation can take place either on-site or off-site.

This chapter discusses available commercial site remediation technologies along the three major groupings of biological, physico-chemical, and thermal treatments. Examined within each group of remediation technologies are *in situ* and *ex situ* technologies. For each technology, the applicability to soil or groundwater contamination is indicated. Because of their significance in site remediation planning, other technologies, such as containment and monitoring, and measurement technologies are also discussed.

The classification scheme offered in this chapter is intended only for streamlining the discussion. In reality, a technology such as incineration may be used for soil remediation as well as groundwater remediation. The term soil in this chapter also includes sediments and sludges. Depending on the site-specific characteristics, more than one technology may be needed to achieve the cleanup goals at a site. An example of a combination of treatment technologies, or treatment train, is thermal desorption, followed by incineration and solidification/stabilization for chlorodiphenyl (PCB) treatment.

Procedurally, before a technology can be commercialized, there must be a demonstration that the technology is proven. Under the Superfund Innovative Technology Evaluation (SITE) program, which was initiated in 1985, EPA enters into cooperative agreements with technology developers to research, demonstrate, and develop innovative or new and promising site remediation technologies. Some programs were created to encourage the development and routine use of treatment technologies and monitoring/measurement technologies.[4] These programs are

- Programs for innovative remediation treatment technologies. As presently structured, the two major programs under this category are the emerging technology program and the technology demonstration program.
- Programs for the advancement of monitoring and measurement technologies. EPA's monitoring and measurement technologies program assesses innovative monitoring, measurement, and site characterization technologies. These technologies are used to investigate the nature and extent of contamination and evaluate the effectiveness of site remediation.

In the emerging technology program EPA provides technical and financial support to developers for bench- and pilot-scale testing and evaluation of innovative technologies that have already been approved in the conceptual stage. Technologies are solicited through requests for preproposals and after a technical review of the preproposals, selected candidates are required to submit a cooperative agreement application and a detailed project proposal that undergoes another full technical review. Projects are considered for either a one- or two-year development effort, providing awards of up to $150,000 per year. The review compares the applicability of particular technologies to site waste characteristics. The review may support technologies that eventually will be evaluated in the technology demonstration program.

In the technology demonstration program, the technology is field-tested on hazardous waste materials; engineering and cost data are gathered so that potential users can assess the technology's applicability for a particular site cleanup. Assessment includes overall performance of the technology, the need for pre- and post-processing of the waste, applicable types of wastes, potential operating problems, and estimates of capital and operating costs.

The demonstration process typically consists of five steps:

1. to match an innovative technology with an appropriate site
2. to prepare a four-part demonstration plan consisting of:
   - a test plan
   - a sampling and analysis plan
   - a quality assurance project plan
   - a health and safety plan
3. to perform community relations activities to gain public acceptance
4. to conduct the demonstration (ranging from days to months)
5. to document results in an application analysis report (AAR) and a technology evaluation report (TER)

The technology developers are responsible for operating their innovative systems at a selected site, removing the equipment from the site, and paying the costs to transport equipment to and from the site. EPA is responsible for project planning, sampling and analysis, quality assurance/quality control, report preparation, and information dissemination. These responsibilities are specified in the cooperative agreements between the technology developers and EPA.

After the completion of the SITE demonstration, EPA prepares reports, videos, bulletins, and project summaries and disseminates the information to provide reliable technical data for site remediation decision making and to promote the technology's commercial use. SITE information is available through the Alternative Treatment Technology Information Center (ATTIC) by calling 301-670-6294, the Vendor Information System for Innovative Treatment Technologies (VISITT) at 800-245-4505, and the Center for Environmental Research Information (CERI) at 513-569-7562.

As of November 1991, 25 technology demonstrations have been completed. Table 11.1 lists the technologies in alphabetical order by the developer's name along with pertinent information on the technology transfer opportunity for the project.

The table shows that only three of the 25 entries are *in situ* treatment technologies. The remaining 22 technologies are designed for *ex situ* treatment. The technologies displayed in the table vary from soil washing to solvent extraction, solidification and stabilization, biological aqueous treatment, thermal destruction, plasma arc vitrification, ultraviolet (UV) radiation and oxidation.

The groupings derived from the table indicate that most of the demonstrated technologies are physico-chemical (64%), followed by thermal treat-

Table 11.1 Completed Technology Demonstrations

Developer	Technology	Site location	Visitors' day and demonstration date	Applications analysis report	Technology evaluation report	EPA contact	Profile page no.
American Combustion Technologies, Inc., Norcross, GA	Pyretron thermal destruction	EPA's combustion research facility in Jefferson, AK Soil from String-fellow Acid Pit Superfund site in Glen Avon, CA	Visitors' Day: January 1988 Demonstration: November 16, 1987–January 29, 1988	American Comustion Pyretron destruction system EPA/540/A5-89/008, April 1989	SITE program demonstration test: The American Combustion Pyretron thermal destruction system at EPA combustion research facility, EPA/540/5-89/008, April 1989	Laurel Staley EPA ORD. 513-569-7863	28
AWD Technologies Inc. San Francisco, CA	Integrated vapor extraction and steam vacuum stripping	San Fernando Valley Groundwater Basin Superfund site in Burbank, CA	Visitors' Day: September 20, 1990 Demonstration: September 1990	EPA/540/A5-91/002	In preparation	Christine Stubbs EPA Region 9 415-744-2248	30
BioTrol, Inc., Chaska, MN	Biological aqueous treatment	MacGillis and Gibbs Superfund site in New Brighton, MN	Visitors' Day: September 27, 1989 Demonstration: July–September 1989	BioTrol, Inc. EPA/540/A5-91/001 May, 1991	In preparation	Rhonda McBride EPA Region 5 312-886-7242	36
BioTrol, Inc., Chaska, MN	Soil washing	MacGillis and Gibbs Superfund site in New Brighton, MN	Visitors' Day: September 27, 1989 Demonstration: September–October 1989	In preparation	In preparation	Rhonda McBride EPA Region 5 312-886-7242	38

Company	Technology	Site	Dates	Report	SITE program report	Contact	Page
CF Systems Corporation, Waltham, MA	Solvent extraction	New Bedford Harbor Superfund site in New Bedford, MA	Visitors' Day: August 26–27, 1988 Demonstration: September 1988	CF Systems organics extraction system, New Bedford, MA EPA/540/A5-90/002, August 1990	SITE program demonstration test: CF Systems Corporation solvent extraction, EPA/540/5-90/002, August 1990	David Lederer EPA Region 1 617-573-9665	44
Chemfix Technologies, Inc., Metairie, LA	Solidification and stabilization	Portable Equipment Salvage Company in Clackamas, OR	Visitors' Day: March 15, 1989 Demonstration: March 1989	Chemfix Technologies, Inc. EPA/540/A5-89/011, May 1991	SITE program demonstration test: Chemfix Technologies, Inc. Solidification/stabilization EPA/540/5-89/011a	John Sainsburg EPA Region 10 206-553-0125	46
Dehydro-Tech Corporation, East Hanover, NJ	Carver-Greenfield process for extraction of oily waste	EPA facility in Edison, NJ	Visitors' Day: August 20, 1991 Demonstration: August 1991	In preparation	In preparation	Laurel Staley EPA ORD 513-569-7863	58
E.I. Du Pont de Nemours and Co., Newark, DE, and Oberlin Filter Co., Waukesha, WI	Membrane microfiltration	Palmerton Zinc Superfund site in Palmerton, PA	Visitors' Day: April 10, 1990 Demonstration: April 1990	In preparation	In preparation	Tony Koller EPA Region 3 215-597-3923	60
Ecova Corporation Redmond, WA	Bioslurry reactor	EPA Test and Evaluation facility in Cincinnati, OH	Visitors' Day: not held Demonstration: October 1991	In preparation	In preparation	Ron Lewis EPA ORD 513-569-7856	64

Table 11.1 Completed Technology Demonstrations (Continued)

Developer	Technology	Site location	Visitors' day and demonstration date	Applications analysis report	Technology evaluation report	EPA contact	Profile page no.
Ecova Corporation (developed by Shirco Infrared Systems), Redmond, WA	Infrared thermal destruction	Rose Township Superfund site in Oakland County, MI	Visitors' Day: November 4, 1987 Demonstration: November 1987	Shirco Infrared Incineration System EPA/540/A5-89/007, June 1989	SITE program demonstration test: Shirco pilot-scale infrared incineration system at the Rose Township Demode Road Superfund site, EPA/540/5-88/007a, Vol. 1, April 1989 Vol. 2, in preparation	Kevin Adler EPA Region 5 312-886-7078	68
Ecova Corporation (developed by Shirco Infrared Systems), Redmond, WA	Infrared thermal destruction	Peak Oil Superfund site in Brandon, FL	Visitors' Day: August, 1987 Demonstration: August 1–5 1987	Shirco Infrared Incineration System EPA/540/A5-89/010, June 1989	SITE program demonstration test: Shirco Infrared incineration system, Peak Oil, Branson, FL, September 1988, EPA/540/5-88/002a	Fred Stroud EPA Region 4 404-347-3931	68

Company/Location	Technology	Site	Dates	Report	SITE program status	Contact	Page
EmTech (formerly Hazcon, Inc.), Oakwood, TX	Solidification and stabilization	Douglassville Superfund site, Berks County, near Reading, PA	Visitors' Day: October 14, 1987 Demonstration: October 1987	Hazcon, Inc., solidification process EPA/540/A5-89/001, May 1989	SITE program demonstration test: Hazcon, Inc., solidification process, Douglassville, PA, EPA/540/5-89/001a, Vol. 1, May 1989	Victor Janosik EPA Region3 215-597-8996	72
Horsehead Research Development, Inc., Monaca, PA	Flame reactor	National Smelting and Refining Company Superfund site in Atlanta, GA	Visitors' Day: March 22, 1991 Demonstration: March 18–22, 1991	In preparation	In preparation	Marta Richards EPA ORD 513-569-7783	90
International Waste Technologies, Wichita, KS, and Geo-Con, Inc., Pittsburgh, PA	In situ solidification and stabilization	General Electric service shop in Hialeah, FL	Visitors' Day: April 14, 1988 Demonstration: April–May 1988	International Waste Tech/Geo-Con, Inc. EPA/540/A5-89/004, August 1989	SITE program demonstration test: International Waste Technologies In situ stabilization/solidification, Hialeah, FL, EPA/540/5-89/004a, June 1989	Mary Stinson EPA ORD 908-321-6683	98
NOVATERRA, Inc., [formerly Toxic Treatment USA, Inc.], Torrance, CA	In situ stream and airstripping	Annex terminal in San Pedro, CA	Visitors' Day: September 7, 1989 Demonstration: September 1989	EPA/540/A5-90/008 June 1991	In preparation	Paul dePercin EPA ORD 513-569-7797	100
Ogden Environmental Services, San Diego, CA	Circulating bed combustor	Ogden's facility in La Jolla, CA	Visitors' Day: Not applicable Demonstration: March 1989	Not available	In preparation	John Blevins EPA Region 9 415-744-2241	102

Table 11.1 Completed Technology Demonstrations (Continued)

Developer	Technology	Site location	Visitors' day and demonstration date	Applications analysis report	Technology evaluation report	EPA contact	Profile page no.
Retech, Inc., Ukiah, CA	Plasma arc vitrification	DOE component development and integration facility in Butte, MT	Visitors' Day: August 8, 1991 Demonstration: July 22–26, 1991	In preparation	In preparation	Laurel Staley EPA ORD 513-569-7863	118
Risk Reduction Engineering Laboratory and IT Corporation, Edison, NJ	Debris washing system	Superfund sites in Detroit, MI; Hopkinsville, KY; and Walker County, GA	Visitors' Day: Not applicable Demonstration: September 1988, December 1989, and August 1990	In preparation	Design and development of a pilot-scale debris decontamination system EPA/540/5-91/006a May 1991	Naomi Barkley EPA ORD. 513-569-7854	124
Silicate Technology Corporation, Scottsdale, AZ	Solidification and stabilization treatment technology	Selma pressure treating site in Selma, CA	Visitors' Day: November 15, 1990 Demonstration: November 1990	In preparation	In preparation	Ed Bates EPA ORD. 513-569-7774	138
SoilTech, Inc.., Englewood, CO	Anaerobic thermal processor	Wide Beach Superfund site in Brant, NY	Visitors' Day: Not applicable Demonstration: May 1991	In preparation	In preparation	Herbert King EPA Region 2 212-264-1129	140
Solidtech, Inc., Houston, TX	Solidification and stabilization	Imperial Oil Company/Champion Chemicals Superfund site in Morganville, NJ	Visitors' Day: December 7, 1988 Demonstration: December 1988	Solidtech Inc. solidification/stabilization process EPA/540/A5-89/005 September 1990	SITE program demonstration test: Solidtech, Inc. solidification/ stabilization process Vol. I EPA/540/A5-89/005a, February 1990 Vol. II EPA/540/5-89/005b	Trevor Anderson EPA Region 2 212-264-5391	142

Company/location	Technology	Site	Visitors' Day/Demonstration	Publication	SITE program	Contact	
Terra Vac, Inc., San Juan, PR	In situ vacuum extraction	Groveland Wells Superfund site, Valley Manufactured Product in Groveland, MA	Visitors' Day: January 15, 1988 Demonstration: December 1987–April 1988	Terra Vac in situ vacuum extraction system, EPA/540/A5-89/003, July 1989	SITE program demonstration test: terra vac in situ vacuum extraction system, Groveland, MA, EPA/540/5-89/003a, April 1989	Robert Leger EPA Region 1 617-573-5734	148
Ultrox International Inc., Santa Ana, CA	Ultraviolet radiation and oxidation	Lorentz Barrel and Drum Company in San Jose, CA	Visitors' Day: March 8, 1989 Demonstration: March 1989	Ultrox International Ultraviolet Radiation/Oxidation Technology EPA/540/A5-89/012 September 1990	SITE program demonstration test: Ultrox International, Inc., ultraviolet radiation/oxidation technology, EPA/540/5-89/012, January 1990	Joseph Healy EPA Region 9 415-744-2231	156
U.S. Environmental Protection Agency	Excavation techniques and foam suppression	McColl Superfund site in Fullerton, CA	Visitors' Day: Not held Demonstration: June and July 1990	In preparation	In preparation	S. Jackson Hubbard EPA ORD. 513-569-7507 and John Blevins EPA Region 9 415-744-2241	158
Wastech, Inc., Oak Ridge, TN	Solidification and stabilization	Robins AFB, Warner Robins, GA	Visitors' Day: August 22, 1991 Demonstration: August 1991	In preparation	In preparation	Terry Lyons EPA ORD. 513-569-7589	160

Source: U.S. Environmental Protection Agency, The Superfund Innovative Technology Evaluation Program: Technology Profile, Washington, DC, 1991, Fourth Edition.

ment (28%). Biological treatment represents only 8%. Please note that the pattern refers to completed treatment technology demonstrations; it does not necessarily reflect the distribution of available technologies in the market.

The following sections provide a review of each group of remediation technologies. The review describes the characteristics of a particular technology, its principle, process, limitations, and applicability to address site contamination. A brief discussion is also provided to examine some innovative technologies. The overall review and discussion is geared towards those technologies that are available in the market or are currently used and tested in site remediation.

1. BIOLOGICAL TREATMENT

Biological treatment or bioremediation converts soil and groundwater toxic contaminants to nontoxic substances or less harmful compounds. This process involves the degradation of organic waste by the action of microorganisms such as bacteria or fungi. When introduced to hydrocarbon contaminated soil, the microorganism creates a biofilm around the hydrocarbon molecule and breaks the molecule down into simpler compounds or oxygen. When the hydrocarbon is consumed, microbial activity is stopped and the microbes die.

Bioremediation is an attractive technology because it is a natural process and relatively inexpensive compared to other technologies. The treatment degrades the target chemical and the residues from the processes, such as water and carbon dioxide, are harmless products. The degradation process alters the molecular structure of the waste's organic compounds. The degree of alteration determines whether complete degradation or partial degradation has occurred.

- Complete degradation refers to the complete breakdown of organic molecules into cellular mass, carbon dioxide, water, and inert inorganic residuals. Complete degradation is also called mineralization.
- Partial degradation refers to the simplification of an organic compound to a daughter compound; this process is commonly known as biotransformation.

Generally all organic wastes can be degraded if the proper microbial population is established and managed. Microorganisms can metabolize organic waste through a series of low temperature enzyme-catalyzed reactions. Metabolism is the collective chemical reactions within an organism to sustain life. For sustaining life, all organisms require both an energy source and a carbon source. When the energy source supplies energy, carbon dioxide must be provided as the carbon source for the production of cellular materials including proteins, carbohydrates, fats, and nucleic acids. The energy source may be sunlight, some reduced inorganic (e.g., ammonia nitrogen) or an organic compound.

In the bioremediation process, metabolism releases energy from the waste and uses the carbon to build new cellular material and, as a result, the waste is degraded. Functionally, organic waste serves both as a carbon source and energy source to living organisms. Although most synthetic organisms are biodegradable, some specific compounds have resisted degradation or their degradation occurs so slowly as to make bioremediation inefficient. The "resisted" group of organics is called the recalcitrant compounds and the "slow" group is termed the persistent compounds. Because bioremediation promotes and maintains a microbial population that metabolizes a target waste, the treatment must assure that the microorganism grows and flourishes. The rate of biodegradation is critical because it determines whether the treatment cost can be competitive with other treatment alternatives.

1.1. Bioremediation Requirements

In order for biodegradation processes to occur, microorganisms require the presence of certain minerals (referred to as nutrients) and an electron acceptor. Other conditions, i.e., temperature, pH, and moisture content will impact the effectiveness of these processes.

Phosphorus and nitrogen are most commonly required mineral nutrients. Frequently, these two macronutrients are not available in sufficient amounts in hazardous waste and must be added, usually as ammonia and orthophosphate. If all hydrocarbons are connected to cell material, it is assumed that the nutrient requirements of carbon to nitrogen to phosphorus ratios are in the order of 100:10:1.[5]

The most common electron acceptor used in bioremediation is oxygen. The mass of oxygen required can be calculated based on stoichiometry or laboratory determination. Norris et al.[5] estimate that stoichiometrically, approximately 3 lb of oxygen are required to convert 1 lb of hydrocarbon to carbon dioxide.

Temperature has a major influence on the biological degradation rate. Cellular activity responds to heat so that the rate of cell growth increases sharply with increasing temperature until the optimum growth is reached. Increases in temperature just a few degrees above the optimum can slow growth dramatically due to the reduction of reproductive capability.[6]

In the bioremediation, an organic waste must first come into contact with a bacterial cell's outermost coating in order for the degradation to occur. This contact initiates a series of metabolic steps in the degradation of the waste. The catalysts of this process are large protein molecules composed primarily of amino acids twisted into complex shapes by peptide links and hydrogen bonding. Produced by the microorganisms, these molecules are called enzymes and their activities depend on pH. The amount of enzyme that is in a catalytically active state changes as the pH changes. Accordingly, the growth of bacteria depends on pH. LaGrega et al.[7] states that most bacteria grow best in a relatively narrow range around neutrality, i.e., pH of 6 to 8. The die-off typically occurs below a pH of 4 to 5 and above a pH of 9 to 9.5.

Moisture is required for biodegradation. Biodegradation in soils can occur at moisture levels well below saturation. As LaGrega et al.[7] indicate, it is generally accepted that the minimum moisture content necessary for treatment of contaminated soil is 40% of saturation.

The search for effective biological agents to destroy organic contaminants is continuing and consists of an assortment of research activities. The activities ranging from attempts to isolate pure strains from real-life contaminated situations to sophisticated genetic-engineering applications to produce a tailor-made organism capable of degrading a specific compound.[8]

Until recently, the use of bioremediation was somewhat limited by the lack of a thorough understanding of the biodegradation processes, their appropriate applications, their control and enhancement in environmental matrices, and the engineering techniques required for broad application of the technology.[9] Because bioremediation relies on microorganisms, and these microorganisms have a wide range of abilities to metabolize different chemicals, the opportunity to tailor the utilization of bioremediation to address contaminants at specific sites, in specific media, and under specific conditions is available. This selection issue will frequently be encountered during the discussion of *in situ* bioremediation in the following section.

1.2. *In situ* Bioremediation

In situ bioremediation relies on the natural biodegradation that occurs spontaneously in the subsurface. In some instances bioremediation occurs automatically without human intervention when a contaminant is released and the site conditions are such that indigenous microorganisms can degrade the contaminant fast enough to keep it from spreading. This is known as intrinsic bioremediation.

Normally, however, without human intervention the degradation rate under natural conditions is too slow. In this situation, effective degradation requires the construction of engineered systems that deliver growth-stimulating materials to the organism. This is called engineered bioremediation.

In situ bioremediation enhances biodegradation through the stimulation and management of the subsurface microbiological communities. Because remediating a soil-contaminated site is different from remediating a site with groundwater contamination, *in situ* bioremediation technology is reviewed from its application in both soil and groundwater remediation.

1.2.1. *Biodegradation and Bioventing*

In the remediation of contaminated soil, the two most common *in situ* biological treatment technologies are biodegradation and bioventing. *In situ* biodegradation refers to biological treatment in which organic contaminants in subsurface soil are treated using naturally occurring microorganisms or special strains of cultured bacteria, or both.

- The activity of naturally occurring microbes is stimulated by circulating water-based solutions, which contain nutrients, through the contaminated soils to enhance the biodegradation of the organic contaminants. This is done by pumping the liquid microbe solution directly into the soil.
- The contact of microbes, oxygen, and nutrients is facilitated in permeable soil. The proper pH, temperature, and oxygen level should be maintained in order to achieve maximum effectiveness.

Some of the limitations of *in situ* biodegradation are that extensive treatability studies and site characterization may be necessary. Care must be taken not to drive the contaminants downward because the circulation may increase the contaminants' mobility and there is a risk of leaching the contaminants into the groundwater. Remediation times can last five years or more depending mainly on the degradation rates of the site contaminants. *In situ* biodegradation may not be applicable at sites where there are high concentrations of heavy metals, highly chlorinated organics, or inorganic salts.[10]

Bioventing is defined as the delivery of air or oxygen to soil to stimulate the aerobic biodegradation of contaminants. Aerobic biodegradation is a biologically destructive process which occurs in the presence of oxygen. *In situ* bioventing is a modification of the technology referred to as soil vapor extraction (SVE), soil gas extraction, and *in situ* volatilization. The bioventing phenomenon initially was observed during the operation of soil vacuum extraction. Besides the volatile organic compounds (VOCs) removed by the vacuum extraction process, it was found that additional reduction in the VOCs was not readily amenable to the vacuum extraction process. This "secondary" removal has since been attributed to biodegradation.[11]

- Unlike soil vacuum extraction, bioventing employs lower air flow rates that provide only the amount of oxygen necessary for biodegradation while minimizing the transfer of a chemical substance from a liquid phase to a gaseous phase (volatilization) and, therefore, minimizing the release of contaminants to the atmosphere.
- The advantages of gas-phase introduction of oxygen into soils as opposed to liquid-phase (as in the case of *in situ* biodegradation) are that gases diffuse more rapidly than liquids into less permeable subsurface formations and much less gas is required to deliver oxygen at the levels needed to stimulate biodegradation. The use of air as a carrier of oxygen is far more efficient than water.

In the early 1980s, bioventing researchers were mainly concerned with the remediation of soil contaminated with hydrocarbons and realized that venting would not only remove gasoline physically but would also enhance the microbial activity and promote the biodegradation of gasoline. Much of the success of bioventing is due to the use of air as the carrier of oxygen. It is estimated that various forms of bioventing have been applied to more than 1,000 sites worldwide; however, little effort has been given to the optimization of these systems.[12]

Bioventing is most often used at sites with mid-weight contaminants (diesel and jet fuel). Lighter site contaminants (e.g., gasoline) tend to volatilize readily and can be removed more rapidly using the SVE technology. Heavier contaminants such as lubricating oils take more time to biodegrade than the lighter contaminants.[13] A typical bioventing layout using extraction wells is displayed in Figure 11.1.

Norris et al.[12] argue that, in addition to the normal site characterization, soil gas surveys are required to determine the amount of contaminants, oxygen, and carbon dioxide in the vapor phase; these are needed to evaluate metabolic processes under site conditions. An estimate of the soil gas permeability along with the radius of influence of the venting wells is also required. This is needed for designing full-scale systems, including well spacing requirements, and to size blower equipment.

Low soil-moisture content may limit biodegradation and, therefore, impact the effectiveness of bioventing. This is because bioventing tends to dry out the soils. Monitoring of off-gases at the soil surface may also be required. Bioventing, however, does not require expensive equipment and can be left unattended for long periods of time. Only a few personnel are involved in the operation and maintenance of the system. Typically, quarterly maintenance monitoring is conducted.[14]

1.2.2. Air Sparging/Airstripping and Biosparging

For remediating groundwater, bioremediation is typically performed by withdrawing groundwater through extraction wells usually located downgradient of the contaminated site, adding oxygen and nutrients, and reinjecting the enriched water through reinjection wells normally placed upgradient of the site. The extraction and injection of water clearly changes groundwater patterns. The combination of withdrawing groundwater downgradient and injecting water upgradient increases the hydraulic head between the two points, thus increasing the rate of flow of enriched water through the aquifer.

The addition of oxygen and nutrients determines the types of bioremediation. Presented in the following section are two common *in situ* groundwater bioremediation technologies, i.e., oxygen enhancement with air sparging, and oxygen enhancement with hydrogen peroxide (H_2O_2). Air sparging is used for the lighter site contaminants (e.g., benzene, toluene, ethylene, xylene, or BTEX) and less applicable to mid-weight contaminants such as diesel fuel and kerosene. In combination with SVE, air sparging can effectively remove volatile compounds from groundwater particularly when the contaminants are

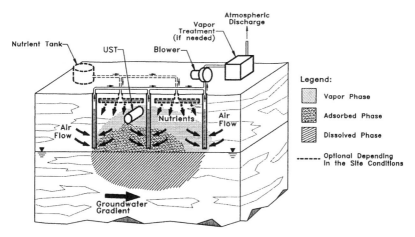

Figure 11.1.   Typical bioventing system using SVE. *Source:* **U.S. Environmental Protection Agency, How to Evaluate Alternative Cleanup Technologies for Underground Storage Tank Sites: A Guide for Corrective Action Plan Reviewers, Washington, D.C., 1994.**

in highly permeable and fairly homogeneous soils. Heterogeneous soils, soils with high organic content and rock formations, and aquifers contaminated with a thick layer of light nonaqueous phase liquids (LNAPLs) restrict the contaminant movement and drastically decrease the efficiency of the system.[15]

Air sparging employs the same concept as bioventing. The difference is that in air sparging, air is injected under pressure below the water table (see Figure 11.2). This increases the groundwater oxygen concentrations and enhances the rate of biological degradation of organic contaminants. Air sparging increases mixing in the saturated zone resulting in increased contact between groundwater and the soil.

- Due to the presence of less permeable strata, sparging can push contaminated groundwater away from the injection point, therefore, a groundwater recovery system may be needed.
- Because air sparging increases pressure in the vadose zone, vapors may rise through the vadose zone and be released into the atmosphere.
- The cost of air sparging is relatively inexpensive. Although the system is relatively new, the related technology, bioventing, is rapidly receiving increased attention from remediation consultants.

Air sparging was initially designed to enhance the *in situ* volatilization of chlorinated solvents from groundwater. When the processes are designed to enhance biodegradation, this technique is called biosparging. The biosparg-

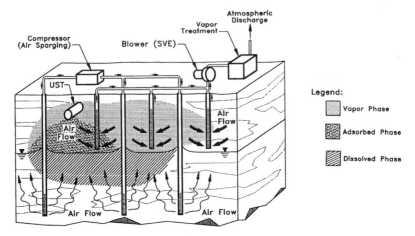

Figure 11.2.   Air sparging system with SVE. *Source:* **U.S. Environmental Protection Agency, How to Evaluate Alternative Cleanup Technologies for Underground Storage Tank Sites: A Guide for Corrective Action Plan Reviewers, Washington, D.C., 1994.**

ing process is similar to air sparging. While air sparging removes constituents primarily through volatilization and limited biodegradation, biosparging promotes the biodegradative process by using lower flow rates than those employed in air sparging. Air sparging is also known as airstripping.

Oxygen enhancement with hydrogen peroxide, H_2O_2, refers to *in situ* bioremediation in which a dilute solution of hydrogen peroxide is circulated throughout a contaminated groundwater zone. The purpose is to increase the oxygen content of the groundwater and, therefore, enhance the rate of aerobic degradation of the organic contaminants by naturally occurring microbes.

- Operation and maintenance costs of the system can be expensive because a continuous source of H_2O_2 must be provided to the contaminated groundwater.
- A groundwater circulation system must be established so that contaminants do not escape from zones of active biodegradation.
- Part of the extracted groundwater can be treated before reinjection, part of it can be discharged elsewhere if permitted by the regulating agencies. A surface treatment system, such as air stripping or carbon adsorption, may be used for treating extracted groundwater.

1.2.3.   Nitrate Enhancement and Co-Metabolism Process

The emerging *in situ* bioremediation technologies are those treatment technologies that are currently developed to address major barriers in bioremediation. In this regard, there are three site characteristics that dictate whether certain substances can be added to stimulate bioremediation, and if so, which ones. These three site factors are[16]

- native microbial population
- types of contaminants
- site hydrogeologic conditions

Native microbes utilize a metabolic process to degrade the contaminants. The process requires an electron donor so that the organisms can oxidize to produce energy. It also needs an electron acceptor that serves as the repository for the electrons taken from the donor. MacDonald and Rittman[16] assert that historically an inadequate supply of electron acceptors, e.g., oxygen, has been the barrier in bioremediation; therefore, engineered bioremediation systems usually are designed to increase the concentration of electron acceptors and when necessary they also supply the elemental nutrients, e.g., nitrogen and phosphorus, that are required for cell synthesis.

Emerging bioremediation technologies that address the inadequacy of electron acceptors include nitrate enhancement in which solubilized nitrate is circulated throughout the groundwater contamination zone. The nitrate solution provides added electron acceptors for biological activity and as a result enhances the rate of degradation.

- The applicability and effectiveness of nitrate enhancement is dependent on homogeneity of the subsurface. In a heterogeneous subsurface, it is very difficult to circulate the nitrate solution throughout every part of the contaminated zone.
- This technology effectively remediates BTEX-contaminated groundwater.

Under laboratory conditions, numerous contaminants have been biodegraded successfully and some have been successfully bioremediated in the field. MacDonald and Rittman[16] state that *in situ* bioremediation is well established for certain classes of petroleum hydrocarbons but is still in the development stage for most other contaminants. The co-metabolism process represents an example of emerging bioremediation technologies that addresses nonpetroleum hydrocarbons, in which water containing dissolved methane and oxygen is injected into groundwater to enhance methanotrophic biological degradation. The solution is used to degrade chlorinated solvents, such as TCE, by co-metabolism.

Co-metabolism means that during the degradation of certain compounds, enzymes are produced and these enzymes fortuitously transform the contaminants. The key requirements for co-metabolism are, therefore, the presence of a substance that, when metabolized, results in the transformation of a contaminant.

- Similar to nitrate enhancement, the applicability and effectiveness of a co-metabolism process also depends on the homogeneity of the subsurface because under heterogeneous conditions it is very difficult to circulate the methane solution throughout every part of the contaminated zone.
- *In situ* application of a co-metabolism process is still at the pilot scale although the development of *ex situ* reactors for methanotrophic TCE biodegradation is progressing well.

The site hydrogeologic conditions under which *in situ* bioremediation works best depend on whether a project is intrinsic or engineered. Because engineered bioremediation uses technology to manipulate site conditions, MacDonald and Rittman[16] argue that natural site characteristics are less important for these projects than for intrinsic bioremediation projects.

Weymann[17] reveals that bioremediation may be the only viable alternative for addressing contaminants such as fuel oils, semi-volatile PNAs, and highly soluble compounds that are not amenable to the traditional pump-and-treat. In spite of its acknowledged potentials and successes, many aspects of bioremediation remain poorly understood. Experts continue to debate issues such as oxygen availability, alternate electron acceptors, and the importance of nutrients and pH.

Brubaker[18] observed that with all the debate and promising advances in *in situ* bioremediation technology, the following three areas are likely to lead to dramatic changes in the next 10 years:

1. *In situ* bioremediation will be increasingly used to manage contaminated groundwater through containment, and less for sitewide contamination removal.
2. Anaerobic biodegradation (or a combination of anaerobic and aerobic processes) will be frequently used to remediate nonchlorinated sites.
3. Although costly and complicated, *in situ* bioremediation of chlorinated solvents will become commercially available, probably utilizing anaerobic processes.

1.3. *Ex situ* Bioremediation

As the name implies, *ex situ* bioremediation is a biological treatment requiring excavation of contaminated soil or extraction of contaminated groundwater before the excavated or extracted media can be treated. Because of its properties, a highly permeable soil can facilitate the flow of contaminants to depths that cannot be treated by *in situ* methods. In such cases, the site must be remediated *ex situ*. The contaminated soil is removed with a backhoe until clean soil is exposed. During the excavation, the lighter hydrocarbons, such as BTEX compounds, may volatilize through aeration. This will result in a slight reduction of the BTEX concentrations.

In anticipation of the *ex situ* bioremediation, a cement slab may be necessary to house the excavated soil in order to prevent leachate runoff from contaminating the clean soil beneath the accumulation of the contaminated soil pile. A simple drainage system or a sprinkler system (for arid environments) may be necessary to maintain proper moisture levels. During the process, the remediated soil can be easily placed back to the pits if *ex situ* bioremediation is performed on-site. The following sections discuss some of the *ex situ* bioremediation technologies.

1.3.1. Landfarming, Slurry, and Solid Phase Biodegradation

There are a number of *ex situ* soil bioremediation technologies. Three of the common technologies discussed in this section are slurry biodegradation, solid phase biodegradation, and landfarming. Slurry biodegradation requires the creation of an aqueous slurry by combining excavated soil or sludge with water and then the slurry is biodegraded aerobically using a bioreactor. The slurry is mixed to keep solids suspended and microorganisms in contact with the contaminants. To enhance the biodegradation process, pH, nutrients, and oxygen in the bioreactor are properly maintained. Once the process is completed, the slurry is dewatered and the treated soil is disposed. Slurry biodegradation is applicable mostly to soils and sludges containing soluble organic contaminants. The treatment results in volume reduction and contaminant destruction.

- A slurry biodegradation is more complex than a solid phase biodegradation. Sizing of materials prior to putting them in the hopper can be difficult and dewatering soil fines after treatment is expensive.
- Ross[19] reveals that this technology degrades waste at a faster rate and requires less land area than solid phase biodegradation.
- The effectiveness of slurry phase treatment depends on the specific soil and chemical properties of the contaminated media. This technology is applicable where the amount of material containing recalcitrant compounds is small.

Solid phase biodegradation means that excavated soils are mixed with soil amendments and staged in aboveground enclosures. The system consists of prepared treatment beds, biotreatment cells, soil piles, and composting.

- A controlled solid phase biological treatment requires a vast area.
- The remediation processes take more time to complete than the slurry phase processes.
- The technology is simple and relatively inexpensive. It can permanently destroy selected organic contaminants.

Landfarming, also known as land treatment or land application, is based on the principle of aerobic decomposition for organic contaminants. It requires that contaminated soil be excavated and placed in a layer of predetermined thickness. This procedure will allow on-site aerobic digestion to occur through aeration and the addition of minerals, nutrients, and moisture.

Tilling or plowing is conducted to facilitate the aeration process and to maximize contact among the contaminants, indigenous microbes, and chemical additives or nutrients. The depth to which the soil is tilled is about 4 to 12 in. and is called the plow zone. The area where treatment occurs is called the treatment zone. The treatment zone extends as far as 5 ft below the soil surface.

Volatile organic compound (VOC) emissions (from the lighter hydrocarbons) may need to be controlled by capturing the vapors before they are emitted to the atmosphere, passing the captured vapors through an appropriate treatment process, and then venting them to the atmosphere. A berm surrounding the landfarm is to be constructed and the groundwater monitoring wells are to be installed at some locations surrounding the farm. If necessary, a leachate collection and leachate treatment system may be provided.

- Landfarming provides better control over operating conditions than the *in situ* options. Compared to bioreactor, the landfarming costs are lower.
- The disadvantages of landfarming are that it is land and management intensive. In landfarming, the organic fraction must be biodegradable at reasonable rates, in order to minimize environmental problems related to migration of contaminants. Proper management of the treatment zone and monitoring of the unsaturated zone are keys to the effectiveness of landfarming and protection of air and water resources.[20]
- Additionally, air emissions and odor are sources of nuisance and probably health hazards.

Land treatment also removes waste contaminants other than organics through physical settling and filtration on the soil that result in the removal of suspended solids. Heavy metals are also removed from solution by adsorption onto soil particles and by precipitation and ion exchange in the soil.

1.3.2. Bioreactor

Bioreactor represents an *ex situ* groundwater biomediation system which typically consists of technologies originally developed for the treatment of industrial wastewater. In this scheme, the contaminated groundwater is fed to the system as influent. The bioreactor, operated under aerobic or anaerobic conditions, contains either suspended or attached biomass of highly active and acclimated microorganisms.

In suspended systems (e.g., activated sludge) contaminated groundwater is circulated in an aeration basin where the microbes aerobically degrade organic matter and produces new cells. In the attached systems (e.g., rotating biological contactors and trickling filters) microorganisms are established on an inert support matrix to aerobically degrade groundwater contaminants.

Before and after the biological treatment, the liquid waste receives several treatment processes. Depending upon the types of groundwater contaminants fed to the system, the pretreatment can be in the form of:[21,22]

1. storage and equalization, i.e., to dampen hydraulic surges and variable organic loadings in the flow system
2. physical/chemical treatment such as precipitation of toxic metals, oxidation, and hydrolysis
3. physical separation including sedimentation of metallic precipitates and removal of floating materials
4. chemical conditioning, i.e., to supply nutrients and adjust pH for optimum results

The post-treatment processes include:

1. physical separation, such as sedimentation, filtration, and flotation
2. final treatment which encompasses adsorption, ion exchange, and microfiltration

In the bioreactor, the dissolved organics are metabolized by the biomass through a biodegradation process. Some organics, however, may not be metabolized but absorbed with colloidals onto the biomass. Some organics volatilize in response to the aeration. This all means that the removal of an organic waste occurs by biodegradation as well as non-biodegradation processes, with no reduction of the toxic nature of the waste. The latter removal is called abiotic loss. In this removal process, the contaminants have been simply transferred to other media which may require further treatment.

2. PHYSICO-CHEMICAL TREATMENT

Physical treatment technologies refer to the processes which are designed to separate and concentrate various components of site contaminants without chemically altering the form of these components. Sedimentation is an example of this process. Chemical treatment, such as oxidation, refers to a process which brings about a chemical transformation of one or more components of contaminants as the primary means of either removing a particular component or reducing its toxicity.[23] Physico-chemical treatment methods integrate both physical and chemical processes as the major features of the technologies. The *in situ* and *ex situ* types of physico-chemical treatments are briefly described in the following paragraphs.

2.1. *In situ* Treatment

An *in situ* physico-chemical treatment technology can be characterized by its application to site remediation for a particular environmental medium, i.e., soil or groundwater. The first part of this section introduces several technologies appropriate for soil remediation. They include soil flushing, solidification/stabilization, and SVE. An *in situ* treatment technology for groundwater remediation, i.e., steam stripping, is addressed after the SVE discussion.

2.1.1. *Soil Flushing, Solidification/Stabilization, and Soil Vapor Extraction*

In utilizing soil flushing technology, water is applied to the soil or injected into the groundwater to raise the water table into the contaminated soil zone. Contaminants are forced to leach into the groundwater. Compatible surfactants may be added to the water to enhance contaminant solubility. The groundwater is extracted and the leached contaminants captured, treated, and removed before the groundwater is recirculated.

- This technique is effective for removing metals and inorganics from coarse soils. It is used effectively only for some organics.
- Soil flushing is appropriate where flushed contaminants and flushing fluid can be contained and recaptured.
- Soil characteristics and uniformity are critical decision-making parameters. Soil porosity and the mobility of contaminants and soil-flushing fluids must be demonstrated before undertaking remediation.

Solidification is a technique that encapsulates hazardous waste into a solid material of high structural integrity.[24] Solidifying fine waste particles is termed microencapsulation; macroencapsulation solidifies wastes in large blocks or containers. Stabilization technologies reduce a hazardous waste's solubility, mobility, or toxicity.

Solidification and stabilization are effective for treating soils and sludges containing metals, asbestos, radioactive materials, inorganics, corrosives and cyanides, and semi-volatile organics. This technique is not effective for treating volatile organics. Solidification eliminates free liquids, reduces hazardous constituent mobility by lowering waste permeability, minimizes constituent leachability, and provides stability for handling, transport, and disposal.[25]

Soil vapor extraction (SVE) is a technology that has been proven effective in reducing concentrations of VOC and certain semi-volatile organic compounds. Principally, a vacuum is applied to the soil matrix to create a negative pressure gradient that causes movement of vapors toward extraction wells (see Figure 11.3). Volatile contaminants are readily removed from the subsurface through the extraction wells. The collected vapors are then treated and discharged to the atmosphere or where permitted, reinjected to the subsurface.

SVE is more successful in removing the lighter (more volatile) compounds; heavier, less volatile compounds are not readily treated by SVE. Bioventing may remove these heavier compounds. SVE, however, is not applicable to all locations. Its efficiency decreases at sites with low soil permeability and at sites with high groundwater tables.

2.1.2. Steam Stripping

The steam stripping method is based on a mass transfer concept which is used to move volatile contaminants from water to air. Steam is injected through an injection well into the aquifer to vaporize volatile and semi-volatile contaminants. The contaminated vapor steam is removed by vacuum extraction, and the contaminants are then captured through condensation and phased separation processes. The vapors from the condenser are treated via activated carbon filters and discharged to the atmosphere. Condensed liquids are transferred to the separators for treatment.

The steam stripping method is relatively more expensive than air stripping. The technology is effective for removing VOCs. It is less effective for treating contaminants that are less volatile or more water soluble, such as acetone. Soil type will significantly affect the effectiveness of this *in situ* groundwater treatment technology.

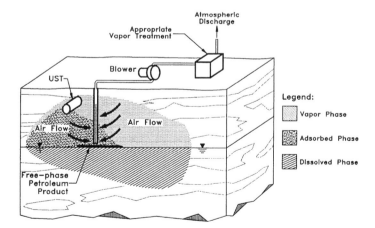

Figure 11.3.   Typical SVE system. *Source:* **U.S. Environmental Protection Agency, How to Evaluate Alternative Cleanup Technologies for Underground Storage Tank Sites: A Guide for Corrective Action Plan Reviewers, Washington, D.C., 1994.**

2.2. *Ex situ* Treatment

In the following section, *ex situ* physico-chemical treatment technologies are categorized into those technologies appropriate for soil treatment and groundwater remediation. Dehalogenation, chemical reduction/oxidation, and solvent extraction are typical *ex situ* technologies for soil treatment while examples for groundwater remediation are carbon adsorption and ultraviolet (UV) oxidation. An emerging *ex situ* technology called supercritical fluids extraction is briefly described at the end of this section.

2.2.1. *Dehalogenation, Chemical Reduction/Oxidation, and Solvent Extraction*

Dehalogenation of organic compounds is divided between haloalipathic and haloaromatic.[26] The classification means that the halogen (fluorine, chlorine, bromine, or iodine) is covalently bonded to either aromatic or alipathic carbon atoms. This treatment technology is used to dehalogenate halogenated compounds. In the haloaromatic method, contaminated soils and reagents are mixed and heated in a treatment vessel. Potassium polyethylene glycolate (KPEG) is widely used as the reagent. The reaction between KPEG and chlorinated organics causes displacement of a chlorine molecule and results in a reduction of toxicity.

Dehalogenation technology requires the excavation of soils contaminated with halogenated organics. This excavation poses a potential health and safety risk to site workers through air emissions and dermal contact; therefore, appropriate personal protective equipment (PPE) must be planned and used properly.

Glycolate dehalogenation of soils containing chlorinated organics of more than 5% requires an excessive volume of reagent. Large amounts of reagent are also required for a site with a soil water content of more than 20%. The treated soil must be washed with water to remove the excess reagent and reaction products. After remediation, the remediated soil usually has a poor physical appearance. Dehalogenation of soils has been successfully field tested in treating chlorinated solvents such as PCBs, dioxin, and furans in soils.

Chemical reduction/oxidation is a chemical conversion of hazardous contaminants to nonhazardous or less toxic compounds. The result is a more stable, less mobile and/or inert material. The agents used to reduce/oxidize soil contaminants are ozone, hydrogen peroxide, hypochlorites, chlorine, and chlorine dioxide. A combination of these chemicals is often necessary.

The chemical conversion is effective for inorganics. Alkaline chlorination is used to oxidize cyanide wastes. Reduction is used for lead, mercury, and chromium. The chemical reduction/oxidation technology requires soil excavation; therefore, proper PPE must be considered to address the health and safety of site workers due to the possibility of dermal contact and inhalation exposure.

Solvent extraction technology involves mixing the solvent and waste in an extractor. The mixing dissolves the organic contaminant into the solvent. The solvent and dissolved organics are then placed in a separator where the organics and solvent are separated and treated.

After the treatment, traces of solvent may remain in the treated solids. Some soil types and moisture content levels may adversely impact the process. Because solvent extraction separates the contaminants without destroying them, solvent extraction is usually used in combination with other treatment technologies.

2.2.2. Carbon Adsorption, UV Oxidation

Carbon adsorption removes organic contaminants from aqueous or vapor phase wastes. Adsorption is a process in which a soluble contaminant is removed from water by contact with an adsorbent. The adsorbent most frequently used is activated carbon, carbon that has been processed to increase the internal surface area. Activated carbon is available in powdered and granular form. Granular activated carbon (GAC) is the popular adsorbent for removal of toxic organic compounds from groundwater.

The principle of carbon adsorption technology is that the targeted waste stream is directed to flow through a series of beds packed with granulated activated carbon which adsorbs the volatile and semi-volatile organics. This technology is appropriate for removing organics from groundwater.

The effectiveness of carbon adsorption is affected by temperature and humidity; therefore, these two factors must be controlled. When the carbon has been saturated with contaminants it can be regenerated on-site, regenerated off-site, or disposed.

Ultraviolet (UV) oxidation technology uses UV radiation, ozone, or hydrogen peroxide to destroy or detoxify organic contaminants as water flows into a treatment tank. The reaction products are dechlorinated materials and chlorine gas. UV light cannot penetrate pollutants effectively in turbid solutions or soil.

2.2.3 Supercritical Fluids Extraction

Supercritical fluids (SCFs) are materials at elevated temperature and pressure that have properties between those of a gas and a liquid. SCF processes represent emerging technologies in the site remediation field. Few full-scale applications of SCF are currently in existence.

In the supercritical fluid extraction technology, the contaminated water is fed into an extraction column. The fluid is pressurized and heated to the critical point. Under these conditions, the organic contaminant dissolves in the SCF. The SCF is then directed to a separator through a pressure reduction valve. In the SCF separator, the organic contaminant is separated from the extracting fluid. The SCF is recompressed and recycled back to the extraction column.

3. THERMAL TREATMENT

Thermal treatment technologies are destruction and removal types of treatment. Most of the technologies are *ex situ*. They include thermal desorption, vitrification, pyrolysis, and incineration. An example of *in situ* thermal treatment (discussed later in this section) is the thermally enhanced SVE technology.

3.1.   Thermal Desorption, Vitrification, Pyrolysis, Incineration

Thermal desorption technology is based on a physical separation system. The process desorbs (physically separates) organics from the soil without decomposition. Volatile and semi-volatile organics are removed from contaminated soil in thermal desorbers at 200 to 600°F for low-temperature thermal desorption (also called soil roasting), or at 600 to 1,000°F for high-temperature thermal desorption.

To transport the volatilized organics and water to the gas treatment system, the process uses an inert carrier gas. The gas treatment units can be condensers or carbon adsorption units which will trap organic compounds for subsequent treatment or disposal. The units can also be afterburners or catalytic oxidizers that destroy the organic constituents.

The bed temperatures and residence times of the desorbers are designed to volatilize selected contaminants, not to oxidize them. Certain less volatile compounds may not be volatilized at low temperatures. When using soil roasting technology, some pre- and post-processing of soil is necessary. The preprocessing activity includes screening of the excavated soils to remove large objects. The large objects may be crushed or shredded and introduced back into the feed material. In the post-processing situation, i.e., after leaving the desorber, soils are cooled, dewatered, and, if necessary, stabilized prior to disposal or reuse. A typical soil roasting technology is provided in Figure 11.4.

In vitrification technology, contaminated soil and sludges are fed into the system. The combustor preheats suspended waste material with glass forming additives to vitrify the materials. Upon the completion of the process, the preheated soil materials are tranferred to the melter, where they form a molten glass. This vitrified, molten material and attendant exhaust gas exit through a channel into a glass- and gas-separation chamber. Proper disposition of the vitrified slag is required. Vitrification is a complex process and requires a high degree of specialized skill and training

Pyrolysis means a chemical decomposition or change due to heating in the absence of oxygen. In practice, it is not possible to achieve a completely oxygen-free atmosphere. Because some oxygen will be present in any pyrolysis system, nominal oxidation will occur.[27] If volatile or semi-volatile materials are present in the waste, thermal desorption will also occur..

Application of pyrolysis to remediation technology involves a two-step process.[28] First, wastes are heated, separating the volatile components (i.e.,

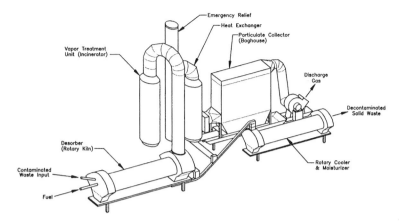

Figure 11.4. Parallel flow rotary LTTD system. *Source:* U.S. Environ-
mental Protection Agency, How to Evaluate Alternative
Cleanup Technologies for Underground Storage Tank Sites:
A Guide for Corrective Action Plan Reviewers, Washington,
D.C., 1994.

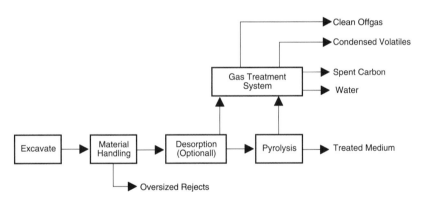

Figure 11.5. Schematic diagram of pyrolysis. *Source:* U.S. Environmental Protection Agency, Engineering Bulletin: Pyrolysis
Treatment, Washington, D.C., 1992.

combustible gases and water vapors) from the nonvolatile char and ash. Then
the volatile components from the first step are burned under the proper conditions to assure incineration of all hazardous components. This two-step
process occurs in the pyrolysis chamber at temperatures of 800 to 2,100°F.

A general schematic of a pyrolysis technology is displayed in Figure 11.5.
The figure shows that after the excavation of the contaminated soil, sludge, or
sediment, the materials are screened for large objects before transfer to the
pyrolysis unit. Depending on the technology design, a desorption unit may be
included in the system.

Incineration technology is intended to permanently destroy organic contaminants. Incineration is a complex system of interacting pieces of equipment and is not just a simple furnace.[29] It represents an integrated system of components for waste preparation, feeding, combustion, and emissions control. An incineration system concept flow diagram is displayed in Figure 11.6. Central to the system is the combustion chamber, or the incinerator. There are four major types of incinerator: rotary kiln, fluidized bed, liquid injection, and infrared.

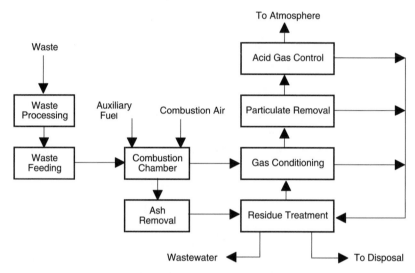

Figure 11.6. **Incineration system concept flow diagram.** *Source:* **U.S. Environmental Protection Agency, Experience in Incineration Applicable to Superfund Site Remediation, Cincinnati, OH, 1988.**

A rotary kiln incinerator is a long, inclined tube that is mounted so that it can be slowly rotated. They are used primarily for the combustion of solids, but liquids and gases may be co-incinerated with solids. A fluidized bed incinerator uses a tubular bed of inert granular material to improve the transfer of heat to the liquids and sludges to be incinerated. A liquid injection incinerator is simply a furnace with fly ash removal that is used to incinerate liquids (with only 10% solid content). Infrared incineration is designed to treat contaminated soils. It is composed of a rectangular carbon steel chamber lined with layers of ceramic fiber blanket. Energy is provided in the form of infrared radiation from silicon carbide resistance heating elements.

3.2.   Thermally Enhanced SVE

The thermally enhanced soil vapor extraction process uses hot air injection or electric heating to increase the mobility of volatile organics and facilitate extraction. This technology includes a system of handling of gases. One of the problems is that debris or large objects buried in the soil can cause operating difficulties. The use of the technology is dependent upon the specific soil and chemical properties of the contaminated site.

4. OTHER TECHNOLOGY

Site remediation technology is certainly not limited to the various types of treatment techniques described in the preceding sections. Remediation related technology also includes those used for sampling, monitoring, and containing a contaminated site. The next section briefly examines these technologies.

4.1. Monitoring and Measurement Technology

Some of the commonly used monitoring and measurement technologies have been elaborated in Chapter 8. As far as EPA's technology development program is concerned, the current preferences are for those technologies that provide cost effective, faster, safer, and better methods than conventional technologies for producing real-time (or near-real-time) data.

EPA is interested in innovative technologies that can detect, monitor, and measure hazardous substances in the subsurface, air, surface waters, and biological tissues, as well as technologies that characterize the physical properties of sites. The types of technologies under this description are:[30]

- chemical sensors for *in situ* measurement
- groundwater sampling devices
- soil and core sampling devices
- soil gas sampling devices
- fluid sampling devices for the vadose zone
- *in situ* and field portable analytical methods
- expert systems that support field sampling or data acquisition and analysis

As of November 1991, 14 technologies actively participated in the monitoring and measurement technology program (MMTP). Table 11.2 lists these technologies in alphabetical order. The table shows that the demonstration or evaluation of 6 of the 14 technologies has been completed. These technologies are identified with asterisks.

As shown in the table, one of the innovative technologies in the monitoring and measurement area is immunoassay. Immunoassay relies on the binding between an antibody and a target analyte. Antibodies or immunoglobulins are proteins that bind to a specific foreign substance. An antibody for immunoassay is developed for a specific analyte. Antibodies are incompatible with typical extraction solvents such as methylene chloride, most oily matrices, and certain reactive waste.[31]

The resulting analyte–antibody compound can be detected through chromogenic reactions, radioactivity, fluorescence, or biosensors. One common immunoassay that uses chromogenic reactions is the enzyme-linked immunosorbent assay (ELISA). In this technique, the selectivity of the antibody for the analyte and the resultant antibody–analyte complex are the basis for the specificity of immunoassays.[32]

ELISA relies on color changes in a solution to indicate the presence or absence of a contaminant. This immunoassay method has been successful in detecting PCB, pentachlorophenol (PCP), and BTEX compounds. An antibody is developed for a specific analyte (e.g., PCP). The antibody is then immobilized at appropriate concentration levels inside a test tube. When a field sample is introduced into the test tube, the target analyte PCP in the sample binds to unoccupied antibody sites. An enzyme–conjugate reagent is developed. Conjugate is a compound (protein) usually covalently bound to an enzyme. This reagent consists of an enzyme and the target analyte.

The bound enzyme–conjugate will react with an added substrate (e.g., tetramethylbenzidine and hydrogen peroxide) to produce a colored product. The darker the color, the lower the contaminant concentration in the sample. The cross-reactivity of the antibody between the target analyte and other similar compounds can confuse the analysis. This problem is similar to interference or lack of chromatographic resolution in classical analytical chemistry.[33]

4.2. Containment Technology

Various containment technologies are used to contain waste or contaminated groundwater, or to keep uncontaminated water from entering the waste. This can be accomplished by employing one of the three most commonly used containment methods:[34]

- groundwater pumping
- establishing a subsurface drainage system
- installation of low permeability barriers

Groundwater pumping is used to manage the groundwater to contain or remove a plume, or to adjust groundwater levels to prevent the formation of a plume. Subsurface drains refer to any type of buried conduit used to convey and collect aqueous discharges by gravity flow. The discharge collection is performed by a sump which is installed as part of the system. The drainage system can only be used in shallow depths.

Low permeability cut-off walls or diversions are installed below ground. They are used to contain, capture, or redirect groundwater flow in the vicinity of a contaminated site. The types of barriers are soil-bentonite slurry walls, cement-bentonite slurry walls, grouted barriers, and sheet piling cut-offs.

Table 11.2 Monitoring and Measurement Technologies Program Participants

Developer	Technology Contact	Technology	EPA project manager	Waste media	Applicable waste Inorganic	Organic
Analytical and Remedial Technology, Inc.,* Menlo Park, CA	Automated volatile organic analytical system	D. MacKay 415-324-2259	Eric Koglin/ Stephen Billets 702-798-2432	Water, air streams	Not applicable	VOCs
Binax Corp., Antox Division, South Portland, ME	Equate® immunoassay	Roger Piasio 207-772-3544	Jeanette Van Emon 702-798-2154	Water	Not applicable	Volatile PAHs
Bruker Instruments,* Billerica, MA	Bruker mobile environmental monitor	John Wronka 506-667-9580	Eric Koglin/ Stephen Billets 702-798-2432	Air streams, water, soil, sludge, sediment	Not applicable	VOCs and SVOCs
CMS Research Corp., Birmingham, AL	MINICAMS	H. Ashley Page 205-773-6911	Richard Berkley 919-541-2439 FTS: 629-2439	Air streams	Not applicable	VOCs
EnSys, Inc.* [developed by Westinghouse Bio-Analytic Systems], Research Triangle Park, NC	ELISA	Stephen Friedman 914-941-5509	Jeanette Van Emon 702-798-2154	Water	Not applicable	PCPs
Graseby Ionics, Ltd., Watford, Herts, England and PCP, Inc.,* West Palm Beach, FL	Ion mobility spectrometry	John Brokenshire/ Martin Cohen 01-44-923-816166/ 407-683-0507	Eric Koglin 702-798-2432	Air streams, vapor, soil, water	Not applicable	VOCs, chloroform, ethylbenzene
HNU Systems, Inc., Newtown, MA	Portable gas chromatograph	Clayton Wood 617-964-6690	Richard Berkley 919-541-2439 FTS: 629-2439	Air streams	Not applicable	VOCs, aromatic compounds, halocarbons

Table 11.2 Monitoring and Measurement Technologies Program Participants (Continued)

Developer	Technology Contact	Technology	EPA project manager	Waste media	Applicable waste	
					Inorganic	Organic
MDA Scientific, Inc., Norcross, GA	Orman Simpson 404-242-0977	Infrared spectrometer	William McClenny 919-541-3158 FTS: 629-3158	Air streams	Nonspecific inorganics	Nonspecific organics
Microsensor Systems, Inc., Springfield, IL	N. L. Jarvis 703-642-6919	Portable gas chromatograph	Richard Berkley 919-541-2439 FTS: 629-2439	Air streams	Not applicable	VOCs
Microsensor Technology, Inc.,* Fremont, CA	Gary Lee 415-490-0900	Portable gas chromatograph	Richard Berkley 919-541-2439 FTS: 629-2439	Air streams	Nonspecific inorganics	Nonspecific organics
Photovac International, Inc., Deer Park, NY	Mark Collins 516-254-4199	Photovac 10S PLUS	Richard Berkley 919-541-2439 FTS: 629-2439	Air streams	Not applicable	VOCs, volatile aromatic compounds, chlorinated olefin compounds
Sentex Sensing Technology, Inc., Ridgefield, NJ	Amos Linenberg 201-945-3694	Portable gas chromatograph	Richard Berkley 919-541-2439 FTS: 629-2439	Air streams, water, soil	Not applicable	VOCs
SRI Instruments, Torrance, CA	Dave Quinn 213-214-5092	Gas chromatograph	Richard Berkley 919-541-2439 FTS: 629-2439	Air streams	Not applicable	VOCs
XonTech, Inc., Van Nuys, CA	Matt Young 818-787-7380	XonTech sector sampler	Joachim Pleil 919-541-4680 FTS: 629-4680	Air streams	Not applicable	VOCs

Source: U.S. Environmental Protection Agency, The Superfund Innovative Technology Evaluation Program: Technology Profiles — Fourth Edition, Washington, D.C., 1991.

* Demonstration or evaluation complete.

A slurry wall is the most common and least costly technique. The walls are constructed in a vertical trench which is excavated under a slurry condition. Soil bentonite and cement bentonite are the types of materials used to backfill the slurry trench. Grouting occurs when a particular fluid is injected into a soil mass to reduce water flow and strengthen the formation. Sheet piling cut-off uses either wood, steel, or precast concrete.

Barrier technologies are not actual remediation technologies, but their use does not preclude and may advance remedial activities. The use of cut-off walls to contain groundwater contamination close to the source, for example, can reduce drastically the pump-and-treat groundwater cleanup time by cutting pumping needs as much as 90%.

Pump-and-treat systems are widely used in site remediation and are effective for containing contaminant plumes and removing contaminant mass. The methods use one or more extraction wells. The groundwater removed during pumping is treated at the surface and discharged. The discharge may occur either to a point where it does not supply any recharge to the aquifer, or to a point in the aquifer itself where it does not influence the velocity field around the pumping wells or enhance natural biodegradation.[36] The two most common methods to treat groundwater are with GAC adsorption and airstripping in conjunction with a carbon adsorption.

The pump-and-treat technique, however, is limited in its effectiveness for complete aquifer restoration.[37] Some contaminant source areas simply cannot be removed and therefore the aquifer cannot be remediated properly. Adequate characterization and identification of the contaminant source area are the keys to successful pump-and-treat groundwater remediation.[38]

In a site contaminated with petroleum hydrocarbons, for example, the hydrocarbons usually behave as LNAPLs, i.e., they drain through the unsaturated zone under the influence of gravity, then accumulate near the water table. Pumping activity in this type of site is normally discontinued when the hydrocarbons can no longer be detected in the extracted water. Unfortunately, after pumping ceases, the concentrations of hydrocarbons rebound in the water. These problems are common with the pump-and-treat approach to aquifer remediation.[39]

As a result of the limitations in pump-and-treat, the systems are being supplanted at many sites by other technologies.[40,41] A coordinated, phased implementation of various technologies in the Savannah River site, near Aiken, SC, has resulted in rapid cleanup progress.[42] One of the largest government-built groundwater pump-and-treat systems in the U.S. was constructed in 1994 in Muskegon County, MI, utilizing a combination of various technologies to remediate extremely varied toxic contaminants.

12   TECHNOLOGY SCREENING AND DATA REQUIREMENTS

A brief introductory description of remediation technologies currently available in the market was provided in Chapter 11. The chapter examined the potential applicability of a technology to remediate contaminated sites. Given the variety of alternatives, planners by necessity must reduce the number of technologies to consider and then identify the most appropriate technologies for more detailed evaluation prior to technology selection. This chapter discusses the selection process of those technologies.

Because the use of a technology is dependent upon the characteristics of the contaminated site, it is necessary for a planner to be familiar with the site's physical and chemical data. This chapter also examines the site soil and water characteristics for which observations and measurements should be compiled.

Basically, the application and effectiveness of a technology to remediate a site depends on four major factors:

1. The local factor, i.e., the acceptability of the technology to local community.
2. The regulatory factor, i.e., the federal, state, and local statutes and regulations that apply to the operation of the technology. Examples are those laws that regulate discharges to surface waters and air emissions. Discharges may be necessary as part of the operation of a technology.
3. The technology factor, i.e., the technical, cost, and performance characteristics of the technology.
4. The site factor, i.e., the hydrogeology, soil structure, particle size, organic content, pH, and moisture content of the site.

The site factor, community acceptance, and local regulatory requirements are all site-specific conditions that affect the applicability of a technology to a site. The technical, cost, and performance characteristics of a technology can normally be generalized for reference to the technology screening process. The following section describes that particular aspect.

1. SCREENING OF REMEDIATION TECHNOLOGIES

The technology screening process can be conceptualized as a decision-making flow chart, displayed in Figure 12.1.[43] The first question to be answered in this process is whether an available technology is applicable to a site based on the site's soil and groundwater properties, and the contaminant's physical and chemical characteristics. The applicability of a technology is assessed in terms of its implementability, effectiveness, and cost. The assessment can be conducted through a literature review including contacting the Environmental Protection Agency's (EPA) technical support offices. Computerized on-line information services and contaminant treatability databases are available for public use.

The computerized on-line information system (COLIS) is operated by the Technical Information Exchange (TIX) at the EPA's Risk Reduction Engineering Laboratory (RREL) in Edison, NJ. The system can be accessed by calling 908-548-4636. COLIS houses several files such as:

- underground storage tank (UST) case history files containing information on UST releases
- Superfund innovative technology evaluation (SITE) applications analysis reports which contain performance and cost information on technologies evaluated under the SITE program
- RREL treatability database

Developed by the RREL in Cincinnati, OH, the RREL treatability database provides data on the treatability of contaminants in soil, sludge, sediment, and water. For each contaminant, the database displays the physical and chemical properties as well as treatability data, such as technology types, matrices treated, study scale, and cleanup levels achieved.

If a technology is not applicable to a site based on available information, the technology will be screened out from the process. If it is applicable, the next test is whether the technology is feasible to be implemented at the site. A feasibility test utilizes the current site information. The information must be able to be used for analyzing whether the technology can achieve the following results:

1. the overall protection of human health and the environment
2. the compliance with applicable or relevant and appropriate requirements
3. the long-term effectiveness and permanence
4. the reduction of toxicity, mobility, and volume of contaminants
5. the short-term effectiveness

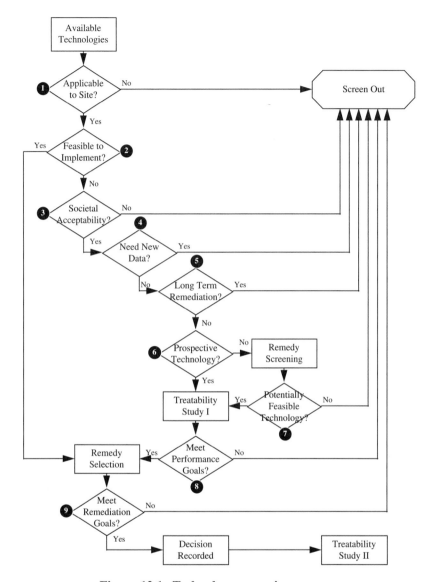

Figure 12.1. Technology screening process.

6. no significant barriers in implementation
7. relatively cost effective
8. compliance with state regulations and local ordinances
9. community acceptance

If the technology passed the feasibility screening, a detailed technology analysis can be conducted; otherwise it will be screened out, unless the technology is socially accepted, does not require additional data for further remedy screening, and the technology can offer short-term effectiveness. If these three conditions apply, the technology is examined whether it is truly a prospective technology. The examination focuses on a qualitative evaluation or comparative performance rating.

In 1993, EPA and the Air Force developed a matrix that summarizes the strengths and limitations of innovative, as well as conventional technologies for the remediation of soils, sediments, and sludges; groundwater; and off-gases. The matrix was developed with intensive input from 30 technical experts representing researchers, technology developers, and users.[44] The matrix can be used to evaluate the qualitative performance of prospective technology. This matrix will be described later in this chapter.

At the remedy screening phase, the technology evaluation is performed in the laboratory. Some technologies may require additional small-scale field tests. The evaluation will generate qualitative data for use in assessing whether the technology is potentially feasible for treating a contaminant-matrix combination. No cost data or design information are generated. In the test, a single contaminant is used as an indicator if a reduction in toxicity, mobility, or volume is occurring. The test is normally performed in a few days at a cost of $10,000 to $50,000.

A prospective technology with an acceptable qualitative performance based on a comparative rating using the technology performance matrix, or through remedy screening, is then subject to the first treatability study. A treatability study's contents and process are described later in this chapter. The findings of a treatability study will determine whether a technology can meet the performance goals, i.e., to achieve a certain cleanup level, and the associated cost. Once the technology is qualified, it will be evaluated further in a detailed technology analysis, also called a remedy selection study.

In this evaluative study, operating parameters of a technology are investigated. The choice of parameters to be investigated is based on the cleanup goals and other waste-specific performance goals. The analysis is performed in the laboratory or pilot-scale field treatment test. Equipment should be designed to simulate the basic operations of the full-scale treatment.

The results of the evaluation are quantitative data that can be used to determine whether the tested technology can meet the cleanup goals and at what cost. When this remediation goal is achieved, the decision to choose the technology is recorded in the Superfund program in a record of decision (ROD). In the Resource Conservation and Recovery Act (RCRA) program, the decision is contained in the format of the remediation permit. Depending on its scale and complexity, a remedy selection study can be conducted at a total cost of $50,000 to $250,000.

The final step in the technology screening process is the execution of the second treatability study. The purpose of the study is to generate the detailed design, cost, and performance data for optimizing the implementation of the selected remedy.

1.1. Remedy Screening Matrix

Intended to be used for the remedy screening process, a matrix was developed to evaluate the qualitative performance of prospective technologies. In the matrix, 48 technologies are evaluated against 13 evaluation factors. Most of those 48 technologies have been introduced in Chapter 11. The 13 evaluation factors address performance, technical merits, cost, development status, and institutional issues (see Table 12.1).

Eight of the 13 factors involve a comparative rating. In this process, technologies are assigned one of three possible ratings: better, average, or worse. In the matrix, the rating levels are indicated by a shape, where a square represents a better rating, a circle means average, and a triangle denotes the worst rating level. In some instances, the rating cannot be conducted due to the inadequacy of information.

The eight comparative rating factors are

1. Overall cost. This includes the design, construction, and operation and maintenance (O&M) costs of the main process that defines a technology. Excavation costs for *ex situ* soil technologies are assumed to be $55.00/metric ton ($50.00/ton) and pumping costs for *ex situ* groundwater technologies are assumed to be $0.07/1,000 l ($0.25/1,000 gal). Excluded in the total costs are mobilization and demobilization, and pre- and post-treatment costs.
2. Commercial availability. This is the number of technology vendors that can design, construct, and maintain the technology.
3. Minimum contaminant concentration available by the technology. This is measured in milligrams/kilogram (mg/kg) for soil technologies, micrograms/liter (μg/l) for groundwater, and mg/kg or μg/kg for off-gases.
4. Time to complete remediation of a "standard" site using the technology. The standard is 18,200 metric tons (20,000 tons) for soil and 3,785,000 l (one million gal) for groundwater.
5. Degree of system reliability and maintainability when using the technology.
6. Level of awareness of remediation consulting community regarding the availability of the technology.
7. Degree of regulatory and permitting acceptability of the technology.
8. Level of community acceptability to the technology.

The remaining five factors demonstrate performance-related questions. Direct answers to these questions are displayed in the matrix. The five technology performance factors are

1. Is the technology capital intensive, i.e., with significant costs for design and construction; or O&M intensive, i.e., with significant costs for labor, operation, maintenance, and repair?

Table 12.1 Remediation Technologies Screening Matrix

REMEDIATION TECHNOLOGIES SCREENING MATRIX

NOTE: There are factors that may limit the applicability and effectiveness of any of the technologies and processes listed below. These factors are discussed in the *Remediation Technologies Screening Matrix Reference Guide*. This Matrix should always be used in conjunction with the *Reference Guide*, which contains additional information that can be useful in identifying potentially applicable technologies.

Technology	Reference Guide Page Number	Status — (F)ull-scale or (P)ilot-scale	Contaminants/Pollutants Treated [a]	Overall Cost	Capital (Cap) or O&M Intensive	Commercial Availability	Typically Part of a Treatment Train (excludes off-gas treatment)	Residuals Produced — (S)olid, (L)iquid, or (V)apor	Minimum Contaminant Concentration Achievable	Addresses (T)oxicity, (M)obility, or (V)olume	Long-Term Effectiveness/ Permanence	Time to Complete Cleanup	System Reliability/ Maintainability	Awareness of Remediation Consulting Community	Regulatory/Permitting Acceptability	Community Acceptability
SOIL, SEDIMENT AND SLUDGE																
In Situ Biological Processes																
Biodegradation	21	F	3,4,5 1,2,6	○	O&M	■	No	None	○	T	Yes	▲	▲	○	▲	■
Bioventing	23	F	3,4,5 1,2,6	■	Neither	■	No	None	○	T	Yes	○	○	○	○	■
In Situ Physical/Chemical Processes																
Soil Vapor Extraction (SVE)	25	F	1,3,5	■	O&M	■	No	L	○	V	Yes	○	■	■	■	■
Soil Flushing	27	P	1,3,7 2,4-6	I	O&M	■	No	L	▲	V	Yes	▲	○	○	○	○
• Solidification/Stabilization	29	F	7 2,4,6	■	Cap	■	No	S	M	I	■	■	■	○	○	
Pneumatic Fracturing (enhancement)	31	P	1-7	■	Neither	▲	Yes	None	NA	M	Yes	NA	■	▲	I	I
In Situ Thermal Processes																
Vitrification	33	P	7 1-6	▲	Both	■	No	L	NA	M	Yes	■	▲	○	▲	▲
Thermally Enhanced SVE	35	F	2,4,6 1,3,5	○	Both	○	No	L	○	V	Yes	■	○	○	○	○
Ex Situ Biological Processes (assuming excavation)																
Slurry Phase Biological Treatment	37	F	3,5 1,2,4,6	○	Both	○	No	None	○	T	Yes	○	○	○	○	○
Controlled Solid Phase Bio. Treatment	39	F	3,5 1,2,4,6	■	Neither	■	No	None	○	T	Yes	○	■	■	■	○
• Landfarming	41	F	3,5 1,2,4,6	■	Neither	■	No	None	○	T	Yes	▲	■	■	○	○
Ex Situ Physical/Chemical Processes (assuming excavation)																
Soil Washing	43	F	2,4,5,7 1,3,6	○	Both	○	Yes	S,L	○	V	Yes	■	○	○	○	○
• Solidification/Stabilization	45	F	7 2,4,6	○	Cap	■	No	S	NA	M	I	■	■	○	○	
Dehalogenation (Glycolate)	47	F	2,6 1	▲	Both	○	No	L	■	T	Yes	▲	■	○	○	○
Dehalogenation (BCD)	49	F	2,6 1	▲		○	No	V		T	Yes	I	I	▲	I	I
Solvent Extraction (chemical extraction)	51	F	2,4,6 1,3,5	▲	Both	○	No	L	○	V	Yes	▲	○	○	○	○
Chemical Reduction/Oxidation	53	F	7 3-6	○	Neither	■	Yes	S	NA	T,M	Yes	■	■	○	■	○
Soil Vapor Extraction (SVE)	55	F	1,3	■	Neither	■	No	L	○	V	Yes	○	■	■	○	○
Ex Situ Thermal Processes (assuming excavation)																
Low Temperature Thermal Desorption	57	F	1,3,5 2,4,6	■	Both	■	Yes	L	■	V	Yes	○	■	■	○	○
High Temperature Thermal Desorption	59	F	2,4,6 1,3,5	○	Both	■	No	L	■	V	Yes	○	■	■	○	○
• Vitrification	61	F	7 1-6	○	Both	○	No	L	NA	M	Yes	○	■	■	▲	▲
• Incineration	63	F	2,4,6 1,3,5	▲	Both	■	No	L,S	■	T	Yes	○	■	○	▲	▲
Pyrolysis	65	P	2,4,6 1,3,5	▲	Both	▲	No	L,S	■	T	Yes	■	I	○	○	○
Other Processes																
• Natural Attenuation	67	NA	3,4,5 1,2,6	■	Neither	■	No	None	○	T	Yes	■	■	○	▲	■
Excavation and Off-Site Disposal	71	NA	1-7	▲	Neither	■	No	NA	NA	M	No	■	■	■	■	■
GROUNDWATER																
In Situ Biological Processes																
• Oxygen Enhancement with H₂O₂	73	F	3,4,5 1,2,6	○	O&M	○	No	None	■	T	Yes	○	▲	■	○	■
• Co-metabolic Processes	75	P	1,2 3-6	○	O&M	▲	No	None	■	T	Yes	○	▲	▲	I	I
Nitrate Enhancement	77	P	3,4,5 1,2,6	○	Neither	▲	No	None	■	T	Yes	○	▲	▲	I	I
Oxygen Enhancement with Air Sparging	79	F	3,4,5 1,2,6	■	Neither	○	No	None	■	T	Yes	○	○	■	○	○
In Situ Physical/Chemical Processes																
• Slurry Walls (containment only)	81	F	1-7	■	Cap	■	NA	NA	NA	M	I	▲	■	○		
Passive Treatment Walls	83	P	1,2,7 3,4,5	▲	Cap	▲	No	S	I	T	I	▲	I	▲	○	I
Hot Water or Steam Flushing/Stripping	85	P	2,4,5 1,3	○	O&M	■	Yes	L,V	○	V	Yes	■	○	■	○	○
Hydrofracturing (enhancement)	87	P	1-7	○	Neither	I	Yes	None	NA	M	Yes	■	▲	○	■	■
Air Sparging	89	F	1,3,5	■	Neither	■	Yes	V	○	V	Yes	○	○	■	○	■
Directional Wells (enhancement)	91	F	1-7	○	Neither	I	Yes	L,S	NA	NA	Yes	■	○	○	■	■
Dual Phase Extraction	93	F	1,3,5	○	O&M	■	No	L,V	○	V	Yes	■	○	○	○	○
Vacuum Vapor Extraction	95	P	1,2,5 3,4,6,7	○	Cap	▲	No	L,V	○	V	Yes	■	○	▲	○	○
• Free Product Recovery	97	F	4,5	■	Neither	■	No	L	○	V	Yes	○	■	○	○	■
Ex Situ Biological Processes (assuming pumping)																
Bioreactors	99	F	3,4,5 1,2,6	■	Cap	■	No	S	■	T	Yes	NA	○	○	■	○
Ex Situ Physical/Chemical Processes (assuming pumping)																
• Air Stripping	101	F	1,3 2,4,5	■	O&M	■	No	L,V	■	V	Yes	NA	○	■	▲	○
• Carbon Adsorption (liquid phase)	103	F	2,4 1,5-7	■	O&M	■	No	S	■	V	Yes	NA	■	■	○	○
UV Oxidation	105	F	1,2,6 3,5	○	Cap	■	No	None	■	T	Yes	NA	○	○	○	○
Other Processes																
• Natural Attenuation	107	NA	3,4,5 1,2,6	■	Neither	■	No	None	■	T	Yes	▲	■	○	▲	■
AIR EMISSIONS/OFF-GAS TREATMENT																
• Carbon Adsorption (vapor phase)	111	F	1-6	■	Neither	■		S	■	V	Yes	NA	■	■	○	○
• Catalytic Oxidation (non-halogenated)	113	F	3,4,5	■	Neither	■		None	■	T	Yes	NA	■	■	○	○
• Catalytic Oxidation (halogenated)	115	F	1,2	■	Neither	■	NA [b]	None	■	T	Yes	NA	■	■	○	○
Biofiltration	117	F	3,5 1	■	Neither	○		None	■	T	Yes	NA	○	▲	I	I
• Thermal Oxidation	119	F	3,4,5	■	Neither	■		None	■	T	Yes	NA	■	■	○	○

[a] The listing of contaminant groups is intended as a general reference only. A technology may treat only selected compounds within the contaminant groups listed. Further investigation is necessary to determine applicability to specific contaminants.

[b] The definition of this factor is not applicable to these technologies. They are, by design, the final step in treatment processes.

• Conventional technologies/processes

Contaminant Codes
1 - Halogenated volatile organics
2 - Halogenated semivolatile organics
3 - Non-halogenated volatile organics
4 - Non-halogenated semivolatile organics
5 - Fuel hydrocarbons
6 - Pesticides
7 - Inorganics

Target contaminants are listed first and in bold type

Rating Codes
■ Better
○ Average
▲ Worse
I Inadequate information
NA Not applicable

Source: U.S. Environmental Protection Agency, *Remediation Technologies Screening Matrix and Reference Guide,* Washington, DC, 1993.

2. Is the technology typically part of a treatment train, i.e., is additional treatment to remediate the contaminated media necessary after the use of the technology?
3. What is the physical form of the residuals generated when the technology is used? Is it solid, liquid, or vapor?
4. What parameter of the contaminated media is the technology primarily designed to address? Is it toxicity, mobility, or volume?
5. What is the long-term effectiveness of the technology? Does the use of the technology maintain the protection of human health and the environment over time after the remediation objectives have been met?

The matrix provides a screening of prospective technologies in a rapid manner. Assume a technology is sought for remediating soils contaminated with fuel hydrocarbons. Nine technologies are listed in the matrix. They are bidegradation, bioventing, SVE, slurry phase bioremediation, solid phase bioremediation, landfarming, soil washing, LTTD, and natural attenuation.

From the nine candidates, two technologies (biodegradation and natural attenuation) can be screened out because of the poor regulatory acceptance and the poor remediation completion. Natural attenuation also received the worst rating for community acceptance. Slurry phase bioremediation and soil washing can also be deleted from the list because they are just the "average" technologies. Of the remaining five technologies, bioventing and SVE represent the better candidates, followed by LTTD.

- Bioventing is the better candidate if the remediation goal is primarily to address the toxicity of soil contaminants.
- SVE is the better choice if the goal is to achieve the optimum volume of remediation; however, the technology is O&M intensive.
- LTTD is similar to SVE and will achieve remediation sooner although it requires capital intensive equipment.

1.2. Treatability Study

A treatability study evaluates a technology and determines whether the technology can achieve a specified cleanup goal and at what level of cost. Treatability studies should be performed in a systematic way to ensure that the data generated can support remedy selection and remediation. EPA has developed a protocol for conducting treatability studies. The protocol consists of the following elements:[45]

1. Establish data quality objectives (DQOs). The DQOs developed for remedy screening are stated in qualitative terms such as "go" or "no-go" statements. For remedy selection, the DQOs are primarily quantitative in nature, e.g., the reduction of PCB to <30 ppm in soils, which is the target cleanup goal established for the site.

2. Identify sources for treatability studies. The sources are qualified remedial contractors or technology vendors with experience and technical expertise to perform the study. For remedy screening, no special equipment is required. Vendor-specific technology may be necessary for remedy selection. More complex tests involving the use of specialized equipment can be required for Treatability Study II.

3. Issue a work assignment that outlines the scope of work to be provided by the contractor. The work assignment describes the site, the objectives of the study, the approach used to perform the study, deliverables, and scheduling.

4. Prepare a workplan to ensure that the data collected in this study are useful for evaluating the performance of a technology. The workplan, prepared by the selected contractor, contains a technical approach for implementing the tasks outlined in the work assignment. The workplan includes, among other things:
   - treatment technology to be tested
   - test design and procedures
   - sampling, data analysis, and interpretation
   - health and safety
   - residual management
   - community relations
   - scheduling

5. Prepare a sampling and analysis plan. The goal is to ensure that samples collected for the treatability study are representative and that the quality of the analytical data generated is known.

6. Prepare a health and safety plan. The plan covers hazard analysis, employee training, personal protective equipment (PPE), medical monitoring, site control, spill containment program, decontamination procedures, confined space entry procedures, and an emergency response plan.

7. Conduct community relations activities. This will provide interested persons an opportunity to comment on and participate in decisions concerning site remediation in a timely manner.

8. Comply with regulatory requirements such as the Comprehensive Environmental Response, Compensation, and Liability Act (CERCLA) and RCRA provisions; also air quality regulations from local ordinances and DOT shipping requirements for off-site treatability testing may apply to the study.

9. Implement the study. This includes sampling the waste, testing, collecting, and analyzing samples of the treated waste and residuals.

10. Analyze and interpret the data. Data interpretation must be based on the test objectives formulated prior to testing.

11. Report the results. Complete and accurate reporting is critical because the remedial decisions will be based partly on the findings of the study.

2.   SITE-SPECIFIC DATA REQUIREMENTS

The screening of remediation technologies conducted through the framework outlined in the preceding paragraphs has shown that the selection of a technology to remediate a site is determined by site-specific data. How this site data affects a technology is the subject of this section's discussion. As a general rule, the applicability of a technology for treating contaminated soil and groundwater is determined by the type, concentration, and distribution of site contaminants as well as the physical and chemical characteristics of the soil and groundwater of the contaminated property.

This means that knowledge of the site contaminants and soil and groundwater characteristics of the site can be used to facilitate the screening of promising technologies. Other kinds of information that can be equally helpful in evaluating the potential success of a technology include site waste handling history (how and where wastes were disposed), site stratigraphy, and topographic and hydrologic details.

2.1.   Soil Data

Soil data required for assessing a technology consists of the traditional engineering properties of soils (e.g., particle size, permeability, density) and soil chemistry including contaminants and oxygen demands. Analytical data required for groundwater are chemistry, oxygen demand, and pH. In the chemical characterization of both soils and groundwater, it is important to know the existence of metals and organic chemicals, whose presence or absence reflects the historical waste-handling activities of the site.

In 1992, EPA published a document that shows the relationship between soil and groundwater characteristics and specific treatment technologies.[46] The relationship is displayed in two tables, one for soil contamination and the other for groundwater. The premise that underlies the development of the tables is that some properties of the soils or groundwater may support the implementation of a technology, some may inhibit a particular process.

Table 12.2 displays the relationship between soil data and each treatment technology group. In the table, the technologies are classified into five groups: physical, chemical, biological, thermal, and solidification/stabilization. The relationship between soil parameters and the technology group is identified in terms of ratings. The rating values are assigned as "higher" and "lower" in showing the tendency of the soil parameters to enhance or to inhibit particular technologies. Where no rating is shown for a characteristic, the effect on the associated technology is considered inconsequential. The rating scheme introduced in the table applies to a technology group (e.g., physical treatment), not to a particular technology within that group (e.g., soil washing).

Some soil conditions are process-limiting. Process-limiting characteristics such as pH or moisture content may sometimes be adjusted.[47] Moisture may hinder the movement of air through the soil in vacuum extraction systems.[48]

**Table 12.2 Soil Characteristics that Assist
in Treatment Technology Preselection**

Characteristic	Treatment technology group				
	Physical	Chemical	Biological	Thermal	S/S
Particle size	■	▼	▼	■	■
Bulk density	▼			■	
Particle density	■				
Permeability	■		■		
Moisture content	▼		■	❑	❑
pH and Eh		▼	▼	▼	
Humic content	❑	❑	❑	▼	❑
Total organic carbon (TOC)		▼	■	■	▼
Biochemical oxygen demand (BOD)			■		
Chemical oxygen demand (COD)		■	■		
Oil and grease	▼	❑			❑
Organic contaminants					
Halogenated volatiles	▼	▼	❑	■	❑
Halogenated semi-volatiles	▼	▼	❑	■	❑
Nonhalogenated volatiles	▼	▼	▼	■	❑
Nonhalogenated semi-volatiles	▼	▼	▼	■	❑
PCBs	▼	▼	▼	■	❑
Pesticides	▼	▼	▼	■	❑
Dioxins/furans	▼	▼	▼	■	❑
Organic cyanides	▼	▼	▼	■	❑
Organic corrosives	▼	▼	▼	❑	❑
Light nonaqueous phase liquid	▼		▼	■	❑
Dense nonaqueous phase liquid	▼		▼	▼	❑
Heating value (Btu content)				■	
Inorganic contaminants					
Volatile metals		▼		❑	
Nonvolatile metals	■	▼	❑	❑	■
Asbestos				❑	■
Radioactive materials	▼	▼	❑	❑	▼
Inorganic cyanides		▼		▼	▼
Inorganic corrosives		▼		❑	▼

Table 12.2 Soil Characteristics that Assist
in Treatment Technology Preselection (Continued)

Characteristic	Treatment technology group				
	Physical	Chemical	Biological	Thermal	S/S
Reactive contaminants					
Oxidizers		▼			
Reducers		▼			

■ = Higher values support preselection of technology group.
❏ = Lower values support preselection of technology group.
▼ = Effect is variable among options within a technology group.
Where no symbol is shown, the effect of that characteristic is considered inconsequential.
Source: U.S. Environmental Protection Agency, Engineering Bulletin: Technology Preselection Data Requirements, Cincinnati, OH, 1992.

Soil moisture affects the application of vitrification and other thermal treatments by increasing energy requirements, thereby increasing costs. On the other hand, increased soil moisture favors *in situ* biological treatments.[49]

Many treatment technologies are affected by the pH of the waste being treated. Low pH can interfere with chemical oxidation and reduction processes. High soil pH can improve the feasibility of applying chemical extraction and alkaline dehalogenation processes.[50] Eh represents the oxidation reduction potential of a material. The Eh of the solid phase must be greater than that of the organic chemical contaminant for oxidation in soil to occur. Maintaining a low Eh in the liquid phase will enhance decomposition of certain halogenated organic compounds.[51]

Particle size distribution in soils affects many technologies. Sands and fine gravels are generally easiest to deal with. Where the soil is composed of large percentages of silt and clay, soil washing may not be effective because of the difficulty of separating fine particles from each other and from washing fluids.[52] *In situ* technologies dependent on the subsurface flowability of fluids (e.g., soil flushing, steam extraction, vacuum extraction biodegradation) will be negatively influenced by the impending effects of clay layers.[53] Heterogeneous soil and waste composition may produce nonuniform feed streams for incineration that result in inconsistent removal rates.[54]

Soil bulk density and particle size distribution are major factors in determining if proper mixing and heat transfer will occur in fluidized bed reactors. Soil particle density is important, for example, in determining the settling velocity of suspended soil particles in flocculation and sedimentation processes.

Soil permeability is one of the controlling components in the effectiveness of *in situ* treatment technologies. The ability of soil flushing fluids to remove contaminants can be limited by low soil permeability or by variations in the permeability of different soil layers.[55] Low permeability also reduces the effectiveness of *in situ* vitrification by slowing vapor releases.

The impacts of high soil humic content (humus) on the effectiveness of a technology are usually unfavorable due to the strong adsorption of the contaminant by the organic material. High humic content can exert an excessive oxygen demand, adversely affecting bioremediation and chemical oxidation.[56]

Total organic carbon (TOC) includes the carbon from organic chemical contaminants and the carbon from naturally occurring organic material and is often used as an indicator of the amount of waste available for biodegradation. Biochemical oxygen demand (BOD) represents the oxygen consumption by the organic material which is readily biodegraded and can be used as an estimate of the biological treatability of the soil contaminants. Chemical oxygen demand (COD) is a measure of the oxygen equivalent of the organic content in a sample that can be oxidized by a strong chemical oxidant. COD and BOD sometimes can be correlated.

Oil and grease will coat the soil particles when they are present in a soil, resulting in weaker bonds between the soil and cement in cement-based solidification. Oil and grease can also interfere with the chemical oxidation and reduction reactions, thus reducing the efficiency of those reactions.

In evaluating technology performance, understanding the site organic and inorganic contaminants is critical. Soils may contain organic chemicals that are not miscible with water and float on top of the water table. Liquid that is lighter than water is known as light nonaqueous phase liquid (LNAPL), and liquid that is heavier than water is called dense nonaqueous phase liquid (DNAPL). Both types can be physically separated from water within the soil particularly if they are not adsorbed to soil particles.

The organic chemical content of the contaminated soil is related to the fuel value and a high fuel value favors thermal treatment. The problem with soil having high concentrations of halogen is that the halogen content will lead to the formation of corrosive acids in the incineration systems. Volatile metals, on the other hand, generate emissions that are difficult to remove, and nonvolatile metals remain in the ash.[57]

A specific technology can be selected because of the presence of one or more of the chemical groups. When radioactive materials are found at a site, treatment options are limited to volume reduction and permanent containment. In the presence of notable amounts of dioxin or furans, regardless of the concentration of other organics, thermal treatment is preferable.

2.2. Groundwater Data

In many sites, the contaminated groundwater and soil may have the same type of contaminants. This is not uncommon because the contaminants normally came from the same previous waste management activities. Groundwater data that affect the performance of a remediation technology are tabulated in Table 12.3.

Table 12.3 Water Characteristics that Assist in Treatment Technology Preselection

Characteristic	Treatment technology group			
	Physical	Chemical	Biological	Thermal
pH, Eh		▼	▼	▼
Total organic carbon (TOC)		▼	■	■
Biochemical oxygen demand (BOD)			■	
Chemical oxygen demand (COD)		■	■	
Oil and grease	▼	❏		
Suspended solids	▼	▼	▼	
Nitrogen and phosphorus			▼	
Organic contaminants				
Halogenated volatiles	▼	▼	❏	■
Halogenated semi-volatiles	▼	▼	❏	■
Nonhalogenated volatiles	▼	▼	▼	■
Nonhalogenated semi-volatiles	▼	▼	▼	■
PCBs	▼	▼	▼	■
Pesticides	▼	▼	▼	■
Dioxins/furans	▼	▼	❏	■
Organic cyanides	▼	▼	▼	■
Organic corrosives	▼	▼	▼	❏
Light nonaqueous phase liquid	▼		▼	■
Dense nonaqueous phase liquid	▼		▼	▼
Inorganic contaminants				
Asbestos				❏
Radioactive materials	▼	▼	❏	❏
Metals (drinking water standards)	▼	■	❏	❏

■ = Higher values support preselection of technology group.
❏ = Lower values support preselection of technology group.
▼ = Effect is variable among options within a technology group.
Where no symbol is shown, the effect of that characteristic is considered inconsequential.
Source: U.S. Environmental Protection Agency, Engineering Bulletin: Technology Preselection Data Requirements, Cincinnati, OH, 1992.

As with soils, water characteristics displayed in the table are important in determining the applicability of many treatment processes. BOD in contaminated water, for example, indicates the biodegradability of the organic contaminants. COD measurements in groundwater provide an indication of the chemically oxidizable contaminants. Processes such as carbon adsorption, ion exchange, and flocculation can be impacted by pH changes in groundwater.

The suspended solid parameter is an important water characteristic because suspended solids can cause resin binding in ion exchange systems and clogging of reverse osmosis membranes, filtration systems, and carbon adsorption units. If the concentration of suspended solids is greater than 5%, analysis of total and soluble metals should be made.[58]

13   PUBLIC PARTICIPATION

The terms public participation and community relations are commonly used in the regulatory arena. The terminology, however, may have negative connotations. People may perceive that the two terms refer to one-way communication by which a communicator speaks and the public just listens. A more appealing term used to reflect two-way communication between a communicator and the public is public involvement, or community involvement. This book defines the term public participation, community relations, community involvement, or public involvement as having the same meaning.

Public participation or community relations programs are intended to promote active communication between communities affected by discharges or releases and those responsible for the response actions. Specific requirements for public participation under CERCLA are codified in 40 CFR 300.415 for removal actions, 40 CFR 300.430 for remedy selection, and 40 CFR 300.435 for remedial activities. The major components of public participation in the Superfund program are

- public notice requirement
- provisions regarding public comments and response to public comments
- establishment of information repository
- availability of technical assistance
- development of a Community Relations Plan (CRP)

A community relations program provides interested persons an opportunity to comment on and participate in decisions concerning site actions and ensures that the community is provided with accurate and timely information about site activities.[59] Guidance for preparing a CRP and conducting community relations activities is provided by Environmental Protection Agency (EPA).[60] The suggested organization of CRP is as follows:

1. Overview of community relations plan. The overview states the purpose and the distinctive features of the CRP.

2. Capsule site description. This part of the CRP provides the historical, geographical, and technical details necessary to show why the site needs to be remediated.
3. Community background. The background description relates to the community and its involvement with the site.
4. Highlights of the community relations program. This part provides concrete details on approaches to be taken and a strategy for communicating the project with a specific community.
5. Community relations activities and timing. A matrix format may be suitable for this task.

   Appendices:
   A. contact list of key community leaders and interested parties
   B. suggested locations of meetings and information repositories

Under the Resource Conservation and Recovery Act (RCRA) statute, the law considers public involvement as extremely important to the success of the RCRA program. The RCRA public participation framework covers a number of major areas. Section 3006, for example, provides requirements that the public be given the opportunity to comment before the issuance of decisions regarding RCRA authorization to a state and the permitting of a treatment, storage, or disposal (TSD) facility. Section 7002 allows a citizen to bring a civil suit against any person or government agency, including EPA, alleged to be in violation of any permit, standard, regulation, or provision of RCRA.

The Hazardous and Solid Waste Amendments (HSWA) of 1984 expand this citizen suit to include imminent and substantial endangerment actions against past and present hazardous waste handlers.[61] Citizen suits, however, are prohibited in the following circumstances:

- in the siting and permitting of TSD facilities
- where EPA is prosecuting under RCRA or Superfund authority
- while EPA or a state is engaged in a Superfund removal or has incurred costs to engage in a remedial action
- where the responsible party is conducting a removal or remedial action pursuant to an EPA order

Citizen suits can have a very substantial effect on a company's operation. Some suits may convince the court to impose large penalties and to order permanent closure of a noncomplying RCRA facility. The above discussion is to illustrate the power that Congress has vested in citizens under RCRA provisions.

RCRA public participation provisions related to site remediation are codified in 40 CFR 124.10, which states that public notice must be given for all important permits (e.g., closure permit), including the preparation of a draft permit. Under 40 CFR 124.11 provision, a 45-day comment period is allowed during which time a person may submit written comments on the draft permit and request a public hearing. When there is a written notice of opposition to the permit and requests for a hearing, or when there is "significant degree of public interest" in a draft permit, a public hearing will be held by the regulatory agency.

The purpose of public participation is clearly to facilitate a consensus between the public, the project owner, and the agency on an acceptable remediation approach for the site. The public has valid concerns on how a site should be remediated, i.e., whether odor, noise, dust, and other public nuisance components would be generated by the activities. The public also has concerns on how clean the site would be after remediation, and how the remediation residuals (e.g., treated groundwater) would be disposed. If these concerns are unsatisfactorily addressed, the public has the right to oppose the project.

In many respects the relationships among the business community, the government, and the general public have become less confrontational and more cooperative in recent years than they were a decade or two ago. While the level of trust among the three groups may not have increased, they have begun to realize that working together often provides greater benefits than maintaining adversarial relationships.

Public participation is essentially about building and maintaining communication links with all parties involved. Public participation concentrates on communication. The goal of public participation planning is to establish effective and efficient communication among all of the stakeholders in the site remediation: the project, the government, and the public for the purpose of building trust, and establishing credibility. The biggest challenge to public participation is to get the fiercest opponents of the project to go along with the proposal, to consent to its implementation. What is obtained here is an informed consent, i.e., the willingness of opponents to go along with a course of action that they were opposed at the beginning of the process, but changed their position during the process.[62]

1. FUNDAMENTALS OF ENVIRONMENTAL COMMUNICATIONS

An effective environmental communication can be viewed as a balance between credibility, sound decision making, positive actions, and community value sensitivity.[63] Of these four attributes of effective communication, developing and maintaining credibility is the core of environmental communications. Building credibility is an ongoing process. It is a commitment that all promises, direct or indirect, will be kept and exceeded wherever possible. Lukaszewski and Serie[63] admit that credibility can be very hard to achieve and maintain because this attribute represents quality expectations which are difficult to define and measure. It is a judgment formed in the hearts and minds of the public.

Credibility is a perception of how believable and how trustworthy a project owner is based on the evaluation of words and actions from the community's perspective. Harrison[64] states that if a project desires to build a public partnership, it must be based on the three fundamental components, i.e., the project must (1) open operations for observation, (2) provide open dialog, and (3) be open to options. Partnership is built on trust.

People normally distrust and fear a project on the basis of what they do not know. When the subject is the long-term operation of a hazardous waste treatment facility and its associated risk, the perception is presumably about something very dangerous and about an uncontrollable situation in their neighborhood. With the right planning, a project tour, which provides the community with an opportunity to examine the project, can build up community acceptance and good will. Harrison[64] argues that in some cases, these kinds of "come-look-us-over" events can help to overcome negative situations.

While the cost-effectiveness interest of a project is obvious, the project must make it clear that it also stands on the side of the public interest. The message must be concise, positive, and nondefensive and the community must be convinced that its interests come first. The key to this, according to Harrison, is open dialog with the community and an attitude of caring and honesty.

Strengthening a partnership can also be achieved by providing the community a list of options for solving environmental problems. The options for a proposed project, such as alternative plant designs, provisions for monitoring environmental change, and financial factors, can all be put on the table for the community to discuss. While this approach may reduce the project's level of control, it still is a logical approach because a community will usually reject the proposal that is presented in finished form. Harrison emphasizes that presentation becomes as important as intention. The goal is to make the community feel that it has some control over its own destiny.

As more and more information flows into the community, the key to public relations today is to involve the public directly in the decision-making process, to communicate clearly, honestly, and constantly about issues that must be faced together. Harrison believes that a true partnership which is based on those three fundamental concepts is the only viable option for sharing the same environment in a community. Working together in a mutually beneficial environment is the goal of a community relations program. An effective program needs hard work and dedication and demands good planning and intensive preparation. An effective program requires a strategy.

2. STRATEGIZING COMMUNITY RELATIONS

Strategizing a community relations program is done for two basic reasons: the public will become involved in the process because a project appeals to their interests or because the project threatens their interests.[65] Relationships that occur because of mutual interests are usually voluntary. Relationships that grow from a perceived threat to the public's interest are usually involuntary, forced by circumstances, and must be identified early and worked on most aggressively.

A community relations program must consider its response to this differing motivation. Programs to address public concerns must be planned delicately. McDermitt[66] provides 10 major steps to successful community relations. Although his strategy is written to win project approval, it can also be used for general community relations.

The first element of the strategy is to plan early. A community relations plan crafted early in project development can help shape public support because nearly every project has a good story to tell, such as increased local revenue, jobs, and environmental improvements. These benefits will help win local support if the right messages are properly communicated. Failure to properly transmit the information will lead to rumors and speculation and the spread of misinformation and opposition.

The second element is to prepare properly. Identify who the community residents respect the most. This action is necessary because the project will want those individuals on its side. There are a number of research tools that can be used. One of them is a public opinion research poll that can provide information on key factors about a community, its attitudes, concerns, and credible local opinion leaders.

The development of a comprehensive plan is considered the third element. The plan clarifies the existing situation, defines the project objectives in clear and concise terms, analyzes barriers to implement the plan and specifies the techniques to address each barrier, outlines the activities to achieve each objective, determines resources required, and schedules plan implementation including assignments of responsibilities for each task.

Allocating the budget, i.e., dedicating sufficient finances to conduct plan implementation represents the fourth element, followed by being proactive. McDermitt[66] emphasizes that the residents of most communities are not radical, antidevelopment fanatics. Most have genuine concerns about their community and the impact a facility may have on their lives. It is the responsibility of the project to demonstrate sensitivity to community concerns and a willingness to address them.

The sixth element of the strategy is to package the project. The community relations project is normally complex to the average observer. A good tool to relay the information to the community is a project-information package. Begin by describing the objectives of the project, avoid using the plant's technology and confusing statistics, explain how the project will protect the environment, but do not become involved in a discussion of environmental issues. The information kit may contain a question-and-answer section.

Handling the politics is the next step in the strategy. This involves the identification of those officials who carry the real power in the host community and show them how the project benefits them and their constituencies. This is followed by identifying supporters who can assist the project, for example, at public hearings. The success of a community relations program will depend largely on how effective a project is in finding, cultivating, and mobilizing local supporters.

The ninth element is to manage the media. The project goal is to achieve fair and accurate coverage of the project by the media. McDermitt emphasizes that it is not only what you say, but how you say it, that will leave a lasting impression on the public. A project spokesperson should be professionally trained to perform this task.

The tenth element of the strategy is to manage the hearings. Proper planning can help the project win public hearings. If possible, avoid any hearing at all. If not, ensure that the hearing is fair and orderly. McDermitt suggests working with the media at public hearings by providing them with the information they need to write their story because most residents' impressions of the hearing will be what they read in the newspaper the following day. The message that underlines the above strategy is that as the public becomes more actively involved in environmental issues that affect their lives, the remediation industry will be required to focus more attention on building and maintaining good relationships with the public. The same message holds true for governmental agencies.

3. DEVELOPING PUBLIC PARTICIPATION PLAN

The term public participation entails a broad scope of activities designed to help citizens participate in decisions, convey information, solicit citizen concerns, heighten public awareness, and motivate participation in various programs.[67] Successful public participation programs are the result of careful planning. By developing a realistic plan, decision makers can evaluate the situation, problems, and issues, and know where best to direct their efforts and resources.

A public participation plan can be defined as consisting of the following elements:

1. Identification of issues that need to be addressed. An effective program should be planned with the community's needs in mind. It is important to understand the different audiences that exist within the community and determine how these diverse groups obtain information. For example, what are citizens concerned about, will they read written information, will they respond well to public notices? Answers to these kinds of questions ensure that the appropriate messages are sent in the appropriate format.

2. Formulation of goals to be reached. The objectives must be concisely stated. Clearly communicate to the public how and when they will have an opportunity to participate. Include enough details in the plan so that everyone involved in implementing the plan knows what is expected of them and when.

3. Description of activities to accomplish each goal. The public participation plan should be positive and provide simple instructions on how to participate. Planned activities should satisfy the information needs of the community and be within the budgetary and resource limits.
4. Availability of resources for each activity. Public participation planning is relatively inexpensive. However, it does require a commitment of funds and staff time to plan and coordinate a successful program.
5. Program scheduling, coordination, and monitoring. Opportunities for communicating with and involving the public should be initiated early in the planning process. Communication with the public and promotion of the program should be ongoing because a public participation program is a dynamic and continuous planning effort.

There is a broad range of possible activities that can be included in a public participation plan. The activities mostly represent educational undertakings. For this reason, public participation planning is often described as public education and involvement planning. In essence, public participation implies integrating public concerns and values at every stage of the decision-making process. It is a dialog, a two-way communication that involves both getting information out to the public and getting back from the public ideas, recommendations, issues, and concerns.

To improve public participation planning, EPA provides a list of information and involvement techniques for use in developing the plan.[68] Information techniques can be used to get information to the public, while involvement techniques are used to get information from the public. These two types of techniques are, respectively, discussed in the next two sections.

3.1. Information Techniques

A successful public participation program requires a well-planned public information strategy. The public needs to know the rationale of why a project or an activity is needed and the consequences if it is not implemented. They need information about alternative technology or options that are available and other facts that they can use to evaluate the project or activity appropriately.

EPA's list of the techniques for communicating information to the public include:

1. Provide briefings. The goal is to keep key elected officials informed on any new event. This can be done simply by a personal visit or a phone call.
2. Assist reporters in writing feature stories. Sending a news release to a newspaper is one way to get media attention but personal contact with an editor is more likely to get someone interested.

3. Mail out reports directly to key individuals. When constructing a mailing list, code the names of key individuals to accommodate this mail-out. A summary of the report can be sent to the remaining names on the mailing list with a note that a copy of the full report is available upon request or at various locations for review.

4. Arrange news conferences. This technique stimulates the interest of the media in developing news stories. The key to a successful news conference is that the topic of the conference or the person conducting it must be newsworthy or no one will show up.

5. Develop newsletters. This medium provides interested individuals with far more information than the news media. The effectiveness of a newsletter often depends on its format, i.e., attractiveness, pictures and graphics, and the use of simple, daily language.

6. Provide newspaper inserts. As long as the insert meets the newspaper's specifications, newspapers can deliver an insert for a moderate cost. Similar to a newsletter, the insert must be attractive and easy-to-read in order to attract readers.

7. Issue news releases. This text should interest the media in doing a news article on the issue. The name and number of the contact person dealing with the issue must be included in news releases.

8. Arrange for paid advertisements. The public is normally appreciative of paid advertisements announcing public meetings or presenting information; however, any advertisement paid with public funds, especially large ads, is also subject to criticism.

9. Make group presentations. Preparing a slide show should be considered because a visual presentation is more interesting to the audience and can be done in a short time.

10. Develop press kits. A kit summarizes key information from a number of documents and reports. Once a press kit is finalized, deliver the kit to interested editors and be prepared to answer questions that arise.

11. Utilize public service announcements. This service is provided free of charge by radio and television stations. The stations are very likely to broadcast the program and invite the public to participate. Through this program the audience will not only listen, but also respond to the issues.

Communication is a two-way process: giving the information to and receiving the concerns, comments, opinions, and suggestions of the public. Overall, the techniques described in the preceding paragraphs are designed to provide the public with specific information. Once the public has been informed, the remaining part of the communication task is to provide mechanisms by which the public can express their concerns and give opinions and suggestions. The mechanism represents a forum of public involvement.

3.2. Involvement Techniques

A public involvement process is necessary for the exchange of information. The process can be conducted in a number of ways. EPA's list of involvement techniques for the effective exchange of information includes the following:

1. Establish task forces. Task forces and advisory groups can serve a number of purposes such as providing communications between key constituencies, a forum for reacting to the more technical aspects of the discussion, and achieving a consensus decision. It must be decided early in the process what mandate the group has in the decision-making process, i.e., whether they are purely advisory, or have some sort of authority in giving a decision.
2. Establish focus groups as an alternative to polls. These are small discussion groups and are used to explore people's reactions to issues. A trained moderator to lead the groups is essential to the success of using this technique.
3. Establish a hotline. The hotline enables people to call to ask questions as well as to comment and express opinions. The key to an effective hotline is the right person receiving the calls.
4. Conduct interviews. Interviews are powerful because some people will provide more information in a one-on-one conversation than they will in an open forum such as a public meeting.
5. Facilitate workshops. Workshops or meetings are widely used public involvement techniques. Ensure that everyone gets a chance to be heard.
6. Conduct a plebiscite. The ultimate test of whether a community supports a project or activity is a direct vote on it.
7. Conduct polls. The purpose of conducting a poll is to assess community viewpoints quantitatively. Polls, however, do not always predict outcome. Polls only provide a snapshot of one moment in time. Polls are helpful and informative but do not replace the need for some of the other involvement techniques described previously.

3.3. Public Participation Pitfalls

The effectiveness of public participation does not result from using a single public involvement technique, but from combining many techniques of public involvement and public information into a unified program strategy. The way the techniques are combined must be geared toward the characteristics of the community. A few cautions, however, should be observed in developing a public participation plan. The pitfalls are

1. Be cautious with public hearings. Public hearings tend to exaggerate differences rather than bring people together. Workshops provide more satisfactory results.

2. Be aware of task force limitations. The most obvious one is the extent of its authority. The second limitation stems from the fact that advisory groups and task forces can spend more time agreeing on procedures rather than on substance.
3. Objectivity of public information. The public information document is designed to provide the information needed to be involved in an informed manner. It is not aimed at selling a particular point of view; therefore, the document must be objective, balanced, and credible. Clarity will help objectivity of information presented in the document.
4. Be honest with media. Provide all the information objectively and factually. The media have two important characteristics for communicating information to the public. They are the most accessible sources of information and are seen by the public as the most credible sources of information.
5. Provide feedback loops. This means that if we ask the public to participate, we should always get back to them promptly with updated information. The primary motivation for participation is the sense that everyone in the community can have an impact. Feedback loops strengthen this motivation.
6. Never surprise elected officials. Keep them informed on the status of activity first-hand.

4. RISK COMMUNICATION IN PUBLIC PARTICIPATION PLANNING

In general terms, site remediation issues are considered complex because they are not only technical, but also societal, economic, and political in nature. One of the most difficult issues in citizen participation planning associated with site remediation is risk communication. The issue is difficult because in the field of environmental management, contamination is legally permitted. As long as the level of the contaminant left in the property's soil and groundwater is below the remediation threshold limit, such contamination is allowed. These allowable concentrations of site contaminant represent a risk to the public, no matter how small and insignificant the risk is. The public may consider that they are forced to take this risk involuntarily.

The public generally evaluates the risk in terms of its perception of the risk. For involuntary risk, the public often demands a zero risk level, something that is impossible to achieve; in reality the perceived risk is always greater than the actual risk. EPA[69] observed that the environmental risks considered most serious by the general public are different than those considered by the technical professionals charged with reducing environmental risks. Although desirable, it is difficult or almost impossible to change public perception instantly.

Traditionally, risk has been conceived by experts as the probability of injury, disease, or death due to exposure to a hazard. Sandman,[70] however, has redefined risk as the sum of the traditional concept of risk and the outrage factors. Examples of the outrage factors are fatal disasters, uncertain risks, untrustworthy sources, and unfair and unethical risks. Sandman observes that the experts tend to focus almost exclusively on hazard and pay no attention to public outrage, while the public has too little regard for the hazard and pays too much attention to outrage. Sandman's model is useful in understanding why risk communication programs often fail.

Many studies have found that risk communication is closely associated with the level of public trust and the credibility of the organization, particularly the credibility of the communicator. The public will examine the dedication and competence of the risk communicator closely. They will look at the communicator and wonder how much this person really cares. Sidhu and Chadzynski[71] reveal that the key factors that are essential to risk communication are clarity and accuracy of the message, and a fair and open discussion process. They argue that risk communication efforts by risk communicators are facilitated when the communicator is perceived as being knowledgeable, objective, and acting in the best interest of the public.

EPA has developed the seven cardinal rules of risk communication. The rules are[72]

1. Accept and involve the public as a legitimate partner.
2. Plan carefully and evaluate efforts.
3. Listen to the public's specific concerns.
4. Be honest, frank, and open.
5. Coordinate and collaborate with other credible sources.
6. Meet the needs of the media.
7. Speak clearly and with compassion.

The overall scope of risk communication is very broad. This chapter does not attempt to discuss this topic in great detail, but suffice it to say that risk communication is a delicate subject and that communicating risk must be carefully planned. There are a number of areas that can be targeted for effective risk communication. Burke[73] suggests the following efforts to enhance a risk communication program:

1. The development of educational approaches to improve public understanding of the concepts and uncertainties of risk.
2. The improvement of methods to characterize health and environmental risks to policy makers.
3. The improvement of understanding of the risk information needs of communities.
4. The development of methods to incorporate community values into the risk communication and decision-making process.

Good risk communication programs must be proactive, ongoing, and cooperative.[74] A recent national survey revealed, however, that the risk communication programs as currently practiced in the U.S. tend to be more reactive than proactive.[75] Oleckno[76] suggests that risk communication programs must be continually evaluated and improved and input from the public is essential if the programs are to remain vital and effective and achieve the goals for which they are designed.

Although risk communication by itself will not eliminate conflicts, it is a significant element in a site remediation program. By combining risk communication with a genuine understanding of community values and interests, an important component of improving the decision-making process to address site remediation issues will be provided.

5. ASSISTANCE IN DEVELOPING PUBLIC PARTICIPATION PLANNING

Developing an effective public participation planning process involves assessing who could provide support on public involvement activities and where to obtain information. Most EPA regional offices have one person assigned as the Public Involvement Coordinator (PIC) who serves as a liaison between community members and permit writers, enforcement personnel, facility owners and operators, and other appropriate individuals or groups involved in implementing public participation programs.[77]

The PIC should be able to provide assistance in obtaining specific manuals, guidance documents, and memoranda that elaborate on regulations and principles of public participation, and to give helpful tips on implementing successful programs. EPA also provides training in a variety of areas including public participation, risk communications and community outreach.

PART IV
REFERENCES

1. U.S. Environmental Protection Agency, *Treatment Technologies — Second Edition*, Government Institutes Inc., Rockville, MD, 1991.
2. Apte, Nitin, Hazardous Waste Treatment Technologies — An Inventory, *Hazmat World*, Vol. 6, No. 8, 1993, 27–32.
3. Mills, C. Hamilton, Bioremediation Comes of Age, *Civil Engineering*, Vol. 65, No. 5, 1995, 80–81.
4. U.S. Environmental Protection Agency, *The Superfund Innovative Technology Evaluation Program: Technology Profiles — Fourth Edition*, Washington, D.C., 1991.
5. Norris, R.D. et al., *In-Situ Bioremediation of Groundwater and Geological Material: A Review of Technologies*, U.S. Environmental Protection Agency, Ada, OK, 1993.
6. Rochkind, M.L., Blackburn, J.W., and Saylor, G.S., *Microbial Decomposition of Chlorinated Aromatic Compounds*, U.S. Environmental Protection Agency, Cincinnati, OH, 1986.
7. LaGrega, M.D., Buckingham, P.L., and Evans, J.C., *Hazardous Waste Management*, McGraw-Hill, New York, 1994.
8. Sferra, P.R., Biodegradation of Environmental Pollutants, in Harry M. Freeman, ed., *Standard Handbook of Hazardous Waste Treatment and Disposal*, McGraw-Hill, New York, 1988, 9.53–9.60.
9. U.S. Environmental Protection Agency, *Bioremediation of Hazardous Wastes*, Washington, D.C., 1992.
10. U.S. Environmental Protection Agency, *Remediation Technologies Screening Matrix and Reference Guide*, Washington, D.C., 1993.
11. Ross, D. and Sudano, P.L., Back to Nature, *Industrial Wastewater*, Vol. 2, No. 6, 1994, 39–41.
12. Norris, R.D., et al., *In-Situ Bioremediation of Groundwater and Geological Material: A Review of Technologies*, U.S. Environmental Protection Agency, Ada, OK, 1993.
13. U.S. Environmental Protection Agency, *How to Evaluate Alternative Cleanup Technologies for Underground Storage Tank Sites: A Guide for Corrective Action Plan Reviewers*, Washington, D.C., 1994.

14. U.S. Environmental Protection Agency, *Remediation Technologies Screening Matrix and Reference Guide*, Washington, D.C., 1993.

15. Abbasi, Rafat A., Groundwater Treatments Gain Ground, *Civil Engineering*, Vol. 66, No. 2, 1996, 53–55.

16. MacDonald, J.A. and Ritmann, B.E., *In Situ* Bioremediation: When Does It Work?, *Industrial Wastewater*, Vol. 2, No. 6, 1994, 32–38.

17. Weymann, David, Biosparging and Subsurface pH Adjustment: A Realistic Assessment, *The National Environmental Journal*, Vol. 5, Issue 1, 1995, 32–36.

18. Brubaker, Gaylen R., The Boom in *In situ* Bioremediation, *Civil Engineering*, Vol. 65, No. 10, 1995, 38–41.

19. Ross, D., Slurry-phase Bioremediation: Case Studies and Cost Comparisons, *Remediation*, Vol. 1, No. 1, 1991, 7–11.

20. Phung, Tan, Land Treatment of Hazardous Wastes, in Harry M. Freeman, ed., *Standard Handbook of Hazardous Waste Treatment and Disposal*, McGraw-Hill, New York, 1988, 9.41–9.51.

21. LaGrega, Michael D., Buckingham, Phillip L., and Evans, Jeffrey C., *Hazardous Waste Management*, McGraw-Hill, New York, 1994.

22. Irvine, Robert L. and Wilderer, Peter A., Aerobic Processes, in Harry M. Freeman, ed., *Standard Handbook of Hazardous Waste Treatment and Disposal*, McGraw-Hill, New York, 1988, 9.3–9.18.

23. Canadian Council of Ministers of the Environment, *National Guidelines on Physical-Chemical-Biological Treatment of Hazardous Waste*, Edmonton, Alberta, 1989.

24. U.S. Environmental Protection Agency Engineering Bulletin, *Solidification/Stabilization of Organics and Inorganics*, Cincinnati, OH, May 1993.

25. Wilk, Charles, Portland Cement Gives Concrete Support to Solidification/Stabilization Technology, *Environmental Solutions*, Vol. 8, No. 5, 1995, 47–51.

26. Kornel, Alfred, Dehalogenation, in Harry M. Freeman, ed., *Standard Handbook of Hazardous Waste Treatment and Disposal*, McGraw-Hill, New York, 1988, 7.53–7.63.

27. U.S. Environmental Protection Agency Engineering Bulletin, *Pyrolysis Treatment*, Cincinnati, OH, 1992.

28. Shah, J.K., Schultz, T.J., and Daiga, V.R., Pyrolysis Processes, in Harry M. Freeman, ed., *Standard Handbook of Hazardous Waste Treatment and Disposal*, McGraw-Hill, New York, 1988, 8.91–8.104.

29. U.S. Environmental Protection Agency, *Experience in Incineration Applicable to Superfund Site Remediation*, Cincinnati, OH, 1988.

30. U.S. Environmental Protection Agency, *The Superfund Innovative Technology Evaluation Program: Technology Profiles — Fourth Edition*, Washington, D.C., 1991.

31. Marsden, Paul, Immunoassay Techniques, *Environmental Testing and Analysis*, January/February 1995, 58–61, 85.

32. U.S. Environmental Protection Agency, *Immunochemical Analysis of Environmental Samples*, Las Vegas, 1994.

33. Van Emon, Jeannette M. and Gerlach, Clare L., A Status Report on Field Portable Immunoassay, *Environmental Science and Technology*, Vol. 29, No. 7, 1995, 312A–317A.

34. James, Stephen C. and Rogoskewski, Paul J., Treatment and Containment Technologies, in Harry M. Freeman, ed., *Standard Handbook of Hazardous Waste Treatment and Disposal*, McGraw-Hill, New York, 1988, 12.35–12.50.

35. Mutch, Robert, as quoted by Daniel Shannon and Mary Greczyn, Cutoff Walls, Barriers See Resurgence from Improved Capabilities, *Hazardous Waste News — Special Report*, Vol. 17, No. 2, 1995, 1–4.

36. Marquis, Samuel Jr., Don't Give Up on Pump and Treat, *Soil and Groundwater Cleanup*, August/September 1995, 46–50.

37. Rorech, Gregory J. and Morello, Bridget S., Groundwater Remediation Technology, *Industrial Wastewater*, Vol. 3, No. 6, 1995, 33–39.

38. Uhlman, Kristine, Strategies to Estimate Pump and Treat Cleanup Schedules, *The National Environmental Journal*, Vol. 5, Issue 1, 1995, 45–50.

39. Wilson, J.L. and Conrad, S.H., Is Physical Displacement of Residual Hydrocarbons a Realistic Possibility in Aquifer Restoration?, *Conference on Petroleum Hydrocarbons and Organic Chemicals in Groundwater: Prevention, Detection, and Restoration*, NWWA/API, Houston, 1984, 274–298.

40. Rorech, Gregory J. and Morello, Bridget S., Groundwater Remediation: From Start to Finish, *Industrial Wastewater*, Vol. 3, No. 4, 1995, 28–31.

41. Speight, Alicia E., Batsel, Kurt R., and Harris, Donald R., Go With the Flow, *Industrial Wastewater*, Vol. 2, No. 5, 1994, 47–54.

42. Bergren, Chris L., Looney, Brian B., and Shealy, Michael D., Lessons Learned at Savannah River: Phased-In Approach Lets New Groundwater Cleanup Methods Complement Traditional Ones, *Environmental Solutions*, Vol. 9, No. 1, 1996, 26–28.

43. U.S. Environmental Protection Agency, *Guide for Conducting Treatability Studies Under CERCLA*, Washington, D.C., 1992.

44. U.S. Environmental Protection Agency, *Remediation Technologies Screening Matrix and Reference Guide*, Washington, D.C., 1993.

45. U.S. Environmental Protection Agency, *Guide for Conducting Treatability Studies Under CERCLA*, Washington, D.C., 1992.

46. U.S. Environmental Protection Agency Engineering Bulletin, *Technology Preselection Data Requirements*, Cincinnati, OH, 1992.

47. Breckenridge, R.P., Williams, J.R., and Keck, J.F., *Groundwater Issue: Characterizing Soils for Hazardous Waste Site Assessments*, Washington, D.C., 1991.

48. U.S. Environmental Protection Agency Engineering Bulletin, *In situ Soil Vapor Extraction Treatment*, Cincinnati, OH, 1991.

49. U.S. Environmental Protection Agency, *Handbook on In situ Treatment of Hazardous Waste Contaminated Soils*, Cincinnati, OH, 1990.

50. U.S. Environmental Protection Agency, *Technology Screening Guide for Treatment of CERCLA Soils and Sludges*, Washington, D.C., 1988.

51. Koboyashi, H. and Rittman, B.E., Microbial Removal of Hazardous Organic Compounds, *Environmental Science and Technology*, Vol. 16, No. 1, 1982, 170A–183A.

52. U.S. Environmental Protection Agency Engineering Bulletin, *Soil Washing Treatment*, Cincinnati, OH, 1990.

53. U.S. Environmental Protection Agency, Engineering Bulletin, *Slurry Biodegradation*, Cincinnati, OH, 1990.

54. U.S. Environmental Protection Agency, *Summary of Treatment Technology Effectiveness for Contaminated Soil*, Washington, D.C., 1991.

55. U.S. Environmental Protection Agency, *Handbook for Stabilization/Solidification of Hazardous Wastes*, Washington, D.C., 1986.

56. U.S. Environmental Protection Agency, Engineering Bulletin, *Chemical Oxidation Treatment*, Cincinnati, OH, 1991.

57. U.S. Environmental Protection Agency, *Summary of Treatment Technology Effectiveness for Contaminated Soil*, Washington, D.C., 1991.
58. U.S. Environmental Protection Agency, *A Compendium of Technologies Used in the Treatment of Hazardous Wastes*, Cincinnati, OH, 1987.
59. U.S. Environmental Protection Agency, *Guide for Conducting Treatability Studies Under CERCLA*, Cincinnati, OH, 1992.
60. U.S. Environmental Protection Agency, *Community Relations in Superfund: A Handbook*, Washington, D.C., 1988.
61. Hall, Ridgway M. Jr., *RCRA Hazardous Wastes Handbook — Ninth Edition*, Government Institutes Inc., Rockville, MD, 1991.
62. Institute for Participatory Management and Planning, *Citizen Participation Handbook for Public Officials and Other Professionals Serving the Public*, Monterey, CA, 1993.
63. Lukaszewski, James E. and Serie, Terry L., Public Consent Built on Credibility is the Goal, *Waste Age*, February 1993, 45–54.
64. Harrison, Bruce E., *Environmental Communication and Public Relations Handbook — Second Edition*, Government Institutes Inc., Rockville, MD, 1992.
65. Lukaszewski, James E. and Serie, Terry L., Relationship Built on Understanding Core Values, *Waste Age*, March 1993, 83–94.
66. McDermitt, David, The Ten Commandments of Community Relations, *World Wastes*, Volume 36, No. 9, 1993, 48–51.
67. U.S. Environmental Protection Agency, *Decision-Makers Guide to Solid Waste Management*, Washington, D.C., 1989.
68. U.S. Environmental Protection Agency, *Sites for Our Solid Waste — A Guidebook for Effective Public Involvement*, Washington, D.C., 1990.
69. U.S. Environmental Protection Agency, *Reducing Risk — Setting Priorities and Strategies for Environmental Protection*, Washington, D.C., 1990.
70. Sandman, P.M., Risk Communication: Facing Public Outrage, *EPA Journal*, No. 13, 1987, 21–22.
71. Sidhu, Kirpal S. and Chadzynski, Lawrence, Risk Communication: Health Risks Associated With Environmentally Contaminated Private Wells Versus Chloroform in a Public Water Supply, *Journal of Environmental Health*, Vol. 56, No. 10, 1994, 13–16.
72. U.S. Environmental Protection Agency, *Seven Cardinal Rules of Risk Communication*, Washington, D.C., 1988.
73. Burke, Thomas A., Overview: Refining Hazardous Waste Site Policies, in Richard B. Gammage and Barry A. Berven, eds., *Hazardous Waste Site Investigations — Toward Better Decisions*, Lewis Publishers, Boca Raton, FL, 1992, 3–10.
74. Hance, B.J., Chess, C., and Sandman, P.M., *Industry Risk Communication Manual: Improving Dialogue with Communities*, Lewis Publishers, Boca Raton, FL, 1990.
75. Chess, C., Salomone, K.L., and Sandman, P.M., Risk Communication Activities of State Health Agencies, *American Journal of Public Health*, No. 81, 1991, 489–491.
76. Oleckno, William A., Guidelines for Improving Risk Communication in Environmental Health, *Journal of Environmental Health*, Vol. 58, No. 1, 1995, 20–23.
77. U.S. Environmental Protection Agency, *RCRA Public Involvement Manual*, Washington, D.C., 1993.

PART V
CONCLUSION

This concluding part consists of only one chapter. Chapter 14 examines the U.S. site remediation trends in general and discusses the three components of the trends. The components are the challenges of the brownfields program, the market opportunities for environmental technologies, and the innovative ideas and accomplishments in streamlining and focusing the site remediation program. Efforts to accelerate the cleanup process identified in this chapter include lender liability exemptions, covenants not to sue the purchaser, risk assessments with deed restrictions, state assistance in voluntary programs, and environmental justice initiatives.

14   TRENDS IN SITE REMEDIATION

In an analysis of the trends of U.S. environmental affairs, the editor of *Waste Tech News* commented that after decades of hostility and conflict, the adversarial paradigm pitting environmentalists, business, and government against each other is now shifting into a common process widely known as "sustainability".[1] Carleton identifies the components of sustainability to include facilitating circumstances for companies to stay in business, the opportunities for people to have jobs, the channeling of waste into revenue streams, refurbishment of resources for availability to consumers, and the elimination of health hazards. Government agencies, businesses, industrial plants, activists, and academia are all embracing sustainability as a way to break out of the inertia, to forge ahead, and to achieve an ideal world in which resources are not depleted, waste does not poison people, everyone makes money, and the quality of life is comfortably attainable for everyone.

The Environmental Protection Agency (EPA) is credited with redirecting its *modus operandi* away from confrontation to collaboration with industry and citizens. Government is beginning to look upon business and industry as a constituency. Industry which used to fight every regulation placed upon it may now be changing. While recognizing that some pockets of dedicated opposition still exist, the majority of business and government agencies have moved away from sending out press releases announcing environmental violation news to press releases removing sites from the Superfund lists or promoting simplified environmental testing kits.

It is on this philosophy of sustainability through environmental management that the future of site remediation rests. Within this new paradigm, the future of the U.S. remediation program is greatly affected by the implementation of the brownfields program, the capitalization of the growing market of remediation technologies, and the streamlining and focusing of the site remediation process. These three aspects are examined in this chapter.

1. THE BROWNFIELDS PROGRAM

Many industrial sites in the U.S. which once housed business enterprises are now abandoned because of environmental contamination. Various legal and financial obstacles prevent the sites from being rehabilitated and placed back into productive use. These contaminated sites are mostly located in strategically developed areas, close to town centers served by public transportation, often in depressed areas of the older and larger municipalities. Redeveloping these "brownfields" sites for urban revitalization will create jobs and tax revenue, reduce the degradating urban blights, and prevent suburban sprawl. The actual number of these brownfields is unknown; however, it is estimated that more than 500,000 sites nationwide show evidence of at least some contamination.[2]

There is strong national interest to aggressively restore the brownfields because of the potential for improving the environment, upgrading the urban landscape, fostering the inner city employment and economic development. While the opportunities exist, there are some identifiable challenges that confront the brownfields program. Categorically, the challenges are[3,4]

1. The liability issue. In this case, lenders may refuse to lend money to an owner or developer of a contaminated property. The current laws generally have no liability exemption provisions for innocent parties such as lenders, owners, and prospective purchasers as related to buying and selling a contaminated site.
2. The cleanup standard. Technical barriers make it difficult to restore sites to the pristine condition required by EPA. In general, remediation standards are difficult to achieve at complex sites. What is needed is a standard that is less stringent, site specific, and still protective of human health and the environment.

1.1. The Lender and Purchaser Liability

As part of EPA's program to streamline Superfund to encourage reuse of brownfields properties, in April 1992 EPA promulgated a lender liability rule to address potential lender liability for cleanup costs at CERCLA sites. This rule was vacated by the U.S. Court of Appeals for the District of Columbia Circuit in February 1994 because EPA lacked authority to issue the rule as a binding regulation. In a policy released in October 1995, EPA and the Department of Justice (DOJ) recognized that the court decision did not preclude EPA and DOJ from following provisions of the rule as enforcement policy.[5]

The policy restates that lenders are not liable if they hold property merely to protect their security interest. It asserts that the federal government will not use its enforcement action against lenders who do not actively participate in the management of a Superfund site.[6] EPA and DOJ specify that the policy does not address lender liability under any statutory or regulatory authority other than CERCLA. Specifically, the policy does not cover lender liability determinations as they relate to the hazardous waste and underground storage tank (UST) programs under RCRA.

The lender liability rule for USTs was promulgated by EPA on September 7, 1995.[7] Under the rule, a lender is exempt from the federal UST requirements as an operator provided that there is a current operator at the site that can be held responsible for compliance with Subtitle I regulatory requirements or that the tanks are emptied within 60 days after foreclosure and the lender either temporarily or permanently closes the tanks. A lender is eligible for an exemption as a UST owner and operator if the lender holds property with USTs as collateral, and does not participate in the management of the tank system nor engage in petroleum production, refining, and marketing. EPA emphasizes that the exemption does not apply to nonpetroleum hazardous substance underground tanks.

A lender who chooses to participate in management is not eligible for the exemption and would face potential responsibility for corrective action in the event of a release. Management participation is defined as the actual involvement in the management or control of decision making related to the operational aspects of day-to-day operations of the UST system, not the financial, administrative, or environmental compliance aspects of the system or property on which the UST system is located.[8]

In order to obtain protection from liability, the prospective purchaser of contaminated property can enter into an agreement with EPA. Under new guidance, which supersedes the agency's 1989 guidance, EPA basically offers a covenant not to sue to purchasers provided that there are direct and substantial benefits of the sales to the government, as well as other justifiable and less direct benefits such as improving the quality of life, or expanding employment opportunities.[9]

Some states have already enacted legislation that offers similar agreement provisions for prospective purchasers. Under new amendments to the state law in Washington, for example, the first agreement was finalized in 1994 between the Washington State Department of Ecology (DOE) and the Northlake Shipyard. In the agreement, the company will pay $400,000 into a cleanup fund at the time of purchase and will contribute a portion of its profit into the fund for up to 15 years.[10] The fund is expected to grow over that time to the estimated cleanup cost amount ($1.1 million). As part of the agreement, the company will not be liable for any further costs related to the cleanup of the past contamination.

Wesolowski and Antol[11] caution that a covenant not to sue obtained from a state does not protect a purchaser or developer against third party actions or the action of other government agencies. They argue that another way to encourage brownfields revitalization is to proactively create economic and other incentives that make the sites attractive to developers. The incentive can be an investment tax credit, allowing cleanup to be financed through recaptured federal and state tax revenues over several years. The incentive can also be cleanup cost recovery in the form of local real estate tax abatements spread over a specified period. Loans and grants from the state or federal governments could help defray costs.

1.2. The Cleanup Standard

The selection of appropriate cleanup standards will continue to move toward the use of risk-based criteria. Developing scientifically defensible, risk-based decision-making capabilities is essential and must be accompanied by an effective political, social, and educational framework to bring key issues before the public.[12] Knotek[12] states that to prevent today's solutions from becoming tomorrow's problems, cleanup decisions must be science-driven and implemented with foresight.

As part of the cleanup decision-making process, the risk assessment is calculated based on the land use of the contaminated property. Because brownfields were and would always be industrial properties, the risk is assessed in relation to this industrial usage. This category of risk assessment requires that the assessed properties be deed-restricted. Risk-based cleanup criteria with deed restriction provisions attached to the remediated properties will be a common practice in the brownfields program.

2. THE DEMAND OF REMEDIATION TECHNOLOGIES

Over the next 20 to 30 years, federal, state, and local governments and private industry will continue to commit billions of dollars to cleanup sites contaminated with hazardous waste and petroleum products. This commitment can be translated into increased use of all types of site remediation services. While existing remediation technologies have been successful, EPA predicts that the investment in site cleanup offers new opportunities for the development of less expensive and more effective solutions.[13]

The demand for remediation technologies is large and growing and can be grouped into eight market segments:

1. Superfund
2. RCRA
3. UST
4. Department of Defense (DOD)
5. Department of Energy (DOE)
6. Other federal agencies
7. States
8. Private parties

Most of these cleanup programs will cost tens of billions of dollars. The DOE program alone will cost hundreds of billions. Contaminants common to most programs are solvents, petroleum products, and metals. Although it is difficult to forecast the usage of specific technologies, EPA asserts that historical trends in the Superfund program can provide some insight.

Table 14.1 lists the types of technology that will be used by five different remediation programs. Included in the list is the estimated remediation cost for each program and the types of contaminants requiring remediation. For the Superfund program, the cost is estimated for sites currently listed on the National Priorities List (NPL). For the DOE program, the cost represents the planned funding through 1998.

Sixteen other federal agencies not listed in the table, identified approximately 350 sites in need of remediation. This need requires a budget of $1 billion for fiscal years 1991–1995. With 11,000 sites in the program, state remediation can easily cost billions of dollars. The number of private party sites is difficult to estimate. The cleanups in this category result from efforts of the private parties to limit their potential future liabilities. EPA roughly estimates that the cleanup costs for companies is $1 billion.

Table 14.1 Technology Used by Different Remediation Programs

Program	Est. cost (in billion $)	Contaminant	Technology
Superfund	16.5[a]	Chlorinated VOCs, nonchlorinated VOCs, PCBs, PAHs, phenols, lead, chromium, arsenic, cadmium	SVE, thermal desorption, bioremediation, incineration, S/S
RCRA	7.4–41.8	Corrosive wastes, ignitable wastes, heavy metals, organic solvents, electroplating wastes, used oil	Incineration, bioremediation, landfilling
UST	30.0	Gasoline, diesel fuel, other petroleum products, small percent of hazardous constituents	SVE, thermal desorption, bioremediation, landfilling
DOD	14.0	Petroleum products, solvents, metals, pesticides, paints, low level radioactive wastes	Bioremediation, SVE, soil washing
DOE	12.3[b]	Hazardous wastes and mixed radioactive wastes with hazardous constituents	SVE, air sparging, soil washing

[a] For sites currently listed on the National Priorities List.

[b] Planned through 1998. The total cleanup cost is estimated to be in billions of dollars.

Source: U.S. Environmental Protection Agency, Cleaning Up the Nation's Waste Sites: Markets and Technology Trends, Washington, D.C., 1993.

With this potential market, the future of environmental technology will still depend on the U.S. economic policy and incentives to promote the use of such technologies.[14] Grant Ferrier claims that stronger incentives for using better environmental technologies would occur in a society that places a monetary value on natural resources and requires the costs of environmental harm from manufacturing to be internalized, or reflected in a product's price. Just as there are tax incentives to promote the housing industry, there should be economic incentives for protecting natural resources.

3. THE STREAMLINING AND FOCUSING OF SITE REMEDIATION

While EPA is struggling to accelerate the pace of its site remediation, the states are very active in streamlining their remediation programs. A rough estimate suggests that EPA spends about $1 billion annually working on about 1,000 sites while the states are spending about $700 million annually working on 11,000 sites.[15] State remediation programs have proven to be effective due to the fact that contaminated sites are local problems and the states are much closer to these site problems, the people involved and the local environment, economy, and land-use issues.[16] Porter[16] argues that given these advantages, the states have also been more innovative than EPA whose responsibility has been to operate a consistent national program.

In Massachusetts, for example, the state has developed a concept for accelerating cleanups without compromising state environmental standards. Under the Massachusetts Superfund Law, the Massachusetts Department of Environmental Protection was charged with determining the responsible parties and requiring them either to do the cleanup or reimburse the state for doing it. Because of the limited state resources, the state created a licensed private sector professionals program. In this program the professionals act as liaisons between the state and those responsible for cleaning contaminated sites.[17] These experts are hired by the responsible parties to manage the cleanup and free the department from supervising the work at every site. The department audits about 20% of the cleaned sites annually.

This type of innovative program will flourish, modified by other states to fit its own conditions. An innovative program in both Indiana and Arizona, for example, was established to provide voluntary site remediation. The program offers assistance to site owners who voluntarily report the discovery of a contaminated site to the state. Under such a program, the state may provide assistance in cleaning up the site by using a state contractor and recovering the cleanup cost from the site owner. When completed, the owner will receive a statement from the state that declares the site has been satisfactorily remediated.

As a national issue, there are many minority and low-income individuals who live in communities near contaminated sites. Due to the nature of the Superfund ranking system, these environmental justice sites may not qualify for cleanups under CERCLA. Because of that ineligibility, EPA established the Gateway Initiative in 1993 that focuses on the myriad of problems of poverty-stricken East St. Louis and neighboring communities in Illinois and Missouri.[18] While site contamination is a major issue in the communities included in the Gateway Initiative, they also suffer from unemployment and badly deteriorating infrastructures.

An important thrust of the project is to identify other agencies who can provide services to the communities that will alleviate their suffering from the contaminated sites. EPA's goal is to establish interagency workshops in all ten EPA regional offices and community advisory groups in at least ten environmental justice sites to facilitate the early involvement of the affected residents in cleanup decisions.

PART V
REFERENCES

1. Carleton, Bobbie, On the Precipice of Change: Environmental Stewardship Comes of Age, *Waste Tech News*, Vol. 7, No. 5, 1995, 3.
2. National Environmental Policy Institute, *How Clean is Clean Project, White Paper on Brownfields — Second Draft*, Washington, D.C., 1995.
3. Gibson, Jeremy A., Harvesting a Bumper Crop of Brownfields Programs, *The National Environmental Journal*, Vol. 5, Issue 5, 1995, 56–59.
4. Rappaport, Diane, Port of Seattle Breaks the Brownfields Gridlock, in *Environmental Solutions*, Vol. 8, No. 11, 1995, 24–27.
5. EPA, Justice Department Policy on CERCLA Enforcement Against Lenders Released October 6, 1995, *Environment Reporter*, October 13, 1995, 1060–1061.
6. EPA, DOJ Establish Policy Freeing Lending Institutions from Liability, *Hazardous Waste News*, October 16, 1995, 330.
7. U.S. Environmental Protection Agency, Underground Storage Tanks — Lender Liability Final Rule, *60 FR 46692*, September 7, 1995, 46692–46715.
8. Underground Tanks: Final Lender Liability Rule Identifies Criteria for More 'Comprehensive' Exemption, *Environment Reporter*, September 15, 1995, 920–921.
9. U.S. Environmental Protection Agency, *Superfund Administrative Reform Fact Sheet: Prospective Purchaser Guidance*, Washington, D.C., 1995.
10. Washington State Finalizes Deal for Sale and Reuse of Polluted Site, *Hazardous Waste News*, Vol. 16, No. 37, 1994, 295.
11. Wesolowski, Ted and Antol, Sara M., Brownfields Initiatives Offer Few Incentives for Prospective Developers, Purchasers, *Environmental Solutions*, Vol. 8, No. 7, 1995, 32–33.
12. Knotek, Michael L., Cleanup Decisions Must Be Science-Driven, *Environmental Solutions*, Vol. 8, No. 7, 1995, 30–31.
13. U.S. Environmental Protection Agency, *Cleaning Up the Nation's Waste Sites: Markets and Technology Trends*, Washington, D.C., 1993.
14. Ferrier, Grant, quoted in: Future of Environmental Technology Hinges on Economic Policy, Incentives, Industry Says, *Environment Reporter*, November 17, 1995, 1225.

15. U.S. Environmental Protection Agency, *An Analysis of State Supported Superfund Programs, 50-State Study 1993 Update*, Washington, D.C., 1993.
16. Porter, J. Winston, *Cleaning Up Superfund: The Case for State Environmental Leadership*, Reason Foundation, Los Angeles, 1995.
17. Massachussetts Environmental Sense, *State Government News*, Vol. 38, No. 11, 1995, xii–xiii.
18. Superfund Update: EPA is Trying to Assist EJ Areas Currently Ineligible for Cleanups, *Hazardous Waste News*, Vol. 16, No. 48, 1994, 378.

INDEX